THE LIBRARY
ST. MARY'S COLLEGE OF MARYLAND
ST. MARY'S CITY, MARYLAND 20686

D. Heling P. Rothe U. Förstner
P. Stoffers (Eds.)

Sediments and Environmental Geochemistry

Selected Aspects and Case Histories

With 148 Figures

Springer-Verlag Berlin Heidelberg New York
London Paris Tokyo HongKong

Prof. Dr. DIETRICH HELING
Institut für Sedimentforschung, Universität Heidelberg,
Im Neuenheimer Feld 236, D-6900 Heidelberg

Prof. Dr. PETER ROTHE
Lehrstuhl für Geologie, Universität Mannheim,
Postfach 10 34 62, D-6800 Mannheim 1

Prof. Dr. ULRICH FÖRSTNER
Arbeitsbereich Umweltschutztechnik, TU Hamburg-Harburg,
Eißendorfer Str. 40, D-2100 Hamburg 90

Prof. Dr. PETER STOFFERS
Geologisch-Paläontologisches Institut, Universität Kiel,
Ludwig-Meyn-Str. 10, D-2300 Kiel 1

ISBN 3-540-51735-9 Springer-Verlag Berlin Heidelberg New York
ISBN 0-387-51735-9 Springer-Verlag New York Berlin Heidelberg

This work is subject to copyright. All rights are reserved, whether the whole or part of the material is concerned, specifically the rights of translation, reprinting, re-use of illustrations, recitation, broadcasting, reproduction on microfilms or in other ways, and storage in data banks. Duplication of this publication or parts thereof is only permitted under the provisions of the German Copyright Law of September 9, 1965, in its version of June 24, 1985, and a copyright fee must always be paid. Violations fall under the prosecution act of the German Copyright Law.

© Springer-Verlag Berlin Heidelberg 1990
Printed in Germany

The use of registered names, trademarks, etc. in this publication does not imply, even in the absence of a specific statement, that such names are exempt from the relevant protective laws and regulations and therefore free for general use.

Typesetting, Printing and binding:
Brühlsche Universitätsdruckerei, Giessen
2131/3145-543210 – Printed on acid-free paper

Preface

This volume is dedicated to German Müller on the occasion of his 60th birthday. The date coincides with the 25th anniversary of the Institut für Sedimentforschung of the University of Heidelberg. This jubilee motivates us to look back on German Müller's past working period as well as to look forward to future prospects for Sedimentary Petrography and Environmental Geochemistry.

The volume includes contributions from German Müller's colleagues, friends and scholars. The subjects range widely from classical sedimentary petrography (analytical methods, carbonate petrology and facies analysis), to petrology of coal, and environmental geochemistry. The diversity of fields reflects the broad scope of interests of the person celebrating his jubilee – and made it difficult to find a title for the volume. *Sediments and Environmental Geochemistry – Aspects and Case Histories* covers the contents of the volume only somewhat incompletely, and certainly reflects just a fraction of the subjects of German Müller's concern during his past period of professional activities. Much more would have to be included in order to complete the list, for instance celestite studies, black shale petrography, diagenesis of argillaceous sediments, or advanced analytical methods, to mention but a few of his many interests.

Thus, the volume is more a collection of selected subjects, rather than an exhaustive summary of Müller's work, nor should it, by its title, indicate the intention to cover all the different special fields of Sedimentary Petrology or Environmental Geochemistry. Therefore, the restricting subtitle "Aspects and Case Histories" was added.

Now for some facts about German Müller's professional life. His doctoral thesis on "The occurrence of carbonates, particularly of ferruginous carbonates, in the coal seams of the Ruhr area" was presented in Bonn in 1952 (at the age of 23!). (For comparison: Today's average age of leaving PhD candidates is 30.) German Müller was Assistant Professor at the Department of Geology at Cologne University for a short time, and subsequently he went to Ankara to work with the Turkish Institute for Economic Geology. After that position he joined Mobil Oil of Germany as a geologist. Three years later he took the opportunity to go to Ethiopia for Texas Africa Exploration Co. Eventually, he completed his time abroad and returned to Germany to become Assistant Professor at the Department of Mineralogy at the University of Tübingen.

With this decision, he started his career as university teacher. Soon after his habilitation with Prof. W. v. Engelhardt in Tübingen he was called to Heidelberg and appointed to establish the "Laboratory for Research of Sediments". This happened in 1964, exactly 25 years ago. After the Laboratory was merged into the "Institut für Sedimentforschung", German Müller became Director of this Institute, a position he holds to this date. He has given definition to the Institute through his work and personality.

Much has changed in geology in the past quarter of a century: Above all, the regrettable but unavoidable diversification of our science into more and more highly specialized fields, which cannot be mastered completely with full competence by any single scientist.

German Müller recognized early which new scientific fields were potentially capable of development. As early as the 1950's, during his time with the oil industry, he understood intuitively the possibilities offered by sedimentary petrography for both oil exploration (stratigraphic traps) and exploitation (greater recovery due to producing strategies adjusted to the geology of reservoir rocks). German Müller contributed by introducing new methods, an interest which resulted later in his first book *Methods of Sedimentary Petrology*.

Similarly, during his time at Tübingen, he started the study of Lake Constance, taking the lake as a natural laboratory for investigations of both recent sediments and sedimentation processes. These studies, initially aimed purely to increase our knowledge of sedimentary petrology, soon revealed (in combination with geochemical data) that recent sediments are clearly indicative of the chemistry of the water above of them. Initially, this was a concept of merely scientific interest. Later on, when the water quality of Lake Constance began to gain public interest due to the increasing danger to the drinking water supply, the relationship between water and sediment contamination became the basis of many investigations of the reasons for "eutrophication".

Subsequently, the knowledge of the correspondence of water chemistry and recent lacustrine sediment composition was further extended, and general relations between adsorption, suspended matter, and discharge in fluvial environments were derived.

By this work German Müller has contributed vitally to opening the door to the field of applied environmental geochemistry – a field which he did not subsequently leave – not the least because of its practical relevance and the desire to secure the very basis of our existence. The direct application of scientific results to socially useful initiatives was the main motive for his active work and it often formed a criterion of both its value and satisfaction. Although these goals may not always conform to the classic principles of conservative science, there is wide appreciation of their contribution in changing our views on how to coexist with nature. As a practical consequence, such environmental studies helped built sewage plants all around Lake Constance, thus reducing the contamination load, in particular of the phosphates, to a tolerable level.

German Müller, moreover, made numerous contributions to various fields of sedimentary petrology; for instance, he joined the Deep Sea Drilling Project on Leg 56 and 57; he worked on the mineralogy of the recent sediments of the Black Sea; he did research on dolomitization of biocalcarenites from Fuerteventura, Canary Islands, and many more. Limited space for this Preface does not allow to mention all the activities in which German Müller has participated during his career to date.

Most recently, he has been concerned intensely with sewage decontamination. The subject of research of another working group is soil contamination, which is of pressing importance in the Rhine-Neckar area. The interests of all these subjects are rooted in the attraction exerted by what is found to be of practical usefulness in general social terms.

The character of practical applicability, which is a common factor in all the research at Müller's Institute, has a favourable influence on the training of diploma and PhD candidates, since those graduating have better chances in finding a position if they have been prepared for working in useful, applied fields. Furthermore, the connections with external research laboratories or government institutions, developed during their research on critical problems of common interest, typically helps in starting a career relating to the subjects of their theses.

The vita of German Müller, briefly outlined above, confirms the idea that the so-called dispassionate science cannot be separated from the personality of the scientist: Science is made by individuals, and the individuals are molded by their engagement with science.

The results of German Müller's scientific career to date fill an imposing list: He is the author of two monographs, coauthor of two more books, and editor or coeditor of four other books. His publications to date include 215 titles of original contributions, and he is the editor or coeditor of a number of journals.

In recognition of his many achievements, the Bedford University in Arizona awarded him a "Doctor of Philosophy Honoris Causa". In 1986 he received the Philip Morris Research Award for the field of "Man and Environment", and in 1989 he became an Honorary Member of the Society of Economic Paleontologists and Mineralogists.

The volume presented here is accompanied by the best wishes from the contributing authors and the many friends of German Müller, wishing that many years of successful work, satisfaction, and stable health may lie ahead of him. May his inexhaustible optimism help overcome all unavoidable difficulties.

THE EDITORS

Acknowledgements

Thanks go to all who helped in the preparation of this volume; above all, Springer-Verlag for the completion of the volume in proper time. Mr. C. Reinish helped translate and go over the English versions of some of the papers. Mrs. U. Beckenbach took care of much of the editorial preparation.

Contents

Part I: Geology

Thoughts on the Growth of Stratiform Stylolites in Buried Limestones.
R. G. C. Bathurst 3

Anthracite and Concentrations of Alkaline Feldspar (Microcline) in
Flat-Lying Undeformed Paleozoic Strata: A Key to Large-Scale
Vertical Crustal Uplift. G. M. Friedman 16

Sediments on Volcanic Islands – On the Importance of the Exception.
P. Rothe 29

Fallout Tephra Layers: Composition and Significance. P. R. Bitschene
and H.-U. Schmincke 48

Isotope Geochemistry of Primary and Secondary Carbonate Minerals in
the Shaban-Deep (Red Sea). P. Stoffers, R. Botz, and J. Scholten . 83

Biogenic Constituents, Cement Types, and Sedimentary Fabrics.
R. Koch and B. Ogorelec 95

The Southern Permian Basin and its Paleogeography. E. Plein . . . 124

Integrated Hydrocarbon Exploration Concepts in the Sedimentary
Basins of West Germany. M. Blohm 134

Sedimentological and Petrophysical Aspects of Primary Petroleum
Migration Pathways. U. Mann 152

The Influence of Subrosion on Three Different Types of Salt Deposits.
W. Sessler 179

Particle Size Distribution of Saliferous Clays in the German Zechstein.
R. Kühn 197

Mineralogical and Petrographic Studies on the *Anhydritmittelsalz* (Leine
Cycle z3) in the Gorleben Salt Dome. R. Fischbeck 210

Lacustrine Paper Shales in the Permocarboniferous Saar-Nahe Basin
(West Germany) – Depositional Environment and
Chemical Characterization. A. Schäfer, U. Rast, and R. Stamm . . . 220

"Search for Poyang Lake and China" ... "get 7 Hits" Information
Management in Environmental Sedimentology. W. SCHMITZ and K.
WINKLER 239

Microbial Modification of Sedimentary Surface Structures.
H.-E. REINECK, G. GERDES, M. CLAES, K. DUNAJTSCHIK,
H. RIEGE, and W. E. KRUMBEIN 254

Part II: Environmental Geochemistry

The Pollution of the River Rhine with Heavy Metals. K.-G. MALLE . . 279

Interactions of Naturally Occurring Aqueous Solutions with the Lower
Toarcian Oil Shale of South Germany. H. PUCHELT and T. NÖLTNER . 291

Sediment Criteria Development – Contributions from Environmental
Geochemistry to Water Quality Management. U. FÖRSTNER, W. AHLF,
W. CALMANO, and M. KERSTEN 311

Transport of Matter in Sediments: A Discussion. D. HELING and
R. GISKOW 339

Pathways of Fine-Grained Clastic Sediments – Examples from the
Amazon, the Weser Estuary, and the North Sea.
G. IRION and V. ZÖLLMER 351

Subject Index 367

List of Contributors

AHLF, W., Arbeitsbereich Umweltschutztechnik, Technische Universität Hamburg-Harburg, Eißendorfer Straße 40, D-2100 Hamburg 90, FRG

BATHURST, R.G.C., Derwen Deg Fawr, Llanfair D.C., Ruthin, Clywd, LL15 2SN, United Kingdom

BITSCHENE, P.R., Institut für Mineralogie RUB, Postfach 102148, D-4630 Bochum, FRG

BLOHM, M., BEB Erdgas und Erdöl GmbH, Riethorst 12, D-3000 Hannover 51, FRG

BOTZ, R., Geologisch-Paläontologisches Institut und Museum, Universität Kiel, Olshausenstraße 40–60, D-2300 Kiel, FRG

CALMANO, W., Arbeitsbereich Umweltschutztechnik, Technische Universität Hamburg-Harburg, Eißendorfer Straße 40, D-2100 Hamburg 90, FRG

CLAES, M., Institut für Chemie und Meeresbiologie, Universität Oldenburg, Postfach 2503, D-2900 Oldenburg, FRG

DUNAJTSCHIK, K., Institut für Chemie und Meeresbiologie, Universität Oldenburg, Postfach 2503, D-2900 Oldenburg, FRG

FISCHBECK, R., Bundesanstalt für Geowissenschaften und Rohstoffe, Stilleweg 2, D-3000 Hannover 51, FRG

FÖRSTNER, U., Arbeitsbereich Umweltschutztechnik, Technische Universität Hamburg-Harburg, Eißendorfer Straße 40, D-2100 Hamburg 90, FRG

FRIEDMAN, G.M., Department of Geology, Brooklyn College, Brooklyn, NY 11210, USA

GERDES, G., Institut für Chemie und Meeresbiologie, Universität Oldenburg, Postfach 2503, D-2900 Oldenburg, FRG

GISKOW, R., Hauptstraße 187, D-6500 Mainz, FRG

HELING, D., Institut für Sedimentforschung, Universität Heidelberg, Im Neuenheimer Feld 236, D-6900 Heidelberg, FRG

IRION, G., Forschungsinstitut Senckenberg, Institut für Meeresgeologie und Meeresbiologie, Schleusenstraße 39a, D-2940 Wilhelmshaven, FRG

KERSTEN, M., Arbeitsbereich Umweltschutztechnik, Technische Universität Hamburg-Harburg, Eißendorfer Straße 40, D-2100 Hamburg 90, FRG

KOCH, R., Institut für Paläontologie, Universität Erlangen-Nürnberg, Loewenichstraße 28, D-8520 Erlangen, FRG

KRUMBEIN, W.E., Institut für Chemie und Meeresbiologie, Universität Oldenburg, Postfach 2503, D-2900 Oldenburg, FRG

KÜHN, R., Institut für Sedimentforschung, Universität Heidelberg, Im Neuenheimer Feld 236, D-6900 Heidelberg, FRG

MALLE, K.-G., BASF Aktiengesellschaft DU/A-C 100, D-6700 Ludwigshafen, FRG

MANN, U., KFA/ICH-5, Postfach 1913, D-5170 Jülich, FRG

NÖLTNER, T., Institut für Petrographie und Geochemie, Universität Karlsruhe, Kaiserstraße 12, D-7500 Karlsruhe, FRG

OGORELEC, B., Geološki Zavod Ljubljana, Parmova 33, YU-61000 Ljubljana, Yugoslavia

PLEIN, E., Institut für Sedimentforschung, Universität Heidelberg, Im Neuenheimer Feld 236, D-6900 Heidelberg, FRG

PUCHELT, H., Institut für Petrographie und Geochemie, Universität Karlsruhe, Kaiserstraße 12, D-7500 Karlsruhe, FRG

RAST, U., Bayerisches Geologisches Landesamt, Heßstraße 128, D-8000 München 40, FRG

REINECK, H.-E., Forschungsinstitut Senckenberg, Schleusenstraße 39a, D-2940 Wilhelmshaven, FRG

RIEGE, H., Institut für Chemie und Meeresbiologie, Universität Oldenburg, Postfach 2503, D-2900 Oldenburg, FRG

ROTHE, P., Lehrstuhl für Geologie, Universität Mannheim, Postfach 103462, D-6800 Mannheim, FRG

SCHÄFER, A., Geologisches Institut, Universität Bonn, Nußallee 8, D-5300 Bonn, FRG

SCHMINCKE, H.-U., Institut für Mineralogie RUB, Postfach 102148, D-4630 Bochum, FRG

SCHMITZ, W., Institut für Sedimentforschung, Universität Heidelberg, Im Neuenheimer Feld 236, D-6900 Heidelberg, FRG

SCHOLTEN, J., Geologisch-Paläontologisches Institut und Museum, Universität Kiel, Olshausenstraße 40–60, D-2300 Kiel, FRG

SESSLER, W., Kali und Salz AG, Friedrich-Ebert-Straße 160, D-3500 Kassel, FRG

STAMM, R., Bayerisches Geologisches Landesamt, Heßstraße 128, D-8000 München 40, FRG

STOFFERS, P., Geologisch-Paläontologisches Institut Universität Kiel, Ludwig-Meyn-Str. 10, D-2300 Kiel 1, FRG

WINKLER, K., Informationsvermittlungsstelle, Universitätsbibliothek, Universität Heidelberg, Im Neuenheimer Feld 368, D-6900 Heidelberg, FRG

ZÖLLMER, V., Forschungsinstitut Senckenberg, Institut für Meeresgeologie und Meeresbiologie, Schleusenstraße 39 a, D-2940 Wilhelmshaven, FRG

: Geology**

Thoughts on the Growth of Stratiform Stylolites in Buried Limestones

ROBIN G. C. BATHURST[1]

CONTENTS

Abstract 3
1 Introduction 4
2 Definitions and Processes 6
3 Discussion 10
References 13

Abstract

The factors that determine the vertical distribution of stratiform stylolites in pure (low-clay) limestones are examined and current theories of stylolite formation are reviewed. The conclusion is reached that stylolites that are *actively growing* cannot act as conduits for the flow of fluids.

Whereas the development of spaced cleavage in pure limestones may well be controlled by simple feedback between dissolution, molecular diffusion and precipitation, this cannot have been true for all stylolites formed during uncomplicated burial. In many buried limestones the location of stylolites bears a close relation to vertical lithological change across bedding. Moreover, stylolites that are parallel for hundreds of metres are unlike the typically anastomosing tectonic stylolites. In argillaceous limestones (in which stratiform stylolites do *not* occur) it is generally agreed that the vertical distribution of pressure-dissolution fabrics reflects a stratified cementation (lithification) early in burial history and a subsequent restriction of pressure dissolution to the less cemented layers. While such an explanation may be appropriate for some stylolites in pure limestones, where the stylolites are closely related to lithological bedding, it is not satisfactory for others where such bedding (if any) was destroyed by bioturbation and the only bedded feature is the stylolite itself. In this case it seems worthwhile to consider the possibility that the distribution of stylolites may have been influenced by vertical changes in texture related to ichnofabric (tiering). Generally speaking, in a field where observational data remain scarce, there remains much to be done.

---

[1] Derwen Deg Fawr, Llanfair D.C., Ruthin, Clwyd, LL15 2SN, UK (Honorary Senior Research Fellow, University of Liverpool).

1 Introduction

The factors that determine the distribution of stylolites in limestones are not well understood. Many stylolites show a rather regular vertical spacing and a definite stratiform parallelism and these are examined in this chapter; particular attention is given to stylolites and their relation to limestone bedding. Other stylolites are irregular and anastomosing and their distribution relates to a variety of background lithogical patterns, as for example, in stylobreccias. There is, regrettably, a shortage of careful observations in the field, in core and with the microscope. There is also a need for more experimental work. Here, an attempt will be made to examine some of the unresolved problems.

As the purer (clay-poor) limestones are buried they commonly develop stratiform stylolites which have a rather regular vertical spacing (Figs. 1–3). A similar pattern evolves as stylolitic spaced cleavage in tectonically stressed rocks. Both processes are a means whereby accumulated strain is relieved. The spacing of stylolites in buried limestones, and their siting in one part of a bed rather than another, must bear a direct relation to some past heterogeneous pattern of stress: this is obvious enough. Yet the factors that control the development of the crucial stress pattern, which in turn initiates the stratiform stylolites, are not at all clear. Whereas spaced cleavage (Borradaile et al. 1982) may be regarded, in a monomineralic rock, as a response

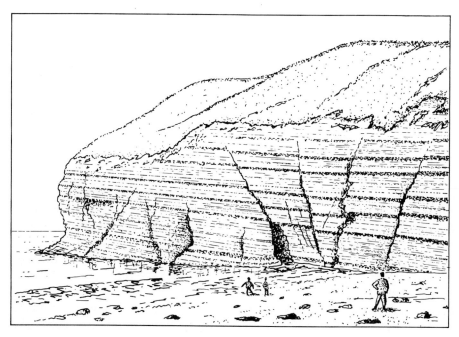

Fig. 1. These bedding planes in the Cretaceous Chalk of Flamborough Head, England, are stylolites as illustrated in Fig. 3

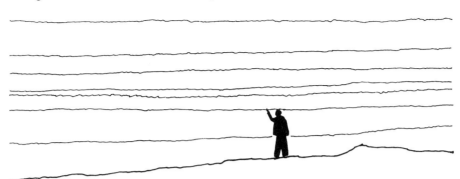

Fig. 2. Stratiform stylolites in the Jurassic Calcaire de Comblanchien of Burgundy: for details, see Fig. 3

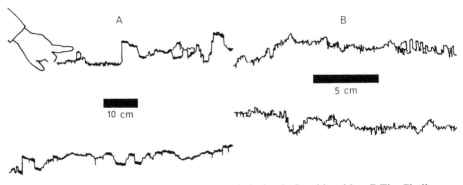

Fig. 3. Details of stylolites in Figs. 1 and 2. **A** Calcaire de Comblanchien; **B** The Chalk

to a simple feedback system involving dissolution-molecular diffusion-precipitation (Merino et al. 1983) in a tectonically controlled stress field (Alvarez et al. 1978), the same cannot be said for the deformation of limestones during simple burial. In these, the stress field is modified by vertical lithological changes related to bedding. Tectonic spaced cleavage, which lies at an angle to bedding, is not controlled by lithology in the same way. Although some stratiform stylolites show no clear relation to lithological bedding, others show a very precise relation. To claim, however, that stylolites follow bedding planes is to oversimplify the matter. Moreover, in many pure limestones the rather regularly spaced stylolites provide the only layered structure to be seen in rocks where the original bedding was totally destroyed syndepositionally during bioturbation. Even where stylolites do follow bedding planes, the causal connection is not immediately obvious.

In argillaceous limestones the controlling factors seem easier to understand. In these rocks (with more than about 8–10% by weight of clay) the role of an early stratified cementation is being increasingly recognized. The vertical distribution of fitted-fabric (interpenetrant grains) and of dissolution-seams in these limestones is regarded by a number of workers as a response to a selective lithification of certain strata early in burial history (Hallam 1964; Campos and Hallam 1979; Eder 1982; Einsele 1982; Ricken and Hemleben 1982; Walther 1982; Alvarez et al. 1985; Bathurst 1987; in press). This stratified cementation gave rise to an alternation of rigid and non-rigid strata. It proceeded sometimes in closed systems, as in pelagic carbonate muds, by dissolution-diffusion-precipitation, or in open systems in shelf sediments with the involvment of meteoric water. Both mechanical and chemical compaction were thenceforth concentrated in the less lithified strata where these abutted against the rigid strata. In a similar way unconsolidated ooids are crushed against chert layers (Chanda et al. 1977) according to the laws governing compression around rigid bodies (Ghosh and Sengupta 1973; Shimamoto 1975). The possibility that some similarly regular spacing of cementation, early in burial history of the purer limestones, may have had a critical influence on the siting of stylolites must now be examined.

2 Definitions and Processes

The stylolite is found only in low-clay limestones (Mossop 1972; Wanless 1979; Marshak and Engelder 1985) and is one of three characteristic fabrics that arise during pressure dissolution in limestones generally. The nomenclature of Buxton and Sibley (1981) is ideally useful, modified slightly with the agreement of the authors. Discussion will be limited to monomineralic limestones, for practical purposes those containing more than about 92% calcite. (In mineralogically heterogeneous rocks the process is more complicated, as in Beach and King 1978; Beach 1979.) *Fitted-fabric* develops by interpenetration of allochems or other crystals or particles. It has been found mainly in grainstones irrespective of clay content. A *dissolution-seam* is a smooth undulose seam of insoluble residue (generally clay) which lacks the sutures of the stylolite. It is restricted to argillaceous limestones. It does not cut through allochems, as a stylolite does, but lies around them, although their margins are commonly corroded. The seams are commonly anastomosing. (The term *microstylolite* has been given so many varied meanings that it is best avoided.) A *stylolite* is a serrated interface between two rock masses (Fig. 3). It appears to be restricted to those limestones that have low clay contents. It has a sutured appearance in sections normal to the plane in which it lies. In three dimensions, columns of one rock mass fit into sockets in the opposed mass. A stylolite transects all fabrics, i. e. allochems, cement, micritic matrix. It normally has an amplitude greater than the local allochem diameter. Insoluble residue may or may not be present. It is equivalent in part to the sutured-seam of Wanless (1979) and is the stratiform stylolite of Purser (1984). The term styolite is not used here to describe sutured contacts between pairs of allochems: these are an aspect of fitted-fabric. The original meaning of stylolite as a stratiform structure (Figs. 1, 2) is retained (as in

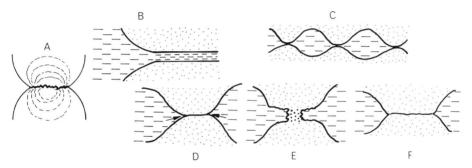

Fig. 4 A–F. Pressure dissolution at rock-rock contacts. **A** two allochems in contact along a slightly serrated interface. Approximate isopleths of elastic strain in each are indicated. **B** Rock-rock contact for the water film diffusion theory: opposing rock masses (*fine dots*) separated by a water film (*dense dashes*) which transmits deviatory stress and allows diffusion of dissolved ions into the open pores (*dispersed dashes*). C-D Theory of plastic deformation plus free-face dissolution. **C** Irregular hummocky interface (high magnification) between opposed rock masses showing rock contacts and labyrinthine water film. **D** Higher magnification free-face dissolution of the more elastically strained rock around the rock-rock interface (*arrows*) leading to **E**. A contact which has become so thin that its yield stress is reached and it deforms plastically (*large dots*) and is dissolved giving rise to new contact at **F**. Here, the cycle of dissolution and failure begins again.

Klöden 1828; Marsh 1867; Wagner 1913) on the grounds that such an important feature requires a specific, unambiguous name.

A general idea of how pressure dissolution works has been developed, although the exact nature of the process is still debated. Two theories are currently important. The *water film diffusion theory* was first propounded by Weyl (1959) and elaborated by De Boer (1977), Robin (1978), Merino et al. (1983) and Guzzetta (1984). The more recent *plastic deformation plus free-face pressure-dissolution* theory has been put forward by Tada and Siever (1986) and Tada et al. (1987). The first process depends for its driving force on the enhanced solubility of elastically strained crystals adjacent to a continuous film of water, which persists between the opposed rock masses (Fig. 4A, B), and the synchronous diffusion of dissolved ions through that film down a chemical potential gradient out into the open pore system. The uniaxial stress is transmitted from one rock mass to the other by the film itself, which obviously does not have the properties of a Newtonian liquid.

The theory of Tada and Siever, on the other hand, is somewhat similar to the undercutting theory discussed by Bathurst (1958) and Weyl (1959). Bathurst noted that, in any sediment made of crystals (single or combined) and water, each crystal lattice is always elastically strained in the vicinity of an intercrystalline boundary owing to misfit of the atomic lattices. The elastic strain is enhanced by the accumulation of overburden stress and is matched by an increase in solubility. Surfaces of strained rock separated by a water film dissolve and the ions diffuse away through the film down a concentration gradient. In order to maintain a water film between the compacting rock masses, Bathurst supposed that the opposed surfaces are hummocky (Fig. 4C), giving rise to numerous point contacts which support the labyrinth of the water film. He added "It would seem that surface contacts [between opposed rock masses] with no intervening solvent, as distinct from point contacts, can never develop because as soon as the smallest surface contact is formed the solvent is excluded from it and further change is impossible. Instead, solution continues at the junction between this interface and the solvent and so destroys it." Bathurst did not explain how the crystalline rock deformed (i. e. how it collapsed) but in 1959 Weyl referred to a similar "undercutting" mechanism and assumed that the point contacts would collapse by mechanical fracture when the undercutting dissolution had proceeded far enough. A similar explanation was implied by Sorby (1908) and is akin to the "house-of-card collapse" of Gray (1981). Weyl nevertheless preferred the hypothesis, elaborated by De Boer and others, of a permanent water film separating the opposed masses which would allow diffusion of dissolved ions and would transmit a deviatory stress.

Tada and Siever have put forward a modified version of the undercutting theory based on laboratory experiments in which they examined the consequence of placing a loaded quartz knife edge on halite in a halite-saturated solution. Instead of fracture collapse of point contacts, they assume that the contacts deform plastically. The driving force of the process is, again, the enhanced solubility of the elastically deformed rock around the point contacts with the addition of a degree of localized plastic strain. Dissolution of elastically strained rock surfaces in the vicinity of the contact (Fig. 4D, arrows) so weakens the rock that brittle fracture or plastic deformation take over and it collapsses mechanically (Fig. 4 E, F). The plastically deformed rock is dissolved and removed. Dissolution of the elastically strained material continues until the contact region is again weakened and once more collapses through plastic strain. The water film does not have to transmit a deviatory stress because this is transmitted via the plastically deforming rock.

The important conclusion is that both processes depend on the simultaneous action of dissolution, molecular diffusion and precipitation. Additional transport of ions in flowing pore water must play a significant role in certain circumstances. A sink into which the transported ions can precipitate is essential. Thus pressure dissolution can only take place in rocks where some residual porosity (and permeability) remains - or new porosity has been created by, say, fracturing. The basic principles are most readily illustrated in the formation of fitted-fabric (Fig. 4A) where dissolution goes on in a water film between the two allochems, ions diffuse into the open pore from where they may be carried away by flowing water to be precipitated on surfaces of unstrained crystals. The allochem framework deforms as the grains interpenetrate and occupy the spaces previously occupied by the dissolved rock. Pores are deformed and reduced in size and water is expelled. Clearly any local precipitation of cement greater than a minimal fringe on allochems will make a rigid framework which will prevent further deformation

(Sorby 1908). Thus removal of ions from any sediment body evolving a fitted-fabric is essential. This is not likely to be difficult because grainstones with fitted-fabric commonly retain considerable porosity and high permeabilities so that pore water can flow easily in them. (They make good hydrocarbon reservoirs.)

In this connection it is important to bear in mind that the growth of stylolites takes place in a rock that combines rigidity with porosity. The process is one whereby deformation is accompanied by the expulsion of both rock and water from the affected regions. If the undeformed rock has a porosity of, say, 25%, then rock and water will be expelled in the ratio water:rock of 1:3. The tiny amount of water held in the films can be neglected. Transport of ions will be dominantly by molecular diffusion. The carriage of ions in flowing water, assuming a reasonable partial pressure of carbon dioxide, will require rock:water ratios of the order of 1:10000 to 1:50000 (Bathurst 1971 p. 440). Such vast volumes of water are rarely available in deeply buried sediments (Land and Dutton 1978; Bjørlykke 1979; Ricken 1986). Because of this restriction, transport distances for solute ions associated with stylolites are likely to be short and of the same order as those involved in the growth of concretions (Berner 1980; Raiswell 1988) which is also controlled largely by molecular diffusion.

While on the subject of process, it is useful to consider the claim often made that the stylolites themselves have been paths for the flow of water or hydrocarbons. This is undeniable. There is evidence not only for the flow of hydrocarbons in stylolites but for the precipitation of wall encrusting cements and even the deposition of internal sediments (Braithwaite 1989). However, flow can only have taken place in inactive stylolites in which pressure dissolution had ceased. This might have happened after the uniaxial lithostatic stress had been reduced until it was less than the hydrostatic stress, by removal of overburden, by generation of hydrostatic overpressure or by tectonic deformation. Flow in active stylolites is impossible because lithostatic pressure then exceeds hydrostatic pressure. Thus, any fluid in the void between two opposed compacting rock masses will continue to flow out of that void until a film of fluid remains in which wall effects dominate and the fluid no longer has the capability to flow. Active stylolites are composed of voids from which any fluid capable of flowing has already been expelled. Thus, active stylolites cannot act as conduits for flow.

To complete the story it should be recalled that the characteristic sutured interface resulting from pressure dissolution was not stimulated by the existence of different solubilities of rock, or of clay content, on either side of some parting (as thought by Stockdale 1926; Bayly 1985 and others). Long ago Dunnington (1954) argued convincingly that such differential solubilities cannot account for the known geometry of stylolite forms. However, neither he nor anyone else was then able to offer an alternative hypothesis. Recently, Guzzetta (1984) has offered a solution. He was able to model all known geometric forms of stylolite on the assumption that dissolution takes place only on one side of the water film between two rock masses at any one time. (He followed the water film diffusion model.) The intervening water film is divided into domains within each of which there is transverse polarity parallel to the principle stress. The domains may shift sideways, expand or reduce and also reverse their polarities. Guzzetta noted that such a polarity is *analogous* to certain properties of ferroelectric crystals, but that its action in rocks remains for the moment hypothetical and unexplained.

In the location of sites of pressure dissolution the initial distribution of clay may have been of some importance in so far as it may have played a role in localizing the stratified cementation (Bathurst 1987). The claim that clay in some way "focusses" hydrostatic pressure, causing it to increase preferentially in clay-rich layers, thus giving rise to local overpressure and an enhanced solubility product for calcium and carbonate ions (as implied in Bayly 1985), is not supported by evidence that water was confined in the clay layers. There seems to be no doubt that the driving force of pressure dissolution is the locally enhanced solubility of portions of elastically or plastically strained rock.

3 Discussion

In summary deformation by pressure dissolution in a monomineralic rock requires:
1. A heterogeneous distribution of solubility in the rock caused by locally enhanced elastic or plastic strain;
2. Selective dissolution at surfaces of the more soluble rock;
3. Diffusion of ions with more or less additional transport in flowing water (especially in fitted-fabric);
4. Precipitation of the dissolved ions.

We can now consider what happens in a clay-free monomineralic sediment. For example, in an unconsolidated carbonate sand under load, the regions of enhanced solubility are situated at the contacts between allochems and the process proceeds as already described for fitted-fabric. If certain strata had been cemented prior to the onset of this process, its action would have been restricted to the uncemented strata. Such alternations have been described by Bathurst (1987) in limestones where he considered that the pressure dissolution took place in bedding-parallel, vertically spaced zones of fitted-fabric. Barrett (1964) described similar development of fitted-fabric along regularly spaced seams (not along stylolites as originally indicated; unpublished observations of Barrett and the author).

Stylolites, on the other hand, grow in rock which has previously been made rigid by some degree of cementation, because stylolites cut across all fabrics including cement. It is, therefore, necessary to discover where, in a lithified sediment, stratiform water films on which stylolites might nucleate are likely to be available. Since cementation occludes voids, it must be supposed that any voids remaining must be those that, for some reason or another, have escaped cementation. (Stylolites typically grow at the margins of concretions). This leads to a consideration of the factors that control the pattern of cementation (precipitation of encrusting calcite crystals). This pattern is influenced by flow of water carrying dissolved ions to the site of crystal growth. This flow is in turn controlled by variations in porosity and permeability (Goldsmith and King 1987). The nucleation of a stylolite should be inhibited, therefore, where cementation is most complete, but encouraged where some residual porosity/permeability remains. For these reasons it seems likely that a pattern of stratified cementation, early in the history of burial, could have influenced the eventual distribution of pressure dissolution in some pure carbonate sediments, just as in argillaceous carbonate sediments. Later, as the stylolite grows, dissolved carbonate will be reprecipitated in pores, the later cements being laid down farthest from the stylolite (Wong and Oldershaw 1981). In this way the distribution of porosity and permeability will be continuously modified until such time as their values fall below critical levels and pressure dissolution halts.

The idea that a pattern of cementation early in burial history could have controlled the distribution of stratiform stylolites fits well with many sedimentary rocks where lithology varies across recognizable bedding planes. In grainstones where bedding-parallel stylolites have amplitudes scarcely greater than the grain diameter, a precise association of stylolites with certain bedding planes or laminar interfaces can commonly be recognized (Fig. 5).

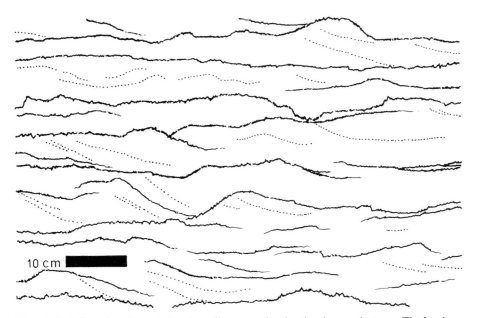

Fig. 5. Stylolites showing relation to sedimentary lamination in a grainstone. The lamination made, evident by colour changes across laminae, is indicated by *dotted lines*. Ordovician Holston Fm., men's washroom, Faculty Club, Louisiana State University

Buxton and Sibley (1981) showed how stylolite positioning can be correlated with the more abrupt lithological transitions in the Devonian Alpena Limestone of Michigan. The concept of control of pressure disolution by stratified cementation is reasonably substantiated for argillaceous limestones. Yet, for the purer stylolitic limestones, it remains a virtually untested working hypothesis. Assertions abound in the literature about the initial siting of stylolites but detailed, carefully argued, case studies scarcely exist.

Moreover, the hypothetical control of stylolite spacing by stratified cementation does not satisfy all cases. Some recent work at the University of Michigan by Finkel and Wilkinson (in press) illustrates this difficulty. The Mississippian Salem Limestone is a grainstone well known for its regularly spaced stylolites lying parallel to the general depositional surface but truncating a pattern of cross-bedded sets. Stylolite amplitudes range up to 25 cm. Despite detailed petrographic observations and analyses of trace elements across the sequence of stylolites, the authors discovered no relation between initial grainstone texture or composition and the spacing of stylolites. There are other limestones in which all layered structures, if any existed, were apparently destroyed within minutes or days of deposition by bioturbation. Nevertheless, in these sediments, now lithified, stylolites are distributed parallel to the original surfaces of deposition with a regularity and a degree of parallelism that suggests a very precise control. Although stylolites show a vertical fluctuation in their positioning on a scale of millimetres and centimetres, on a larger scale of metres or tens of metres they commonly remain essentially parallel. An example is the dense coccolithic Cretaceous Chalk of Flamborough Head, northeast coast of England, where parallel stylolites provide the only layered structure in a bioturbated sediment (Figs. 1, 3) and can be followed laterally for several kilometres. The same remarkable parallelism of stylolites was observed by the author in the Jurassic Comblachien of Burgundy (Figs. 2, 3; Purser 1972), which is a much burrowed oncolitic micrite. In *Amphypora* packstones of the Devonian (Frasnian) Swan Hills Formation, in Alberta, bedding-parallel stylolites cut across the rock fabric without any obvious relation to composition or texture (author's observations). For these rocks it is tempting to assume that the regularity cannot have been dependent on variation in bedded lithology. In tectonically spaced cleavage the feedback mechanism already mentioned and described by Merino et al. (1983) may have been the dominant control. Yet, although this might account for a regularity on the scale of spaced cleavage, it is difficult to imagine the process acting with such precise uniformity where parallelism extends for hundreds of metres or more. It is more likely that such stylolites would have an anastomosing distribution, as in the pressure-dissolution surfaces which are caused by burial but are transverse to syndepositionally folded bedding in the pelagic Jurassic-Tertiary limestones in Umbria (Alvarez et al. 1985). Even if some seafloor-parallel signal was imprinted at regular intervals on the sediments and then escaped total dispersion during bioturbation (as in argillaceous limestones; Bathurst 1987, in press) it is difficult at present to see how this might give rise to such extreme parallelism.

The questions raised by the distribution of stylolites in clay-poor, buried limestones clearly have yet to receive satisfactory answers. Control of the pattern of pressure dissolution by early stratified cementation, as in many argillaceous limestones, may provide an appropriate explanation in some rocks, but even this has yet to be demonstrated. Satisfactory explanations for the disposition of regularly spaced, parallel stylolites in limestones, where original, layered information has apparently been randomized through bioturbation, have yet to make their appearance. A possibility that might be worthy of consideration is some process akin to the tiering of ichnofabric (Bromley and Ekdale 1986). Certainly, in some limestones in the field the precisely defined bedding is, in fact, nothing more than a layering of different discrete ichnofabrics (e. g. Fig. 28 in Bathurst 1987) doubtlessly related to successive stillstands.

Acknowledgements. I am deeply indebted to Dr. A. E. Adams and Dr. J. D. Marshall for their helpful and constructive criticism of an earlier draft.

References

Alvarez W, Engelder T, Geiser PA (1978) Classification of solution cleavage in pelagic limestones. Geology 6:263–266

Alvarez W, Colacicchi R, Montanari A (1985) Synsedimentary slides and bedding formation in Apennine pelagic limestones. J Sediment Petrol 55:720–734

Barrett PJ (1964) Residual seams and cementation in Oligocene shell calcarenites, Te Kuiti Group. J Sediment Petrol 34:524–531

Bathurst RGC (1958) Diagenetic fabrics in some British Dinantian limestones. Liverpool Manchester Geol J 2:1–36

Bathurst RGC (1971) Carbonate sediments and their diagenesis. Elsevier, Amsterdam 620 pp

Bathurst RGC (1987) Diagenetically enhanced bedding in argillaceous platform limestones: stratified cementation and selective compaction. Sedimentology 34:749–778

Bathurst RGC (in press) Pressure-dissolution and limestone bedding: the influence of stratified cementation. In: Einsele G, Ricken W, Seilacher A (eds) Cycles and events in stratigraphy. Springer, Berlin Heidelberg New York Tokyo

Bayly B (1985) A mechanism for development of stylolites. J Geol 94:431–435

Beach A (1979) Pressure solution as a metamorphic process in deformed terrigenous sedimentary rocks. Lithos 12:51–58

Beach A, King M (1978) Discussion on pressure solution. J Geol Soc London 135:649–651

Berner RA (1980) Early diagenesis: a theoretical approach. Univ Press, Princeton, p241

Bjørlykke K (1979) Discussion - cementation of sandstones. J Sediment Petrol 49:1358–1359

Borradaile GJ, Bayley MB, Powell CMcA (1982) Atlas of deformational and metamorphic rock fabrics. Springer, Berlin, Heidelberg, New York, 551 pp

Braithwaite CJR (1989) Stylolites as open fluid conduits. Mar Petrol Geol 6:93–96

Bromley RG, Ekdale AA (1986) Composite ichnofabrics and tiering of burrows. Geol Mag 123:59–65

Buxton TM, Sibley DF (1981) Pressure solution features in shallow buried limestone. J Sediment Petrol 51:19–26

Campos HS, Hallam A (1979) Diagenesis of English Lower Jurassic limestones as inferred from oxygen and carbon isotope analysis. Earth Planet Sci Lett 45:23–31

Chanda SK, Bhattacharyya A, Sarkar S (1977) Deformation of ooids by compaction in the Precambrian Bhander Limestone, India: implications for lithification. Bull Geol Soc Am 88:1577–1585

De Boer RB (1977) On the thermodynamics of pressure solution - interaction between chemical and mechanical forces. Geochim Cosmochim Acta 41:249–256

Dunnington HV (1954) Stylolite development post-dates rock induration. J Sediment Petrol 24:27–49

Eder W (1982) Diagenetic redistribution of carbonate, a process in forming limestone-marl alternations (Devonian and Carboniferous, Reinisches Schiefergebirge, W. Germany). In: Einsele G, Seilacher A (eds) Cyclic and event stratification. Springer, Berlin, Heidelberg, New York, pp 98–112

Einsele G (1982) Limestone-marl cycles (periodites): diagnosis, significance, causes - a review. In: Einsele G, Seilacher A (eds) Cyclic and event stratification. Springer, Berlin, Heidelberg, New York, pp 8–53

Finkel EA, Wilkinson BH (in press) Stylolitization and cementation in the Mississippian Salem Limestone, west-central Indiana. Bull Am Assoc Petrol Geol

Ghosh SK, Sengupta S (1973) Compression and simple shear of test models with rigid and deformable inclusions. Tectonophysics 17:133–175

Goldsmith IR, King P (1987) Hydrodynamic modelling of cementation patterns in modern reefs. In: Marshall JD (ed) Diagenesis of sedimentary sequences. Blackwell, London, pp 1–13

Gray DR (1981) Compound tectonic fabrics in singly folded rocks from southwest Virginia, U.S.A. Tectonophysics 78:229–248

Guzzetta G (1984) Kinematics of stylolite formation and physics of the pressure-solution process. Tectonophysics 101:383–394

Hallam A (1964) Origin of the limestone-shale rythem in the Blue Lias of England: a composite theory. J Geol 72:157–169

Klöden KF (1828) Beiträge zur Mineralogischen und Geonostischen Kenntiss der Mark Brandenburg, I. Dieterici, Berlin

Land LS, Dutton SP (1978) Cementation of a Pennsylvanian deltaic sandstone: isotopic data. J Sediment Petrol 48:1167–1176

Marsh OC (1867) On the origin of the so-called lignilitites or epsomites. Proc Am Assoc Adv Sci 16:135–143

Marshak S, Engelder T (1985) Development of cleavage in limestones of a fold-thrust belt in eastern New York. J Struct Geol 7:3–14

Merino E, Ortoleva P, Strickholm P (1983) Generation of evenly-spaced pressure-solution seams during (late) diagenesis. Contrib Mineral Petrol 82:360–370

Mossop GD (1972) Origin of the peripheral rim, Redwater reef, Alberta. Bull Can Petrol Geol 20:238–280

Purser BH (1972) Subdivision et interprétation des séquences carbonatées. Mem Bur Rech Geol Min 77:679–698

Purser BH (1984) Stratiform stylolites and the distribution of porosity: examples from the Middle Jurassic limestones of the Paris Basin. In: Yahya FA (ed) Stylolites and associated phenomena: relevance to hydrocarbon reservoirs. Abu Dhabi Natl Reservoir Res Found, Abu Dhabi, pp 203–216

Raiswell R (1988) Evidence for surface reaction-controlled growth of carbonate concretions in shales. Sedimentology 35:571–575

Ricken W (1986) Diagenetic bedding. Lecture notes in earth sciences. Springer, Berlin, Heidelberg, New York, Tokyo 210 pp

Ricken W, Hemleben C (1982) Origin of marl-limestone alternation (Oxford 2) in southwest Germany. In: Einsele G, Seilacher A (1982) Cyclic and event stratification. Springer, Berlin, Heidelberg, New York, pp 63–71

Robin P-YF (1978) Pressure-solution at grain-to-grain contacts. Geochim Cosmochim Acta 42:1383–1389

Shimamoto T (1975) The finite element analysis of the deformation of a viscous spherical body embedded in a viscous medium. J Geol Soc Jpn 81:255–267

Sorby HC (1908) On the application of quantitative methods to the study of the structure and history of rocks. Q J Geol Soc London 64:171–233

Stockdale PB (1926) The stratigraphic significance of solution in rocks. J Geol 34:385–398

Tada R, Siever R (1986) Experimental knife-edge pressure solution of halite. Geochim Cosmochim Acta 50:29–36

Tada R, Maliva R, Siever R (1987) A new mechanism for pressure solution in porous quartzose sandstone. Geochim Cosmochim Acta 51:2295–2301

Wagner G (1913) Stylolithen and Drucksuturen. Geol Palaeontol Abh NF 11:101–128

Walther M (1982) A contribution to the origin of limestone-shale sequences. In: Einsele G, Seilacher A (eds) Cyclic and event stratification. Springer, Berlin, Heidelberg, New York, pp 113–120

Wanless HR (1979) Limestone response to stress: pressure solution and dolomitization. J Sediment Petrol 49:437–462

Weyl PK (1959) Pressure solution and the force of crystallization - a phenomenological theory. J Geophys Res 64:2001–2025

Wong PK, Oldershaw A (1981) Burial cementation in the Devonian, Kaybob Reef Complex, Alberta, Canada. J Sediment Petrol 51:507–52

Anthracite and Concentrations of Alkaline Feldspar (Microcline) in Flat-Lying Undeformed Paleozoic Strata: A Key to Large-Scale Vertical Crustal Uplift

GERALD M. FRIEDMAN[1]

CONTENTS

Abstract	16
1 Discovery of Anthracite in Flat-Lying Devonian Sandstone Strata	17
2 Discovery of High Concentrations of Alkaline Feldspar (Microcline) in Flat-Lying Lower Ordovician Limestone and Dolostone Strata	18
3 Significance of Anthracite and Alkali Feldspar (Microcline) in Flat-Lying Paleozoic Strata	22
4 Mississippi Valley-Type (MVT). Mineralization in Flat-Lying Silurian Strata	24
5 Burial Case Histories of Middle Ordovician, and Lower and Middle Devonian Strata	25
6 Relevance of Epeirogeny	25
7 Timing of Epeirogeny	25
References	26

Abstract

On visits to the United States Professor German Müller visited localities in which the presence of anthracite and concentrations of microcline (alkaline feldspar) in flat-lying undeformed strata created problems. These occurrences were puzzling because these strata are exposed at the present Earth's surface and the modern geothermal gradient in the area is low. These occurrences resulted from some combination of pressure and temperature that is not easily explained by the accepted geologic history of the region.

Various techniques of study, including fluid homogenization, vitrinite reflectance, oxygen isotopes, fission-track analysis on fluorite, and $^{40}Ar/^{39}Ar$ analyses on authigenic feldspar provided evidence of former deep burial followed by large amounts of uplift and erosion of these strata. Large-scale Carboniferous to Permian vertical movements of the crust and litho-

---

[1] Department of Geology, Brooklyn College, Brooklyn, NY 11210, USA, and Northeastern Science Foundation affiliated with Brooklyn College of the City University of New York, P.O. Box 746, Troy, NY 12181-0746, USA.

sphere brought deeply buried Ordovician to Devonian strata to shallow burial depth and ultimately to the present land surface.

1 Discovery of Anthracite in Flat-Lying Devonian Sandstone Strata

When Professor German Müller visited the United States in 1966, I took him to the Catskill Mountains in New York State where my then graduate student and PhD candidate Kenneth G. Johnson interpreted facies changes in Middle to Upper Devonian clastic strata (for location, see site, labeled New York, including an outline of New York State, in Fig. 4). In 1989 Prof. Müller returned to this site to reexamine this same material (Fig. 1). In 1966 Johnson and I had shown Prof. Müller the interpreted slip-off slope of a fluvial channel which truncated dark, highly organic pyritic lenses of a marsh facies. These lenses contained abundant, very coarse plant fragments (Johnson and Friedman 1969). On his return to Germany in 1966 Prof. Müller submitted these samples to the Geologisches Landesamt, Nordrhein-Westfalen, where Dr. M. Teichmüller noted that the woody cells of the plant debris were in part converted to anthracite and in part replaced by pyrite. A year later Prof. Müller invited me as a visiting professor to the University of Heidelberg where Won C. Park attended my lecture sequence. On his return to the United States Park worked for the Kennecott Copper Corporation

Fig. 1. Anthracite in flat-lying Devonian sandstone strata. East Windham, Catskill Mountains, New York. (**a**) View of vertical sequence showing fluvial and overbank strata. Prof. German Müller's hammer rests at base of black, recessed layer of marsh facies in which anthracite occurs. (**b**) Prof. Müller holds a sample of marsh facies containing anthracite.

whose facilities he used to determine that the mean reflectivity in oil of the vitrinites in plant debris, using light with a wavelenght of 546 µm, was 2.5%. This indicated that the vitrinite contains 93 to 94% so-called fixed carbon, and about 3.5% hydrogen and other volatile matter. He also reported that the coalified plant chambers contain anhedral pyrite and galena. Outside the plant debris, small amounts of chalcopyrite and possibly sphalerite are present. A vitrinite reflectance of 2.5% implies a LOM of 16 (Hood et al. 1975) and indicates that the specimens have been subjected to conditions in the upper part of the zone of metagenesis (Tissot and Welte 1978).

Discovery of anthracite (Johnson and Friedman 1969) in this plant debris within flat-lying strata created questions (Friedman and Sanders 1978, p.186). This occurrence is puzzling for the following reasons: (1) The amount of overburden in this area today is only 50 m or so. The maximum depth to which these strata may have been buried in the past is not known but is thought to have been only 1000 m or less. (2) The modern geothermal gradient in the area is low. Surely, this Devonian anthracite resulted from some combination of pressure and temperature that is not easily explained by the accepted geologic history of the region. Either these Devonian strata were deeply buried by many thousands of meters of overburden, or the geothermal gradient was high sometime in the past, or both. The Catskill front is located close to the Acadian Mountain chain. The initial interpretation of the presence of anthracite in these strata was attributed to the Acadian orogeny, during which the geothermal gradient might have been raised to a level sufficient to form anthracite, even if burial was not deeper than 1000 m.

2 Discovery of High Concentrations of Alkaline Feldspar (Microcline) in Flat-Lying Lower Ordovician Limestone and Dolostone Strata

On the same American visit in 1966, during which Prof. Müller sampled the plant debris, I introduced him to Lower Ordovician carbonate strata of the Tribes Hill Formation in the Mohawk Valley of New York, studied at that time by the then graduate student, Moshe Braun. Figure 2, photographed in Prof. Müller's presence, shows stromatolites in a dolomite rock. Stromatolites in Ordovician dolostones are not unusual, but the fine laminae consist of alternating dolomite and microcline (alkaline feldspar). The microcline crystals are aligned with their axis of elongation parallel to the bedding. The insoluble residue, for the most part composed of microcline, makes up between 35 and 67% by weight in samples studied. Table 1 shows chemical analyses of insoluble residues in the carbonate samples.

The abundance of feldspar which characterized some of the lithofacies was used in naming them; such as "laminated feldspathic dolomite" for the stromatolitic facies and "mottled feldspathic dolomite" for a lithofacies in which gray-black mottling was prominent (Braun and Friedman 1969). This ubiquitous feldspar, which formed as a replacement of dolomite and as crys-

Fig. 2. Stromatolite in a Lower Ordovician carbonate rock composed of finely alternating laminae of dolomite and microcline feldspar, Fort Hunter, N.Y. Photograph taken in Prof. German Müller's presence

Table 1. Chemical analyses of insoluble residues (in percent) in Ordovician carbonate rocks of the Tribes Hill Formation, Mohawk Valley, New York[a]

Samples	1 %	2 %	2 %	3 %	5 %	7 %
SiO_2	68.36	74.60	73.41	67.22	62.81	67.87
Al_2O_3	16.18	13.31	12.91	15.66	15.74	13.32
Fe_2O_3	0.04	0.05	–	0.93	2.35	2.16
FeO	0.14	0.14	0.21	Nil	0.93	–
Ti_2	0.66	0.58	0.60	0.76	0.79	1.00
CaO	0.42	0.07	0.63	0.53	1.96	0.90
MgO	0.81	0.45	0.55	0.44	1.22	0.56
K_2O	10.79	10.13	10.32	10.83	10.47	7.37
Na_2O	0.16	0.14	0.14	0.12	0.12	0.19
MnO	0.01	Traces	Traces	Nil	0.12	
P_2O_5	0.016	0.026	0.05	0.028	0.02	
SO_3	0.25	0.11	0.14	2.41	2.86	
Cl	0.07	0.03	0.10	0.13	0.31	
Free SiO_2	27.36	34.1	34.2	26.07	23.03	40.0
Free Al_2O_3	4.42	2.17	1.56	3.75	4.23	5.2

[a] The calculations for free SiO_2 and Al_2O_3 are based on the assumption that all the K_2O comes from the mineral microcline, which may be wrong. Some of the K_2O may come from clay minerals or mica. Chemical analyses: Mr. M. Gaon of the Geological Survey of Israel (Braun and Friedman 1969)

tals that grew between the dolomite rhombs, was interpreted as of authigenic origin precipitated from hypersaline brines as a result of the high activity ratio of alkali ions to hydrogen ions (Braun and Friedman 1969; Friedman and Braun 1975).

Further work involving my then graduate student, M. Raymond Buyce, revealed that strata containing abundant authigenic feldspar, which previously had been considered rarities in the geologic column, appear to be characteristic of the Cambrian-Ordovician rocks that formed on the shelf of the proto-Atlantic Ocean. They crop out from the mid-Continent of North America to the present Appalachian Mountain chain and along the paleo-shelf edge through Newfoundland, Scotland, and Greenland (Buyce and Friedman 1975; Fig. 3). Volcanic glass alters successively to various zeolites and finally to feldspar. The tephra at the margin of some Quaternary lakes changes in the center of the lake to potassium feldspar. As occurrences of

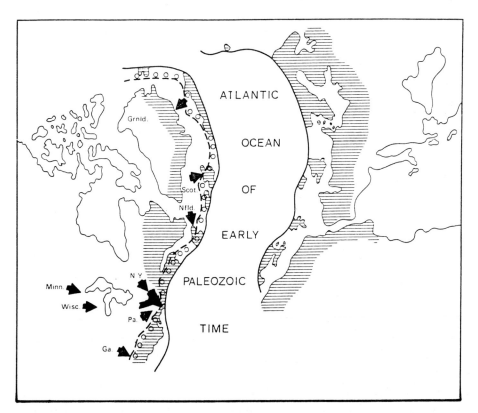

Fig. 3. Reported occurrences of abundant K-feldspar (*arrows*) in Cambrian and Ordovician strata that formed on the proto-Atlantic shelf. Tephra from a proposed zone of island arcs (*circles*) that lay offshore of the North American shelf-edge (*dashed line*) may possibly have contributed to the development of the authigenic K-feldspar (Buyce and Friedman 1975). For location of northern Appalachian Basin and sites of samples discussed in this work, see outline of New York State indicated by notation "*NY*".

Precambrian and lower Paleozoic zeolites are unknown and as feldspar is the stable end product, I think it probable that zeolites that antedated the mid-Paleozoic Era probably have changed to feldspar. The presence of some volcanic rocks of probable Cambrian and Early Ordovician age in New England suggests the possibility that the authigenic K-feldspar of the shelf rocks is the only remaining evidence of older tephra. The feldspar crystals in these carbonate rocks may have replaced wind-transported tephra that accumulated among algal mats on the hypersaline shoals at the margin of the deep ocean in which the volcanoes were active. Intercalations of shaly stringers rich in feldspar and high feldspar content in the carbonate rocks possessing 30 to 50% insoluble residue may have resulted from abundant eruptions and wind transport of debris from ancient deep-sea volcanoes. Chemical analyses of over 1300 carbonate rock samples of Cambrian to Devonian age from the shelf in New York State indicate that Cambrian and Lower Ordovician strata contain seven times more normative K-feldspar than the younger strata (Fig. 4). Petrographic study utilizing cathodoluminescence of Early Ordovician carbonates shows that the K-feldspar is not entirely authigenic but rather, is composed of subequal amounts of detrital K-feldspar as cores of particles and authigenic K-feldspars as overgrowths (Buyce and Friedman 1975). The cores of particles may represent wind-transported debris from ancient deep-sea volcanoes and the authigenic feldspar overgrowths are diagenetic and formed later following burial.

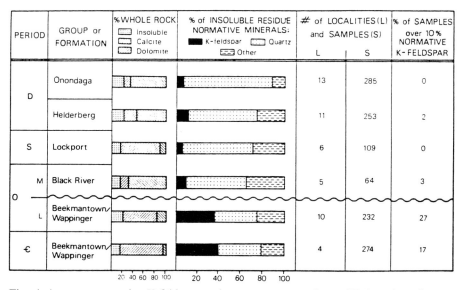

Fig. 4. Average normative K-feldspar and quartz concentrations of Paleozoic carbonate strata of the northern Appalachian Basin. Note that the normative K-feldspar abundance below the Lower Ordovician-Middle Ordovician unconformity is considerably greater than that above. Comparison of the percentage of individual samples with more than 10% normative K-feldspar shows a similar relationship (Buyce and Friedman 1973)

3 Significance of Anthracite and Alkali Feldspar (Microcline) in Flat-Lying Paleozoic Strata

As already mentioned, the initial discovery of the presence of anthracite in flat-lying Devonian sandstone of the Catskill Mountains was attributed to the Acadian orogeny. However, this interpretation was dropped (Friedman and Sanders 1982) through further study of the rocks which revealed that authigenic chlorite and local sericite filled interparticle pores in sandstone and that the anthracite has a vitrinite reflectance of 2.5%. This implies a possible level of organic metamorphism (LOM) of 16. A similar degree of thermal activity is implied by the black color (Staplin kerogen alteration index of 4) of the associated carbonized kerogen and a conodont alteration index of 4 (Harris et al. 1978, sheet 2). The data suggest a former depth of burial of approximately 6.5 km, consistent with the idea that the thick ($\sim$6.5 km) Carboniferous strata of northeastern Pennsylvania extended northeast far enough to bury the Catskill in New York State (Friedman and Sanders 1982).

Initially this interpretation was contested and opponents considered it "misleading" and suggested that various lines of evidence indicate that the rocks exposed at the present Catskill surface were never deeply buried (Levine 1983). However, support for the deep-burial model has spread (Lakatos and Miller 1983; Gale and Siever 1986; Johnson 1986; Gerlach and Cercone 1986; Duddy et al. 1987). Even the opposing school (Levine 1983) reconsidered and claimed a depth of burial ranging from 6 to as much as 9 km (Levine 1986). If these strata were buried to a depth of 6.5 km as the Catskill data imply, unexpectedly large amounts of uplift and erosion must have taken place to bring these strata to the present land surface.

Data obtained from the analyses of fluid inclusions in calcite-healed fractures of the Lower Ordovician carbonate strata, in which microcline crystals are so prominent, indicate higher paleotemperatures and greater depth of burial than have previously been inferred for the rocks of this region (Urschel and Friedman 1984; Friedman 1987a,b). Average fluid-homogenization temperatures range from 96° to 159° C (Table 2). These high paleotemperatures are supported by oxygen-isotope (Table 2) and conodont-alteration data (Harris et al. 1978). A former depth of burial >7 km is implied when a geothermal gradient of 26° C/km (Friedman and Sanders 1982; Sanders 1983) is used.

These data from the northern Appalachian Basin coincide with data obtained from carbonate rocks of the southern Appalachians, where homogenization temperatures range from 100° to 200° C (Hearn et al. 1987). Authigenic feldspars in Cambro-Ordovician carbonates occur along the entire proto-Atlantic (Iapetus) shelf from the southern to the northern Appalachians and farther north and east to Newfoundland, Scotland, and Greenland (Fig. 3) paralleling the Taconic belt.

Table 2. Oxygen-isotope and fluid-inclusion homogenization temperatures for Beekmantown (Lower Ordovician) carbonates

Sample	Location	$\delta^{18}O$	Isotopic temp.[a] (°C)	No. of inclusions	Homogenization temp. (°C)		Freezing temp. (°C)	
					Range	Average	Range	Average
Saddle dolomite	Subsurface core, Mohawk Valley	−13.01	133, 164, 175, 193	10	59–150	96	−12 to −39	−30.4
Dolomite vein	Subsurface core, Mohawk Valley	−11.17	115, 145, 155, 171	—	—	—	—	—
Dolomite vein	Tribes Hill, Mohawk Valley	−9.91	104, 132, 142, 158	8	102–121	112.9	−12 to −17	−14.5
Calcite vein	Smith basin, Hudson Valley	−10.10	74, 96, 104, 117	—	—	—	—	—
Dolomite vein	Ticonderoga, Champlain Valley	−13.98	143, 175, 187, 205	9	100–206	159.4	−12 to −15	−13.4
Dolomite vein	Fort Ann, Hudson Valley	−12.62	129, 160, 171, 188	15	117–190	159.3	—	−5.0
Dolostone	Baldwin Corners, Hudson Valley	−6.80	78, 103, 112, 126	4	123–160	138.6	−8 to −12	−10.0
Dolostone	Ticonderoga, Champlain Valley	−7.38	82, 108, 117, 132	2	140–145	142.5	—	—

After Urschel and Friedman (1984).
[a] Determined from Craig's (1965) equation for calcite and Fritz and Smith's (1970) equation for dolomite.

4 Mississippi Valley – Type (MVT) Mineralization in Flat-Lying Silurian Strata

In the northern Appalachian Basin (Fig. 3) minor concentrations of sulfides occur throughout most of the undeformed strata, especially those composed of carbonate rock. In a host rock of gray, dense, hard, massive, fine-grained dolostone (dolomite), the Lockport Formation (Middle Silurian) contains a remarkable mineral suite filling fractures, geodes, vugs, and cavities. These minerals include saddle dolomite, sphalerite, galena, marcasite, pyrite, fluorite, anhydrite, gypsum, calcite, quartz, barite, and celestite. Fluid inclu-

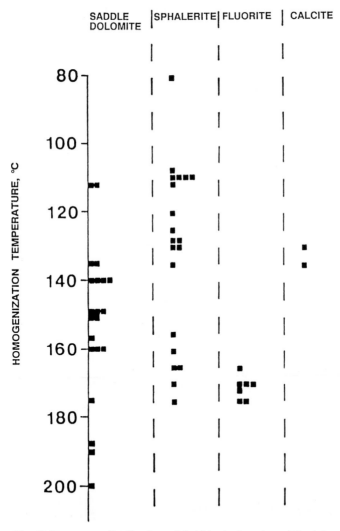

Fig. 5. Frequency distribution of fluid inclusions in saddle dolomite, sphalerite, fluorite, and calcite crystals (not pressure corrected)

sions trapped during the growth of saddle dolomite, sphalerite, fluorite, and calcite indicate that the precipitating fluids were very hot (Fig. 5): saddle dolomite mean homogenization temperatures were ~150° C; sphalerite showed a range of 80–179° C, clustering at 130–140° C; fluorite indicates a temperature range of 167–177° C; and calcite a range of 137–139° C. For a paleogeothermal gradient of 26° C/km (Friedman and Sanders 1982), a former depth of burial of ~5 km is implied for these undeformed strata which are now exposed on the earth's surface. This range of temperature is also supported by the data of Kinsland (1977), Harris et al. (1978), and Harris (1979). The data imply that considerable uplift and erosion have taken place. Because the strata are undeformed, vertical uplift and isostatic unroofing through erosion have brought the inferred, formerly deeply buried strata to the surface.

5 Burial Case Histories of Middle Ordovician, and Lower and Middle Devonian Strata

Fluid homogenization, vitrinite reflectance, and stable-isotope data on calcite cement and vein filling suggest that northern Appalachian limestones were subjected to high paleotemperatures, implying former burial and subsequent uplift of 4.5 to 5 km depending on location and stratigraphic position (Friedman 1987 a,b).

6 Relevance of Epeirogeny

The information just presented implies that strata exposed at the surface in undeformed areas have been heated to temperatures that suggest a great depth of burial. Large amounts of uplift and erosion have reexposed these formerly deeply buried rocks. This indicates that vertical movements have stripped off thick sections of strata, a concept termed epeirogeny (Gilbert 1890). In the past, epeirogeny was commonly explained in terms of glacial rebound. Current interest in episodes of crustal epeirogeny relates to vertical movements of the crust in continental platform regimes and intracratonic basins.

7 Timing of Epeirogeny

Friedman and Sanders (1982) suggested a paleoburial depth of 6.5 km for anthracite-containing Devonian fluvial rocks, consistent with the idea that the thick (~6.4 km) Carboniferous strata of northeastern Pennsylvania extended northeast far enough to bury the Catskill. Johnsson (1986) likewise implied burial by Carboniferous strata. Hearn et al. (1987), using $^{40}A/^{39}A$

analyses for the authigenic feldspar in Cambro-Ordovician carbonate rocks of the central and southern Appalachians obtained Late Carboniferous-Early Permian ages (278–322 Ma). For the northern Appalachian Basin of this study the authigenic alkali feldspar in flat-lying lower Ordovician dolostone strata (Fig. 2) yielded an age of uplift of approximately 320 Ma (Matt Heizler, analyst), likewise Carboniferous in age. The observation that authigenic feldspar in Cambro-Ordovician carbonates occurs along the proto-Atlantic shelf from the Appalachian Basin to Newfoundland, Scotland, and Greenland paralleling the Taconic belt suggests active crustal epeirogeny in Pennsylvanian-Permian time along this entire belt. Paleomagnetic studies have documented that the magnetite carries a well-defined magnetization of Pennsylvanian-Permian age. A two-stage model suggests deep burial and elevated temperatures followed by partial uplift and in-situ growth of diagenetic magnetite (Jackson et al. 1988).

Thus, following subsidence to great depth, Pennsylvanian to Permian epeirogeny uplifted the strata, resulting in deep erosion. This leads to the surprising conclusion that isostatic unroofing following uplift has stripped off thick sections of strata whose presence was previously unsuspected.

Acknowledgements. Thanks are expressed to the Department of Energy which supported this study through Grant No. DE-FG02-84-ER 13322. I thank Vicky Harder of Texas A&M University for the fission-track analysis and Matt Heizler for the $^{40}Ar/^{39}Ar$ analysis on feldspar.

References

Braun M, Friedman GM (1969) Carbonate lithofacies and environments of the Tribes Hill Formation (Lower Ordovician) of the Mohawk Valley, New York J Sediment Petrol 39:113–135

Buyce MR, Friedman GM (1975) Significance of authigenic K-feldspar in Cambro-Ordovician carbonate rocks of the proto-Atlantic-shelf in North America. J Sediment Petrol 45:808–821

Craig H (1965) The measurement of oxygen isotope paleotemperatures. In: Tongiori E (ed) Stable isotopes in oceanographic studies and paleotemperatures. Pisa Cons Nazl Rich, Lab Geol Nucl, pp 161–182

Duddy IR, Gleadow AJW, Green PF, Hegarty KA, Marhallsea SA, Tingate PR, Lovering JF (1987) Quantitative estimates of thermal history and maturation using AFTA (apatite fission track analysis) in extensional and foreland basins–selected case studies: Am Assoc Petrol Geol Bull 71:550–551

Friedman GM (1987a) Deep-burial diagenesis: its implications for vertical movements of the crust, uplift of the lithosphere and isostatic unroofing- a review. Sediment Geol 50:67–94

Friedman GM (1987b) Vertical movements of the crust: case histories from the northern Appalachian Basin. Geol 15:1130–1133

Friedman GM, Braun M (1975) Shoaling and tidal deposits that accumulated marginal to the proto-Atlantic Ocean: the Tribes Hill Formation (Lower Ordovician) of the Mohawk Valley, New York. In: Ginsburg RN (ed) Tidal deposits, a casebook of Recent examples and fossil counterparts. Springer, Berlin Heidelberg New York, pp 307–314 (428 pp)

Friedman GM, Sanders JE (1978) Principles of sedimentology John Wiley & Sons New York, 792 pp

Friedman GM, Sanders JE (1982) Time-temperature significance of Devonian anthracite implies former great (~ 6.5 km) depth of burial of Catskill Mountains, New York. Geology 10:93–96

Friedman GM and Sanders JE (1983) Reply to comment on "time-temperature-burial significance of Devonian anthracite implies former great (~ 6.5 km) depth of burial of Catskill Mountains, New York". Geology 11:123–124

Fritz P, Smith DGW (1970) The isotopic composition of secondary dolomite. Geochim Cosmochim Acta 34:1161–1173

Gale PE, Siever R (1986) Diagenesis of Middle to Upper Devonian Catskill facies sandstones in southeastern New York. Am. Assoc Petrol Geol Bull 70:592–593

Gerlach JB, Cercone KR (1986) Post-Devonian depositional history of northern Appalachian Basin based on regional vitrinite reflectance trends. Geol Soc Am Abstr Progr 18:612

Gilbert GK (1890) Lake Bonneville. US Geol Sur Monogr 1:438 pp

Gurney GG, Friedman GM (1987) Burial history of the Cherry Valley carbonate sequence, Cherry Valley, New York. Northeastern Geology 9:1–11

Harris AG (1979) Conodont color alteration: an organo-mineral metamorphism index and its application to Appalachian Basin geology. Soc Econ Paleontol Mineral Spec Publ 26:3–16

Harris AG, Harris LD, Eppstein JB (1978) Oil and gas data from Paleozoic rocks in Appalachian Basin: maps for assessing hydrocarbon potential and thermal maturity (conodont color alteration isograds and overburden isopachs). U S Geol Surv Misc Invest Map I-917-E, Scale 1:2 500 000, 4 Sheets

Hearn PP Jr, Sutter JF, Belkin HE (1987) Evidence for Late-Paleozoic brine migration in Cambrian carbonate rocks of the central and southern Appalachians: implications for Mississippi Valley-type sulfide mineralization Geochim Cosmochim Acta 51: 1323–1334

Hood A, Gutjahr CCM, Heacock RL (1975) Organic metamorphism and the generation of petroleum. Am Assoc Petrol Geol 59:986–996

Jackson M, McCabe C, Ballard MM, van der Voo R (1988) Magnetite authigenesis and diagenetic paleotemperatures across the northern Appalachian Basin. Geology 16:592–595

Johnson KG, Friedman GM (1969) The Tully correlatives (Upper Devonian) of New York State: a model for recognition of alluvial, dune (?), tidal nearshore (bar and lagoon), and offshore sedimentary environments in a tectonic delta complex. J Sediment Petrol 39:451–485

Johnsson MJ (1986) Distribution of maximum burial temperatures across the northern Appalachian Basin and implications for Carboniferous sedimentation patterns. Geology 14:384–387

Kinsland GL (1977) Formation temperature of fluorite in the Lockport Dolomite in upper New York State as indicated by fluid inclusion studies – with a discussion of heat sources. Econ Geol 72:849–854

Lakatos S, and Miller DS (1983) Fission track analysis of apatite and zircon defines a burial depth of 4 to 7 km for lowermost Upper Devonian Catskill Mountains. Geology 11:103–104

Levine JR (1983) Comment on "Time-temperature-burial significance of Devonian anthracite implies former great (~ 6.5 km) depth of burial of Catskill Mountains, New York". Geology 11:122–123

Levine JR (1986) Deep burial of coal-bearing strata, anthracite region, Pennsylvania: sedimentation or tectonics. Geology 14:577–580

Sanders JE (1983) Reinterpretation of the subsurface structure of the Middle town gas well (Crom-Wells, nc. 1 Fee) in light of concept of large-scale bedding thrusts. In: Friedman GM (ed) Petroleum geology and energy resources of the Northeastern United States, Troy, N Y, United States, Oct. 14, 1983. Northeastern Geology 5:3–4:172–180

Tissot BP and Welte DH (1978) Petroleum formation and occurrence. A new approach to oil and gas exploration. Springer, Berlin Heidelberg New York, 538 pp

Urschel S, Friedman GM (1984) Paleodepth of burial of Lower Ordovician Beekmantown Group carbonates in New York State. Compass of Sigma Gamma Epsilon 61:4, 205–215

Sediments on Volcanic Islands –
On the Importance of the Exception

PETER ROTHE[1]

CONTENTS

Abstract	29
1 Introduction	30
2 Mesozoic Sedimentary Rocks	33
3 Neogene to Quaternary Carbonate Sands and Carbonate Rocks	35
3.1 Age and Composition	36
3.2 Diagenesis	37
3.3 Marine Terraces and Associated Eolianites	40
3.4 Caliche (Calcrete)	43
References	44

Abstract

Sediments of the Middle Atlantic Islands are discussed along with their position relative to volcanic events. They can be subdivided into prevolcanic, synvolcanic, and postvolcanic sediments.

Small-scale outcrops of prevolcanic sediments occur on some of the Cape Verde Islands and the Canary Islands. They comprise deep-water suites which must have been uplifted to their present position by volcano-tectonic processes. Their rocks include sandstones, siltstones, claystones, marlstones, limestones, and chert of Upper Jurassic to probably Eocene age on the Cape Verde Islands, and Lower Cretaceous to Oligocene on the Canary Islands, the latter already contain submarine volcanics. Most of the rocks are turbidites which originated at the adjacent African continent.

The synvolcanic sediments are essentially biocalcarenites which occur on every group of the Middle Atlantic Islands. They were formed after the islands had reached a position close to sea-level during the Neogene; most of them seem to belong to the Upper Miocene and Pliocene. They occur within or above lava streams and pyroclastic deposits and were once the only means for dating the island's volcanism. Some of them still expose the original mineralogy of the biogenic components, i. e. aragonite and high-magnesium

---

Universität Mannheim, Geographisches Institut, Abteilung Geologie, D-6800 Mannheim 1, FRG

calcite. Their diagenesis depends on the geologic situation. Meteoric cementation with drusy mosaic calcite and meniscus cements occurs where carbonate sands had been piled up to form dunes. Submarine and beach-rock cements are mainly micritic or even-rim, high-magnesium calcite. Recent dolomitization of some occurrences seems to follow the seepage-refluxion model but an influence from overlying basaltic lava streams is also possible, as well as a mixing of fresh- and seawater. Regressive diagenesis occurs in such carbonates which came again in contact with seawater after having been subject to meteoric influence.

Much of the carbonates belong to marine terraces; their levels cannot be taken as a means of determining their age since eustatic sea-level changes interfere with the vertical ups and downs of the islands. The processes of carbonate formation and diagenesis seem to continue today, including the formation of caliche. Detailed studies are lacking for almost all Middle Atlantic Islands except for some of the Canary Islands and Porto Santo.

1 Introduction

Geologists visiting volcanic islands usually go there to study basalts and the phenomena of volcanism. Sediments never received much attention, although they were of prime importance prior to physical age dating: Sediments were, at that time, the only means to date the volcanic rocks, provided they contained index fossils. More recently, and particularly after the period of intensive research on carbonates which began in the late 1950s, it proved worthwhile to draw more attention to such sediments. German Müller was the first who studied carbonates from the Canary Islands with the modern tools of staining and X-ray diffraction. He first published a study on high-magnesium calcite cements producing beachrock on the Canary Islands (Müller 1964), and two subsequent dissertations were started: Jens Müller began to discover more about the origin and composition of the carbonate sediments which seem to have formed recently on some of the Canarian shelf areas. The work included scuba diving information and received much support from discussions with John Milliman, who came as a visiting scientist, and Robin Bathurst and Gerald Friedman, who joined us as visiting professors during their sabbatical years. G. Tietz tried to unravel the diagenesis of biocalcarenites on Fuerteventura: from carbonate sands to carbonate rocks. During supervision of some of the field-work, I was lucky to detect a whole suite of much older sedimentary rocks which proved to be Mesozoic in age.

It was during this time when we first met in a shabby Fuerteventura Hotel which we then termed "Hilton Rosario", and subsequently, the Canary Islands became one of the institute's prime interest areas.

I shall try to summarize some of the aspects of these studies, including results from the other Middle Atlantic Islands. Obviously, the Canary Islands represent at present the archipelago which has been studied in more detail than any of the others, which will include the Cape Verde Islands, the Madeira Group with Porto Santo, the Selvagens and the Azores. The discussion is faced with the problem that the state of research of the individual island groups is very different. Mitchell-Thomé (1974), who wrote a summary paper on the sedimentary rocks of Makaronesia (another name for the Middle Atlantic Islands, from the Greek *makáron neseu*, which means islands of the fortunate), has already stated that we are in most cases merely in the stage of reconnaissance. His paper deals mainly with the outcrop situations and stratigraphy; this chapter will discuss aspects of mineral composition and diagenesis. Nevertheless, I have taken his table on stratigraphy, which is adjusted according to new results (Table 1).

Sediments are found in three different geologic settings. Below, within, and on top of volcanic sequences. They may thus be differentiated into prevolcanic, synvolcanic, and postvolcanic occurrences; the latter term should not be taken too seriously since volcanism on the Middle Atlantic Islands cannot be regarded as extinct.

Prevolcanic sediments seem to be restricted to single islands of the Cape Verde's and the Canaries. They belong to the Mesozoic and will be treated first, according to the stratigraphic thinking of geologists.

Syn- (or intra-)volcanic sediments once offered the only means to age dating of the volcanic events and even to learn about the age of the islands. Since outcrops of biostratigraphically datable sediments intercalated within volcanics are rare, the time scale of volcanic events was not well established. Until recently, the age of these islands remained open to speculation, but according to some fossil-bearing sediments, it was already clear during the past century that much of the volcanism occurred during the Miocene; this general picture was not drastically changed by applying paleomagnetic and K/Ar-dating during the late 1960s (Abdel-Monem et al. 1967, 1971, 1972). Prior to this, however, the ages of the island's rocks were open to speculation, the extremes ranging even back to the Precambrian. It now seems clear that most of the volcanic island's parts above sea-level were formed over the past 20 million years. The sediments associated with the corresponding volcanic piles are essentially calcarenites which could have formed only after the volcanic edifices hat reached a position close to sea-level. Very often they are associated with basaltic and other volcanic conglomerates which indicates reworking in a high-energy environment.

The formation and diagenesis of such carbonates will be treated in more detail. This field was stimulated by German Müller when the Heidelberg "Laboratorium (now "Institut") für Sedimentforschung" was established under his guidance during the mid-1960s. His first study on cementation of such carbonate sands revealed that high-magnesium calcite was the main cement phase. Much attention was given to this carbonate phase during the early days of the institute. We learned that no low-magnesium calcite could be precipitated from sea-water because of its high Mg/Ca ionic ratio, and consequently Müller and Blaschke (1969a) published a short note about the

32 Peter Rothe

(Stratigraphic chart of volcanic activity in the Macaronesian archipelagos — Azores, Madeira Group, Selvagens, Canary Islands, and Cape Verde Islands — plotted against geological stages from Callovian to Upper Pliocene. Legend: ■ Recognized, • Probable or Possible, ○ Conjectural.)

white layers in the Black Sea sediments consisting of low-magnesium calcite: which was against the rules. And the rules proved to be correct, the white layers turned out to consist of nannoplankton which forms shells of low-magnesium calcite, which was also consequently detected from similar black shales, e.g. the famous Posidonia-Shale of the Lower Jurassic (Müller and Blaschke 1969b).

2 Mesozoic Sedimentary Rocks

There is no need to study quartzitic rocks on Atlantic Islands to obtain general information on the formation of quartzitic rocks. But once they have been found, they should be mentioned since their occurrence in an unusual geologic setting needs special attention for purposes other than the origin of sedimentary rocks.

Karl von Fritsch reported, in 1867, for the first time on sedimentary rocks from Fuerteventura. The rocks were mentioned as occurring among syenite, diorite, gabbro, diabase and similar greenstones of a basement area called "Mittelgebirge" which were attributed later as of "paleozoic appearance" by Gagel (1910) who evidently had seen v. Fritsch's collections:

"Unter den leider nur spärlich erhaltenen Resten seiner Sammlungen finden sich denn auch ganz zersetzte Diabasbreccien und grüne, stark zersetzte, diabasartige Gesteine mit sekundären Calcitausscheidungen, die mit den alten Grundgebirgsgesteinen aus dem Gran Barranco von La Palma auf das beste übereinstimmen, daneben dunkle, paläozoisch aussehende, sehr dünnschiefrige Tonschiefer ohne Fossilien, sehr harte, schwarze, anscheinend kontaktmetamorphe Tonschiefer und marine, bräunliche, krystalline Kalke, die sehr schlecht erhaltene Fossilien, darunter unverkennbare Austernschalbruchstücke und Echinodermentafeln enthalten; nach dem Etikett liegen sie bei La Pena über Diabas und unter Basalt."(Gagel 1910, p. 16).

Strange enough, almost 30 years later, the findings were completely denied.

J. Bourcart and Jeremine (1938, p. 60) stated *ex cathedra*:

"Il faut enfin éliminer de cette description géologique toutes les roches sédimentaires qui ont été citées par erreur, notamment les schistes argileux et les calcaires qui seraient, d'après FRITSCH, intercalés dans les syénites. Les

---

Table 1. Mesozoic and Cainozoic sediments on the Middle Atlantic Islands (After Mitchell-Thomé 1974). The Mesozoic stratigraphy of Maio/Cape Verde Islands and Fuerteventura/Canary Islands was adjusted according to new data from Robertson and Bernoulli (1982). The general stratigraphy was also modified according to the Geological Time Table (Eysinga) which includes the Cretaceous/Tertiary boundary, and boundaries within the Tertiary. For comparison with the literature, however, the Vindobonian was left as it stands. The Pliocene on Tenerife corresponds to the pyroclastics close to Adeje where Burchard and Ahl (1927) found giant landturtles.

calcaires de type liasique, les "marbres", etc. appartiennent tous à la série des travertins quaternaires."

Bourcart was an authority and hence there were no more Mesozoic sediments. Whenever the history of geology is written, it is surprising how advancement has been influenced by strong personalities (Stille might be a good example as well).

When I first published the rediscovery of these sediments (Rothe 1967a, 1968a) I was unaware of the mapping results of my Spanish colleagues who had detected them already a few years earlier (Fuster Casas and Aguilar 1965).

It proved to be a whole suite of different lithologies including limestones, marlstones, shales, sandstones, siltstones and carbonates with chertlenses (Rothe 1968a) which are folded and partly overturned. Duplications could not be ruled out but a minimum thickness of about 1500 m seemed obvious.

The uncertainty came from the fact that most of the sediments are dissected by greenstone dikes which have also cut other basement rocks in Fuerteventura (e.g. gabbros). The greenstones of von Fritsch and Gagel, later on regarded as tilted lava streams by Hausen (1958), proved to be an immense dike swarm of Tertiary age.

The finding of these old sediments raised a general interest of other geologists and subsequently more detailed studies were carried out (Robertson and Stillman 1979; Robertson and Bernoulli 1982). As for the depositional environment, the sediments were regarded as flysch (Rothe 1968a), and this interpretation was further substantiated by comparison with the adjacent Deep Sea Drilling Sites (Robertson and Bernoulli 1982).

Without going into further details here, it should be mentioned, however, that Bernoulli found Early Cretaceous ammonites in the Fuerteventura sediments. Other fossils, particularly foraminifera, suggest an Albian to Early Cenomanian age for the lower part of the series, which is also comparable to sediments cored at DSDP Sites 370/416 and 415 drilled NE of the Canaries. It may generally be stated that the Fuerteventura flysch probably represents the distal part of a deep-water fan complex; the sediments originated from the adjacent African continent – ocean boundary area.

Higher up the stratigraphic section there are also fossiliferous carbonates, sandstones and siltstones and volcaniclastics of Oligocene age but no continuous section can be established because of the density of the dike swarm (Robertson and Bernoulli 1982).

The Mesozoic sediments on Fuerteventura seem to continue to the north and west. Quartzitic xenoliths were found within Miocene basalts on southern Lanzarote (Rothe 1967b) and fine-grained quartzites, carbonates, shales and chert occur on Gomera where they form small remnants of host rocks within a similar dike swarm (Cendrero 1967, 1971).

Mesozoic sediments of a deep-water environment are also known from the Cape Verde Archipelago where a Neocomian age was attributed to limestones with aptychi on Maio by Stahlecker (1934). More recent data on biostratigraphy by various authors (summary see Robertson and Bernoulli 1982) suggest a possible Oxfordian age for the lowermost part of the section. The whole sequence ranges up-section into the Upper Cretaceous but recent work has shown that tectonic repetitions are present and a revision of the stratigraphy is required. The sedimentary rocks comprise shales, marls, pelagic limestones associated with chert, but also graywackes and volcaniclastics; the total thickness is about 400 m. The sequence can be compared, to a certain extent, with adjacent DSDP Sites 367 and 368 (Lancelot et al. 1977). As on Fuerteventura, the Maio Mesozoic sediments are cut by dikes, sills and plutonics of Early to Middle Miocene age (Grunau et al. 1975), thus a similar geologic setting can be observed on two Middle Atlantic islands, each of which is situated at the easternmost edge of their respective archipelago. This has been stated by Rothe (1968a) and, more detailed, by Robertson and Bernoulli (1982). However, on Maio the oldest pillowed tholeiitic basalts suggest a mid-ocean ridge stage during the early phase which cannot be stated for Fuerteventura. The geometry of the Maio sedimentary rocks clearly points to a more central volcano-tectonic uplift (Serralheiro 1970), whereas the Fuerteventura sequence was obviously folded and partly overturned prior to the island's main volcanism. The structure mechanics are still not sufficiently explained.

My own work on the Fuerteventura sediments was carried out during the period when I was an assistent at Prof. German Müller's institute. This was regional geology rather than sedimentology and it demonstrates his open-mindedness to any interesting problem, provided it had only marginally to do with sediments.

As for a detailed discussion on the stratigraphy of the pre-Tertiary sediments on the Cape Verde Archipelago, the reader is referred to Mitchell-Thomé (1964, 1974, 1976), Serralheiro (1970), and Robertson and Bernoulli (1982).

3 Neogene to Quaternary Carbonate Sands and Carbonate Rocks

The second, and main part, deals with Neogene to Quaternary carbonates most of which can be summarized by Grabaus terminus calcarenites, or, to be more specific, biocalcarenites. They are the most abundant sediment types on many of the Middle Atlantic Islands, and they expose almost all lithification stages from unconsolidated carbonate sands to completely cemented, limestones or dolostones. The origin of the carbonates is almost entirely biogenic. Molluscs, algae, echinoderms, bryozoa, corals, foraminifera, serpulids and others contribute to the sands which are widespread on the island shelves. The situation of the Middle Atlantic Islands in subtropical and tropical climatic regimes favors the abundant carbonate production.

The discussion presented is faced with the problem that most of these carbonates have not yet been studied to a similar extent on all the islands mentioned. Regarding present-day standards most of them need reexamination.

3.1 Age and Composition

Some of the more important occurrences were known already long ago:
1. The limestones from Santa Maria/Azores, to which a Miocene, sometimes Helvetian or "Vindobonian" age is attributed (Hartung 1860; Krejci-Graf 1961a);
2. The limestones of the Ribeira S. Vicente area on Madeira and similar occurrences on Porto Santo. Again, they are regarded as Helvetian or Vindobonian in the literature (Hartung 1864; Krejci-Graf 1955);
3. The "Las Palmas terrace" sediments in NE Gran Canaria which were already mentioned by L. von Buch (1825), Ch. Lyell (1865) and many others. They were regarded as Miocene by Rothpletz and Simonelli (1890).

Furthermore, many outcrops of similar sediments are known from the Cape Verde Islands and the Selvagens group (see Figueira 1964; Mitchell-Thomé 1974, 1976).

All consist essentially of biogenic carbonates. For many, a Vindobonian age is attributed to them in the literature but recent studies, e. g. on the Las Palmas terrace sediments of Gran Canaria have shown that the older data must be treated with caution; some of them are based only on lithological comparison or unsuitable fossils. This holds true even for the Miocene dating of the Las Palmas terrace using *Clypeaster altus* Lk. by Rothpletz and Simonelli (1890), who also mentioned this sea urchin from Pliocene strata (for discussion see Rothe 1966). The more detailed study of Lietz and Schmincke (1975) on these sediments, now termed the "Las Palmas Formation", and a comparable "Arguineguin Formation" in southern Gran Canaria (Schmincke 1976) used K/Ar-dating of intercalated lava streams. At least three distinct marine levels of different ages were recognized, and only the oldest of them may correspond to the Helvetian or Tortonian of the original description by Rothpletz and Simonelli (1890). Even detailed paleontological studies of the Santa Maria/Azores limestones present problems of dating. Their stratigraphic connotation to the Helvetian/Tortonian, based on macrofauna studies of many authors (for summary see Krejci-Graf et al. 1958), has been a controversial subject since Colom (in Krejci-Graf et al. 1958) determined foraminifera, which seems in favor of an Upper Miocene or even Pliocene age.

Since some of the calcarenites must be regarded as lithified fossil dunes, the problem of dating by using microfossils was apparent:

After Colom (in Hausen 1958) had determined Miocene foraminifera in such carbonates overlain by or intercalated between basalts on the west coast of Fuerteventura, the question of the sediments age was raised by Hausen (1958).

Hausen regarded the carbonates as Quaternary, and the fossils reworked. The more recent studies of Fuster et al. (1968) regarded the overlying basalts as Quaternary, but again paleontology is now in favor of an Upper Miocene a8ge (Meco and Stearns 1981). This shows that there is a general need to reexamine the ages of the so-called Miocene sediments on all of the Middle Atlantic Islands.

A common factor is that they are associated with volcanics. In most cases basaltic components are mixed with the carbonates, or the carbonates are underlain and/or overlain by basaltic lava streams or pyroclasts. This may have an influence on diagenesis of the carbonates either directly by heating from a lava spreading on a carbonate surface (which may even reach the field of metamorphosis), or the weathering of basalts may yield fluids of unusual composition, e.g. Mg-enriched solutions which may be the reason for dolomitization. The other source of Mg is sea-water. The distance of the carbonates to the coast (in fact, most of the carbonates occur at, or close to, the coast) may play an important role in dolomitization, which is presently in the stage of a reevaluation on the Canary Island of Fuerteventura (Meder, in preparation).

Fragments or entire carbonate shells seem to be the most important components of the mid-Atlantic Neogene carbonates. Detailed studies, allowing a comparison of all the islands groups mentioned, are still lacking. At present, only the sediments on Fuerteventura and Lanzarote of the Canaries and Porto Santo of the Madeira group have been studied in more detail.

According to Mitchell-Thomé (1974, Table 2), most of the sediments on the Middle Atlantic Islands are of Neogene age. This stratigraphic connotation is based mainly on older literature. A further subdivision into various stages seems at present very uncertain and modern studies are urgently needed.

There have been only a few publications in the past 15 years which can be regarded as valuable contributions to the problem of dating of these sediments.

The main problem of dating was, and is, probably the lack of index fossils within such sediments which essentially represent shallow water or even coastal regimes. Attempts to find suitable nannofossils were unsuccessful. Although a costly method of obtaining reliable results physical age dating is probably most promising in all cases where the sediments are overlain by volcanics. This could be the prospectus of an interesting new research program.

3.2 Diagenesis

Formation and deposition of the carbonate sediments in the shelf areas surrounding the islands is also subject to sea-level changes. Eustatic phases, transgressive or regressive, cannot be reliably determined on volcanic islands since they interfere with vertical uplift and/or subsidence (Krejci-Graf 1961b). This means that the sediments can be influenced by both submarine and subaerial diagenesis.

Diagenesis was one of the main areas of interest of German Müller's group when studying carbonates on Fuerteventura and Lanzarote.

Starting from an inventory of the unconsolidated carbonate sediments of the islands shelves (J. Müller 1969), various stages of diagenesis could be recognized. The original sediments consist essentially of biogenic high-Mg calcite and aragonite, and some low-Mg calcite. High-Mg calcite can mainly be referred to lithothamnium-type coralline algae which occur as isolated nodule-shaped balls and crusts, bryozoa, echinoderms and milliolid-type benthic foraminifera. Aragonite stems from corals, gastropods and pelecypods and low-Mg calcite can be referred to both foraminifera and ostracods (J. Müller 1969).

Similar carbonate mineralogies for the Cape Verde Islands can be assumed from findings of aragonite, high-Mg calcite, calcite and scarce dolomite within reworked shallow-water carbonates in sediment cores from the adjacent deep-sea areas. Graded bedding of such sediments and their association with volcanic components makes an origin from the island shelves and transportation by turbidity currents likely (Rothe 1973).

Cementation includes three carbonate phases; high-Mg calcite and aragonite occur in beach rocks (Müller 1964; Tietz and Müller 1971), and low-Mg calcite in eolianites where the former shelf and beach sands have been piled up to form carbonate dunes in near-coastal areas, indicating meteoric diagenesis.

In addition to this mineralogical aspect, various types of morphologically different cements were found. High-Mg calcites occur as micritic coatings or equant crusts on fossils and volcanic fragments, while aragonite forms fibrous crusts (Tietz and Müller 1971).

In addition, calcitic drusy mosaic, dog-tooth, and meniscus cements were described (Tietz 1969). Both morphology and mineralogical composition of cements allowed reconstructions of the diagenetic environments, where marine, mixed marine-freshwater, and carbonates influenced by fresh-water could be distinguished. Of particular interest was the recognition of regressive diagenesis in Pleistocene eolianites from Fuerteventura: The original high-Mg calcite of clasts of red algae was replaced by low-Mg calcite in a subaerial (= influenced by fresh-water) environment, and during a later stage the cell walls of the red algae were reconstituted to become high-Mg calcite again, in cases where the rocks were exposed to sea-water (Müller and Tietz 1975). This is evidently also a function of carbonate ultrastructures, extremely small crystal size, and hence the large specific surface of the red algae.

Micritic envelopes, particularly affecting mollusc shell fragments and foraminifera were observed; the original carbonate of the biogenic particles was dissolved during fresh-water diagenesis but the more stable cements survived and very often such micritic linings are the only remnants reminiscent of the former fossils (Tietz 1969).

For the geologist, the diagenetic history of the carbonates offers clues to the ups and downs of the islands, combined with the rise and fall of sea-level.

Presently, however, we are far from any general picture since no attempt has been made toward a more systematic approach; the few examples have not gone beyond the stage of case studies, which have only intensified the interest.

One of the most interesting results, however, is the occurrence of dolomitic rocks on the Canary Islands. Recent dolomitization was observed within Quaternary biocalcarenites of Fuerteventura (Müller and Tietz 1966) which seems to have a Miocene counterpart in Lanzarote (Rothe 1968b). The recent occurrences at the windward, high-energy coast in southern Fuerteventura can be explained by the seepage-refluxion model of Adams and Rhodes (1960) but contrary to similar dolomites in the Florida and Bahama areas, the Canary dolomitization affected earlier consolidated rocks (Müller and Tietz 1966).

Dolomitization of Miocene calcarenites overlain by several 100 m of subaerial basaltic lava streams and pyroclasts occurs on Lanzarote (Rothe 1968b). The dolomites occur within biocalcarenites which represent a former land surface. These marine-derived carbonate sands consist essentially of foraminifera and mollusc debris but also contain land snails and fossil ostrich eggs (Rothe 1964, 1966; Sauer and Rothe 1972). The carbonates are dolomitized from their former horizontal surface down to about 3–4 m; no dolomite occurs below this depth. Dolomitization can either be explained by downward-percolating brines prior to eruption of the overlying basalts (thus according to the seepage-refluxion model) or possibly by the weathering solutions percolating through the basalts which release the necessary magnesium, as proposed by Rothe (1968b); if this is true, the dolomitization continues even today.

At the time of our studies, during the 1960s, the eastern Canary Islands offered a picture of essentially arid conditions, which seems to have changed over the past few years. Fuerteventura has had episodically high rainfall, and fresh-water percolates through the carbonates in coastal areas, where a mixture with sea-water occurs, thus producing conditions favorable to the Dorag-Model of dolomitization (Badiozamani 1973). We are looking forward to K. Meder's actual studies on dolomitization there, which includes the sampling and geochemical studies of spray. Of particular interest might be a reexamination of all these dolomites by applying the methods of Lasemi et al. (1989) who found evidence of a cement origin of dolomite in the classical environments of Andros Island/Bahamas, thus supporting earlier findings of Gebelein et al. (1980).

Dolomite crystals forming both inter- and intraparticle cements of the even-rim type, but also completely filling the pores between remnants of red algae, were described by Tietz in his dissertation (Tietz 1969). At this time, we regarded the dolomite as a diagenetic product replacing former calcite, particularly high-Mg calcite precursors. Direct precipitation, as now reported by Lasemi et al. (1989), was discarded as a possible mechanism for the formation. It might be worthwhile to reexamine some of the Fuerteventura dolomites under this aspect.

A different mode of dolomitization seems to exist in combination with volcanic heat.

Although not yet adequately studied, calcarenites immediately below basaltic lava streams in Barranco de los Molinos (western-coastal area of Fuerteventura) are completely dolomitized; some of the cavities contain zeolites [phillipsite (?) rosettes] (Tietz 1969). In the uppermost part, we found, years ago, thin-shelled eggs of landturtle (Klemmer and Rothe, in press) which consist of completely recrystallized calcite. Similar findings of Hutterer et al. (in prep.) from this outcrop include a whole nest which indicates that this area was a favorable breeding area before being covered by lava. The carbonate sand must have been unlithified prior to its dolomitization. Thermal metamorphosis of organic compounds of the eggshells seems possible since many of them are dark gray, and are malodorous during treatment for thin sections. A similar interpretation is possible for the Tertiary calcarenites of northern Lanzarote: The ratite eggs, including two types of different pore patterns (Sauer and Rothe 1972), suggest loose carbonate sands similar to the present lowland area of the small island of Graciosa where abundant land snails are scattered throughout. A thermal event may have caused the dark color of the eggshells but not all of them express this phenomenon. The smell is similar to the Fuerteventura turtleshells.

3.3 Marine Terraces and Associated Eolianites

The carbonate rocks on the Middle Atlantic Islands include elevated and drowned marine terraces and in some cases lithified carbonate sand dunes (eolianites) are associated with them. It is possible to discuss their formation by studying the actual conditions at the southern coastal areas of Fuerteventura and Gran Canaria.

Carbonate shells are essentially formed in the shallow shelf areas surrounding parts of the islands, ranging from the intertidal regime to the shelf edge at more than 130 m water depth. Individual groups of carbonate builders seem to follow a belt-like pattern. (J. Müller 1969 for the eastern Canaries, but almost no data from any of the other islands). The shells are reworked and sorted and at the coastal areas they are mixed with the residues of the weathered volcanic rocks.

Preferably during low tide, the sands are blown toward the land areas. They form dune fields as in the famous vacation area around Maspalomas in southern Gran Canaria (where such dunes can reach 10 m in height), or the carbonate sand is simply blown uphill. In southern Fuerteventura, such sands form a thin veil covering the basaltic cordillera there.

A similar situation must have existed there during a higher sea-level stand. In the Morro Jable area of the Jandia Peninsula, such dunes "hang" on the hillside, thinning upwards, and overlie a fossil beach with pebbles and thick-shelled macrofossils (see also Tietz 1969).

Although the formation of such dunes can presently be observed, it must have been much more obvious during lower sea-level when a much broader shelf area was exposed. This was reported by Hausen (1958) for Fuerteventura and later by Lietz and Schwarzbach (1971) for Porto Santo. Torres and Soares (1946) reported such dunes also from the Cape Verde Islands.

Dune formation occurred essentially during the cold periods of the Quaternary, as stated by Lietz and Schwarzbach (1971) for Porto Santo; the dunes were thus termed "regression dunes".

The dunes offer a special field to study carbonate diagenesis above the high tide level. Meteoric cementation by drusy mosaic low-magnesium calcite was found to be the main factor in lithification of Quaternary eolianites South of Puerto Rosario/Fuerteventura, as well as in many other similar occurrences (Tietz 1969).

Sea spray may be responsible for dolomitization phenomena within these eolianites. It may also promote the phenomenon of regressive diagenesis mentioned earlier, indicating a possible rise in sea-level or a lowering of this particular area. It might be interesting to study diagenetic features of carbonate rocks from the terraces presently below sea level.

Although we have only scarce ^{14}C-data on such eolianites, a rather recent age can be assumed. From Barranco de la Monja/Fuerteventura calcarenites were dated at 38–45000 years B.P. with a caliche on top of about 25000 years (Müller and Tietz 1975). From Barranco de los Molinos/Fuerteventura, calcarenites were dated at greater than 39000 years B.P. (Rona and Nalwalk 1970). The only data from other islands (to my present knowledge) are from Porto Santo, where eolianites have been dated at 21570 and 13480 years B.P. (Lietz and Schwarzbach 1971).

Elevated marine terraces have raised much interest even amongst volcanologists since they may be regarded as indicators of vertical ups (and downs!) of individual islands or even parts of islands. There is, however, a problem regarding such an interpretation: Along many of the continental margins, terraces are taken as indicators of eustatic sea-level changes. Particularly among geographers, their situation above sea-level is, at the same time taken as an indicator of their age. In the Mediterranean, a whole system of Upper Tertiary and Quaternary terraces was developed and many type localities are situated there, e.g. Monastir, Sicily and the Tyrrhenian Sea. Studies to find the corresponding stages on the Middle-Atlantic Islands are faced with the problem that on volcanic islands the individual ups and downs interfere with the eustatic sea-level changes. Krejci-Graf (1961b) has criticized the approaches of many researchers under this view. Besides some exceptions (discussed below), most of the marine carbonates and associated volcanic conglomerates slightly dip towards the sea, thus surrounding the islands in their outer parts. Erosion, cutting back such inclined sedimentary layers which are often interpreted as marine terraces, thus determines their height above sea-level: the farther erosion has progressed, the higher the corresponding "strandline".

Other indicators of former high sea-levels stands include caves which were interpreted as sea caves. Krejci-Graf has amply demonstrated that caves are quite common within volcanic rock piles (lava tubes, etc.). Unless there are marine conglomerates associated with such caves, they should not be mistaken as indicators of high sea-level stands.

Among the older literature on marine terraces of the Middle Atlantic Islands, Zeuner (1958) must be mentioned. He found marine terraces at 16.3, 7.7, and 4.25 m NN, respectively, on the Canaries associated with a fauna containing *Strombus*. The 16.3 and 7.7 m terraces were placed by him into the Monastirian, hence the last Interglacial, whereas the lowermost was placed in his Epimonastirian. A later, more detailed study dealing with geomorphology of the Canaries (Klug 1968) found seven groups of marine terraces on the eastern islands, which seemed more stable with regard to volcano-tectonic, vertical movements. Levels of 60–55, 35, 25, 18–25, 7–6, 4–3 and 2–1 m were measured; all were attributed to the Quaternary. The main reason for this stratigraphic connotation was their situation above sea-level, and the main problem remains their dating based on paleontology. The faunas contain abundant mollusc shells, echinoderms, red algae, sometimes bryozoa and corals, but also foraminifera (mainly benthic) which are all of low stratigraphic importance. The stratigraphic connotation was essentially based on a comparison with the system of different height levels established for the African coast and the Mediterranean. The uppermost level was regarded as Sicilian (or Milazzian) by Klug, who did not agree to a Miocene age of such a level as suggested earlier by Hausen (1958) for Fuerteventura. The 35 m and 25–28 m levels were regarded as Tyrrhenian I (Paleo-Tyrrhenian) since they are situated between the 60 m level and a rather constant level at 15 m, which was regarded as Tyrrhenian IIa (Eu-Tyrrhenian). The lower levels at 6–7 and 1–2 m were placed into the Holocene according to their recent faunal aspect (Staesche, cited in Klug 1968), the 6–7 m level is attributed to the Neotyrrhenian. Moreover, a marine platform between 0 and 1 m was recognized from the eastern coastal area of Lanzarote, which was regarded as Intertyrrhenian.

A more complicated situation seems to exist on Gran Canaria, which appears less stable with regard to volcano-tectonics.

Without going into more detail here, it should be mentioned that a more recent study obtained quite different results for the eastern Canary Islands. Meco and Stearns (1981) have presented reasonable arguments for an Early Pliocene age of the oldest terraces on Lanzarote, based on paleontology. The occurrence of the Pliocene at different levels (the "55 m stage" of Driscoll et al. 1965) indicates post-Pliocene tectonic displacements. According to Meco and Stearns, all remaining terraces belong to the Upper Pleistocene, thus representing events of the Last Interglacial.

Evidence of similar displacements of Miocene marine carbonates was reported from Porto Santo (Lietz and Schwarzbach 1970) where they occur at different heights; the highest outcrops were found 350 m above present sea-level.

Miocene and Pliocene high sea-level stands, an elevation of the island since the Miocene, or a combination of both, can be considered for Gran Canaria (Lietz and Schmincke 1975). However, little is known from the other Canary Islands, and data are lacking for most of the other Middle Atlantic Islands.

The main problem is the dating of the terrace material provided it contains fossils. But even fossils might not be an adequate tool since there is generally little change between species from the Miocene to the Quaternary in these areas. Possibly, but this is merely a brainstorming idea, one could try to isolate marine carbonate cements and study their oxygen isotopic composition or, where present, analyze the planktonic foraminifera to determine whether their $\delta^{18}O$-value fits the isotope stages already established.

3.4 Caliche (Calcrete)

Among the soil-forming processes on the Canary Islands, the formation of caliche has always raised major interest. Caliche was mentioned by the authors of the 19th century, and was sometimes confused with calcarenites or carbonate sands (the Spanish *"tosca blanca"* or *"canto blanco"* sometimes stands for both), and its formation is still a matter of debate. Caliche crusts occur on or within sedimentary rocks (mainly the calcarenites described above) and on surfaces of volcanics. Again, the most spectacular outcrops occur on the eastern Canary Islands, but similar occurrences seem to be developed on all of the Middle Atlantic Islands (except for the Azores), particulary the Madeira Group (Porto Santo, Hartung 1864; Krejci-Graf 1961c; Lietz and Schwarzbach 1971) and on the Cape Verde Islands (Torres and Soares 1946; Krejci-Graf 1960, 1961d; Serralheiro 1970).

The formation of caliche (calcrete) is generally discussed along with semi-arid climatic conditions. The most modern German textbook on sediments (Füchtbauer et al. 1988) summarizes that descending solutions are reponsible for the formation of most calcretes; the author does not mention, however, the widespread occurrences of caliche, where capillary Ca^{2+}-transport to the surface is obvious. For the Fuerteventura outcrops, lateral transport was assumed (Hempel 1978) or ascending capillary solutions. Since Yaalon and Ganor (1973) have discussed the contribution of wind-blown dust, which seems also likely for most occurrences of calcrete on the Middle Atlantic Islands, a reexamination including a quantitative approach, together with ^{14}C-dating, is needed. At present, we have little information even on the mineralogical composition of more than a few samples. On many of the islands, caliche occurs principally on two types of bedrock: carbonates (e.g. calcarenites) and volcanic rocks, essentially basalts. In the latter case, an origin is explained by the solution of Ca^{2+} from minerals of the bedrock (e.g. calcic plagioclase, pyroxene) and ascendent capillary transport to the surface with subsequent precipitation as $CaCO_3$ (Hartung 1857; Lyell 1865; Krejci-

Graf 1960, 1961d; Fuster et al. 1968). In fact, caliche is the reason for the whitish aspect of Quaternary volcanoes on Lanzarote and Fuerteventura, which can thus be distinguished from the recent vents and lavas of the 18th century (Hartung 1857; Hausen 1959).

The most abundant caliches, however, are decimeter thick, irregular, subparallel beds on top or beneath the surfaces of carbonate sands. Such limestones were quarried on Fuerteventura, burnt to mortar and shipped to the other islands for construction purposes. Several authors have reported that the caliche was formed under conditions different from the present climate. Hempel (1978) stated that the crusts are now being eroded where they are exposed at the surface. The more recent work of soil scientists (Stahr's group at the TU Berlin) has now presented the first good data on soil formation in Lanzarote. Their calculations, comparing the Ca and Mg concentrations in the volcanic bedrock and caliche, have shown that the amounts released from the bedrock, possibly with additional Ca and Mg from spray, are sufficient for the carbonates in soils. A rather continuous enrichment of carbonate has been assumed for the past 40 000 years. The work of Jahn makes evident that formation of caliche continues under actual climatic conditions (Jahn 1988), in contrast to Hempel's earlier statements.

Acknowledgements. Support from the Deutsche Forschungsgemeinschaft during my fieldwork on the Canary Islands is acknowledged. Further thanks are due to my secretary E. Sagebiel who patiently typed both manuscript and Table 1, and to M. Mitlehner for the drawing.

References

Abdel-Monem A, Watkins NF, Gast PW (1967) K-Ar Geochronology and paleomagnetic studies of the volcanism on the Canary Islands. Abstr Pap Submitted Annu Meet, New Orleans, Lou Geol Soc Am, p 1

Abdel-Monem A, Watkins ND, Gast PW (1971) Potassium-argon ages, volcanic stratigraphy, and geomagnetic polarity history of the Canary Islands: Lanzarote, Fuerteventura, Gran Canaria, and La Gomera. Am J Sci 271:490–512

Abdel-Monem A, Watkins ND, Gast PW (1972) Potassium-argon ages, volcanic stratigraphy, and geomagnetic polarity history of the Canary Islands: Tenerife, La Palma, and Hierro. Am J Sci 272:805–825

Adams JE, Rhodes ML (1960) Dolomitization by seepage refluxion. Bull Am Assoc Petrol Geol 44:1912–1920

Badiozamani K (1973) The Dorag dolomitization model-application to the Middle Ordovician of Wisconsin. J Sediment Petrol 43:965–984

Bourcart J, Jérémine E (1938) Fuerteventura. Bull Volcanol Ser II, 4:51–109

Buch L von (1825) Physikalische Beschreibung der Canarischen Inseln. Königl. Akad. Wissensch. Berlin, 201 S.

Burchard O, Ahl E (1927) Neue Funde von Riesen-Landschildkröten auf Teneriffa. Z Dtsch Geol Ges A Abh 79:439–447

Cendrero A (1967) Nota previa sobre la geología del complejo basal de la isla de La Gomera (Canaries). Estud Geol Inst Lucas Mallada 23:71–79

Cendrero A (1971) Estudio geológico y petrológico del complejo basal de la isla de La Gomera (Canaries). Estud Geol Inst Lucas Mallada 27:3–73
Driscoll EM, Hendry GO, Tinkler KJ (1965) The geology and geomorphology of Los Ajaches, Lanzarote. Geol J 4:2
Figueira AJG (1964) The Salvage Islands: some geographical, geological and historical notes. Bol Mus Municip Funchal 18:132–139
Fritsch K von (1867) Reisebilder von den Canarischen Inseln. Petermanns Geogr Mitt Ergänzungsh. 22:44 pp
Füchtbauer H (ed), Heling D, Müller G, Richter DK, Schmincke H-U, Schneider H-J, Valeton I, Walther HW, Wolf M (1988) Sedimente und Sedimentgesteine, 4th edn. Schweizerbart, Stuttgart, 1141 pp
Fuster Casas JM, Aguilar Tomas MJ (1965) Nota previa sobre la geología del macizo de Betancuria, Fuerteventura (Islas Canarias). Estud Geol Inst Lucas Mallada 21:181–197
Fuster JM, Cendrero A, Gastesi P, Ibarrola E, Lopez Ruiz J (1968) Geología y Volcanología de las Islas Canarias. Fuerteventura. Inst Lucas Mallada, Madrid, 239 pp
Gagel C (1910) Die mittelatlantischen Vulkaninseln. Handb Reg Geol 7, 10, 4:1–32
Gebelein, CD, Steinen RP, Garrett P, Hoffmann EJ, Queen JM, Plummer LN (1980) Subsurface dolomitization beneath the tidal flats of central-west Andros Island, Bahamas. SEPM Spec Publ 28:31–49
Grunau HR, Lehner P, Cleintuar MR, Allenbach P, Bakker G (1975) New radiometric ages and seismic data from Fuerteventura (Canary Islands), Maio (Cape Verde Islands) and São Tomé (Gulf of Guinea). Progr Geodyn 1975, pp 89–118
Hartung G (1857) Die geologischen Verhältnisse der Inseln Lanzarote und Fuertaventura. N Denkschr Allg Schweiz Ges Naturwiss 15/4:1–168
Hartung G (1860) Die Azoren in ihrer äußeren Erscheinung und nach ihrer geognostischen Natur. (Mit Beschreibung der fossilen Reste, von HG Bronn). Engelmann, Leipzig, 350 pp
Hartung G (1864) Geologische Beschreibung der Inseln Madeira und Porto Santo. Engelmann, Leipzig, 298 pp
Hausen H (1958) On the geology of Fuerteventura (Canary Islands). Soc Sci Fenn Comm Phys Math 22/1:211 pp
Hausen H (1959) On the geology of Lanzarote, Graciosa and the Isletas (Canarian Archipelago). Soc Sci Fenn Comm Phys Math 23, 4:116 pp
Hempel L (1978) Physiogeographische Studien auf der Insel Fuerteventura (Kanarische Inseln). Münstersche Geogr Arb 3:51–103
Jahn R (1988) Böden Lanzarotes. Vorkommen, Genese und Eigenschaften von Böden aus Vulkaniten im semiariden Klima Lanzarotes (Kanarische Inseln). Hohenheimer Arbeiten. Ulmer, Stuttgart, 257 pp
Klemmer K, Rothe P (in press) Fossile Landschildkröten-Eier aus pliozänen Kalkareniten von Fuerteventura (Kanarische Inseln, Spanien). Senck Leth 70
Klug H (1968) Morphologische Studien auf den Kanarischen Inseln. Beiträge zur Küstenentwicklung und Talbildung auf einem vulkanischen Archipel. Schr Geogr Inst Univ Kiel 24, 3:184 pp
Krejci-Graf K (1955) Vulkaninseln und Inselvulkane. Madeira. Nat Volk 85/2:40–51
Krejci-Graf K (1960) Krustenkalke. Zur Geologie der Makaronesen 4 Z Dtsch Geol Ges 112/1:36–61
Krejci-Graf K (1961a) Vulkaninseln und Inselvulkane. Santa Maria (Azoren). Nat Volk 91/10:351–358
Krejci-Graf K (1961b) Vertikal-Bewegungen der Makaronesen. Zur Geologie der Makaronesen 7. Geol Rundsch 51:73–122
Krejci-Graf K (1961c) Vulkaninseln und Inselvulkane. Porto Santo. Nat Volk 91/2:33–38
Krejci-Graf K (1961d) "Versteinerte Büsche" als klimabedingte Bildungen. N Jahrb Geol Paläontol 113/1:1–22

Krejci-Graf K, Frechen J, Wetzel W, Colom G (1958) Gesteine und Fossilien von den Azore Sencken. Leth 39, 5/6:347–351
Lancelot Y, Seibold E et al. (1977) Initial Reports of the Deep Sea Drilling Project 41. US Gov Print Office, Washington, DC
Lasemi Z, Boardman MR, Sandberg PHA (1989) Cement origin of supratidal dolomite, Andros Island, Bahamas. J sediment Petrol 59/2:249–257
Lietz J, Schmincke H-U (1975) Miocene-Pliocene sea-level changes and volcanic phases on Gran Canaria (Canary Islands) in the light of New K-Ar ages. Palaeogeogr Palaeoclimatol Palaeoecol 18:213–239
Lietz J, Schwarzbach M (1970) Neue Funde von marinem Tertiär auf der Atlantik-Insel Porto Santo (Madeira-Archipel). N Jahrb Geol Paläontol, Mh 5:270–282
Lietz J, Schwarzbach M (1971) Quartäre Sedimente auf der Atlantik-Insel Porto Santo (Madeira-Archipel) und ihre paläoklimatische Deutung. Eiszeitalter Gegenw 22:89–109
Lyell Ch (1865) Elements of geology, 6th edn. Murray, London, 794 pp
Meco J, Stearns CE (1981) Emergent littoral deposits in the eastern Canary Islands. Quat Res 15:199–208
Mitchell-Thomé RC (1964) The sediments of the Cape Verde Archipelago. Publ Serv Geol Luxembourg 14:229–251
Mitchell-Thomé RC (1974) The sedimentary rocks of Macaronesia. Geol Rundsch 63/3:1179–1216
Mitchell-Thomé RC (1976) Geology of the Middle Atlantic Islands. In: Bender F, Jacobshagen V, de Jong J D, Lüttig G (eds) Beiträge zur regionalen Geologie der Erde, vol 12. Bornträger, Berlin, 382 pp
Müller J (1969) Mineralogisch-sedimentpetrographische Untersuchungen an Karbonatsedimenten aus dem Schelfbereich um Fuerteventura und Lanzarote (Kanarische Inseln). Diss, Ruprecht-Karl-Univ, Heidelberg, 99 pp
Müller G (1964) Frühdiagenetische allochthone Zementation mariner Küsten-Sande durch evaporitische Calcit-Ausscheidung im Gebiet der Kanarischen Inseln. Beitr Mineral Petrogr 10:125–131
Müller G, Blaschke R (1969a) Zur Entstehung des Tiefsee-Kalkschlammes im Schwarzen Meer. Naturwissenschaften 11:561–562
Müller G, Blaschke R (1969b) Zur Entstehung des Posidonienschiefers (Lias ε). Naturwissenschaften 12:635
Müller G, Tietz G (1966) Recent dolomitization of Quaternary biocalcarenites from Fuerteventura (Canary Islands). Contrib Mineral Petrol 13:89–96
Müller G, Tietz G (1971) Dolomite replacing "cement A" in biocalcarenites from Fuerteventura, Canary Islands, Spain. In: Bricker O P (ed) Carbonate cements. Johns Hopkins Univ Stud Geol 19:327–329
Müller G, Tietz G (1975) Regressive diagenesis in Pleistocene eolianites from Fuerteventura, Canary Islands. Sedimentology 22:485–496
Navarro JM, Aparicio A, Garcia Cacho L (1969) Estudio de los depósitos sedimentarios de Tafira a Las Palmas. Estud Geol Inst Lucas Mallada 25/69:235–248
Robertson AHF, Bernoulli D (1982) Stratigraphy, facies, and significance of Late Mesozoic and Early Tertiary sedimentary rocks of Fuerteventura (Canary Islands) and Maio (Cape Verde Islands). In: Rad U von, Hinz K, Sarnthein M, Seibold E (eds) Geology of the northwest African continental margin. Springer, Berlin Heidelberg New York pp 498–525
Robertson AHF, Stillman CJ (1979) Submarine volcanic and associated sedimentary rocks of the Fuerteventura Basal Complex, Canary Islands. Geol Mag 116, 3:203–214
Rona A, Nalwalk AJ (1970) Post-Early Pliocene unconformity on Fuerteventura, Canary Islands. Bull Geol Soc Am 81:2117–2122
Rothe P (1964) Fossile Straußeneier auf Lanzarote. Nat Mus 94/5:175–187

Rothe P (1966) Zum Alter des Vulkanismus auf den östlichen Kanaren. Soc Sci Fenn Comm Phys Math 31/13:80 pp

Rothe P (1967a) Prävulkanische Sedimentgesteine auf Fuerteventura (Kanarische Inseln). Naturwissenschaften 14:366–367

Rothe P (1967b) Petrographische Untersuchungen an Basalten und Trachyandesiten. Ein Beitrag zur Vulkano-Stratigraphie von Lanzarote (Kanarische Inseln, Spanien). N Jahrb Mineral Mh 2/3:71–84

Rothe P (1968a) Mesozoische Flysch-Ablagerungen auf der Kanareninsel Fuerteventura. Geol Rundsch 58/1:314–332

Rothe P (1968b) Dolomitization of Biocalcarenites of Late-Tertiary Age from northern Lanzarote (Canary Islands). In: Müller G, Friedman GM (eds) Recent developments in carbonate sedimentology in central Europe. Springer, Berlin Heidelberg New York, pp 38–45

Rothe P (1973) Sedimentation in the deep-sea areas adjacent to the Canary and Cape Verde Islands. Mar Geol 14:191–206

Rothpletz A, Simonelli V (1890) Die marinen Ablagerungen auf Gran Canaria. Z Dtsch Geol Ges 42:677–736

Sauer EGF, Rothe P (1972) Ratite eggshells from Lanzarote, Canary Islands. Science 176:43–45

Schmincke H-U (1976) The geology of the Canary Islands. In: Kunkel G (ed) Biogeography and ecology in the Canary Islands. Junk, The Hague, pp 67–184

Serralheiro A (1970) Geologia da Ilha de Maio. Junta Invest Ultramar, 103 pp

Stahlecker R (1934) Neokom auf der Kapverden-Insel Maio. Jahrb Mineral Beil 78B:265–301

Tietz GF (1969) Mineralogische, sedimentpetrographische und chemische Untersuchungen an quartären Kalkgesteinen Fuerteventuras. (Kanarische Inseln, Spanien). Diss, Ruprecht-Karl-Univ, Heidelberg, 149 pp

Tietz G, Müller G (1971) High-magnesian calcite and aragonite cementation in Recent Beachrocks, Fuerteventura, Canary Islands, Spain. In: Bricker OP (ed) Carbonate cements. Johns Hopkins Univ Stud Geol 19:4–8

Torres AS, Soares JMP (1946) Formacões Sedimentares do Arquipélago de Cabo Verde, vol 2. Mem Ser Geol 3:397 pp

Yaalon DH, Ganor E (1973) The influence of dust on soils during the Quaternary. Soil Sci 116:146–155

Zeuner FE (1958) Lineas costeras del Pleistoceno en las Islas Canarias. Anu Estud Atlant 4:9–16

Fallout Tephra Layers: Composition and Significance

PETER RENE BITSCHENE and HANS-ULRICH SCHMINCKE[1]

CONTENTS

	Abstract	49
1	Introduction	49
2	Definition of Terms	50
3	Compositional and Textural Characteristics of Fallout Tephra Layers	50
3.1	Methods	51
3.2	Problems	51
3.3	Macroscopic Identification	52
3.4	Thickness and Areal Distribution	52
3.5	Origin of Widespread Fallout Tephra Layers	55
3.6	Physical Characteristics	55
3.7	Chemical Characteristics	58
3.8	Grain Size and Sorting	59
4	Significance of Tephra Layers	61
4.1	Regional Distribution and Plate Tectonic Setting	61
4.2	Correlation and Stratigraphy	61
4.2.1	Intramarine Stratigraphic Correlation	62
4.2.2	Land-Sea Correlation	63
4.2.3	Ash Layers in Special Environments	63
4.3	Physical Dating and Calibration of Time Scale	64
4.3.1	Potassium Argon Method and Calibration of Time Scale	64
4.3.2	Fission Track	67
4.3.3	Other Direct Methods	68
4.3.4	Indirect Dating Methods	69
4.4	Petrogenetic and Geodynamic Implications	70
4.4.1	Source Regions and Petrogenesis	70
4.4.2	Compositional Heterogeneity of Magma Columns	70
4.4.3	Volcanic Episodicity	72
	References	77

---

[1] Institut für Mineralogie RUB, Postfach 102148, D-4630 Bochum, FRG.

Abstract

Tephra layers differ in color, thickness, composition, and origin from their enclosing non-volcanic sediments. They record instant geologic events extending across marine and continental environments, reflect climatic and paleoenvironmental changes and geotectonic activities, and can provide an orthostratigraphic time frame for sedimentary sequences when numerically dated (Schmincke & Bogaard 1990).

Tephra layers reflect episodic volcanism on a global and regional scale and can be used to delineate the chronologic and compositional evolution of long-lived active volcanic centers and regions. Recent work in the North Atlantic has shown that tholeiitic basaltic ash layers can be distinguished by their Ti-concentrations and attributed to Icelandic volcanic centers. Similarly, rhyolitic ash layers can be distinguished by their K_2O concentrations and reflect different source regions in Iceland (Neogene low-K rhyolites), E-Greenland continental margin (Oligocene-Miocene high-K rhyolites) and Jan Mayen high-K province (Quaternary high-K rhyodacites). Compositional differences also reflect source inhomogeneities and evolution of the erupted basaltic ash from N-MORB types to E/P-MORB types through the Neogene. The undisturbed and nearly complete recovery of Neogene ash layer sequences from the Vøring Plateau (ODP Leg 104) do not support the idea of episodic volcanic activity in the North Atlantic region but reflect relatively constant ash layer frequencies of five to ten layers per Ma through the Neogene. On a global scale, however, ash layer frequencies, magma discharge rates, and K-Ar dates may indicate peaks of Cenozoic volcanic activity in the Middle Miocene, Plio-Pleistocene, and Late Quaternary.

Tephra layers are also excellent tools for numerical calibration of the geologic time scale. Recent advances in the K-Ar method, applying single grain laser dating and $^{40}Ar/^{39}Ar$-stepwise heating techniques, have evolved as powerful methods in tephrochronology and enable high-resolution calibration of the geologic time scale, bracketing of terrestrial ice ages, and correlation to the marine record.

1 Introduction

Modern studies on marine fallout tephra layers started with Bramlette and Bradley (1942), who interpreted silicic ash layers from piston cores in the North Atlantic as subaerial fallout, most likely from Iceland or the Jan Mayen islands. Thorarinsson (1944) used Icelandic Quaternary ash layers for correlation and age estimates on Iceland. Until 1970, marine tephra studies were restricted to the upper few meters of the sediment, because the piston core penetration depth is only a few meters. Most tephrochronological studies are still performed on piston cores which allow closely spaced sampling in contrast to deep holes drilled by the Glomar Challenger and Joides

Resolution drill ships of the "Deep Sea Drilling/Ocean Drilling Program (DSDP/ODP)". The international DSDP/ODP project allows recovery of the Cenozoic marine tephra record and evaluation of composition and evolution of volcanic activity worldwide.

The identification of fallout tephra layers and their geologic significance has been reviewed by Fisher and Schmincke (1984) and Kennett (1981, 1983). The volume *Tephra studies* (Self and Sparks 1981) contains many reviews on selected aspects (correlations, case studies, definitions) of tephra. Heiken and Wohletz (1985) show abundant photographs on shard morphology and give relevant definitions of specific terms.

2 Definition of Terms

Fallout tephra layers, hereinafter called tephra (regardless of particle size) or ash (grain size < 2 mm) layers, result from instant subaerial or subaquatic fallout of pyroclastic fragments expelled from volcanic foci through powerful eruptions (eruptive origin) or elutriated from pyroclastic mass flows travelling into the aquatic environment (coignimbrite origin). *Pyroclastic fragments* or *pyroclasts* are particles expelled through volcanic vents. They are called *essential* or *juvenile*, when derived from erupting magma, *accessory* or *cognate*, when derived from disrupted comagmatic rocks from previous eruptions, and *accidental clasts*, when derived from the country rock of any composition (Fisher and Schmincke 1984). *Tephrochronology* is the science of identification, correlation, and dating of tephra layers (Thorarinsson 1981). *Tephra* (Thorarinsson 1944), synonymous with *pyroclastic material*, is a collective term for all types of pyroclasts regardless of grain size, mode of origin, and depositional mechanism. *Volcanic ash* refers to tephra with grain sizes < 2 mm. Most tephra layers have grain size < 2 mm and are therefore correctly termed *ash layers*. *Tuff* is lithified ash. *Bentonite* (when intensely lithified also called *tonstein,* especially in coal-bearing strata) is used for soft to plastic, altered ash consisting essentially of expandable clay minerals and a few crystals such as sanidine, zircon, and quartz. Most bentonites are interlayered with shallow marine sediments that may develop on the slowly foundering continental slopes. In general, the term bentonite used with the appropriate mineral identifier (smectite, kaolinite) circumvents nomenclatorial problems (Fisher and Schmincke 1984).

3 Compositional and Textural Characteristics of Fallout Tephra Layers

Tephra studies include analyses of grain size and clast morphology, petrography, chemical composition, and isotopy of glasses and crystals. Magnetic susceptibility, natural gamma-ray log, conductivity, and acidity are measured in boreholes or ice cores to identify ash layers.

3.1 Methods

Smear slides are used on board research vessels and in field camps, whereas petrographic thin sections are used in the laboratory for identifying ash particles and layers. Polished thin sections are needed for microprobe analysis to determine major and some trace element concentrations. XRF, AA, INAA, and other methods used on bulk ash samples allow one to chemically characterize tephra layers in more detail. Tephra layers are generally mixtures of different types of pyroclasts and may contain detrital components. Chemical compositions of essential pyroclasts are therefore still best determined by microprobe. For age dating and isotope studies, a few 100 mg of pure and clean fractions have to be prepared by conventional mineral separation techniques (heavy liquids, magnetic separation, hand-picking). Analysis of the morphology of ash particles requires binocular and REM studies, and on-line image analyzing techniques, the latter being helpful to determine grain size, modal composition, and morphological parameters. The most useful methods, however, to measure grain sizes are dry or wet sieving and settling columns.

3.2 Problems

Repeated analyses of silicic glasses nearly always show low oxide sums around 95 ± 2 wt%. They nevertheless document the original compositions of the magma and reflect hydration and/or high initial volatile contents in silicic magmas. Initial volatile concentrations in silicic magmas range from ca 2 to 7 wt% of H_2O, around 0.02 wt% CO_2, and up to 0.08 wt% Cl (e.g. Anderson et al. 1989).

Another problem when analyzing volcanic glass shards is Na loss during microprobe analysis due to Na evaporation under the electron beam. Other effects like reaction with seawater (halmyrolysis) can result in Na loss and K gain (Jezek and Noble 1978). The effect of secondary K and Mg uptake and Si, Fe, Ca, and Na loss is reflected in thoroughly altered ash layers (bentonites). By using immobile elements (Zr, Nb, Y, Ti), Ordovician K-rich bentonite (Huff and Türkmenoglu 1981) was identified as originally trachyandesitic, and Carboniferous ash layers as rhyolite (Schmincke and Sunkel 1987). The chemical and mineralogical changes of volcanic glass due to alteration have been reviewed in detail by Fisher and Schmincke (1984, pp 312–345).

The volcanic origin of bentonites (cf. Sect. 2) is reflected in 1) thinness of beds (generally < 10 cm) combined with wide areal distribution; 2) vitroclastic textures or even relic primary volcanic glass; 3) phenocrysts such as clear quartz, high-sanidine and plagioclase, amphibole, pyroxene, biotite, zircon, and their distinct morphologic features; 4) clay minerals like montmorillonite/nontronite, kaolinite, and paligorskite, or zeolites like phillipsite, clinoptilolite, or analcite, all derived from altered volcanic glass (e. g. Kennett 1981; Fisher and Schmincke 1984). Geochemical indicators for a volcanic origin of suspicious layers may be (1) high concentrations of Cr (characteristic of basaltic ash) or (2) high concentrations of immobile incompatible elements such as REE and Zr (felsic, especially alkali-rhyolitic ash). Müller (1961) reported on in situ devitrification and alteration of volcanic ash in the Gulf of Naples (Italy) and formation of clay minerals, analcite, and silica. This devitrification and alteration scheme of tephra may be interpreted as a possible precursor of bentonite formation.

3.3 Macroscopic Identification

Ash layers are thin and easily overlooked. Moreover, they are commonly discontinuous because ash can become quickly eroded immediately after deposition. Ash layers are best identified in drill cores within undisturbed and continuously accumulated sediments. *Discrete ash layers* have a sharp base and a sharp to gradational top. They can be zoned in color and composition due to sorting by wind and settling through the water column, may comprise several eruptions, or result from successive emptying of zoned magma columns. Ash layers without a sharp base and top and with appreciable detrital components (up to 80%), are called *disseminated* and then are often reworked. Ash layers commonly become reworked and mixed into the sediment (Ruddiman and Glover 1972; McCoy 1981), or entrained into the dwellings of burrowing animals, where they can survive erosion or reworking as *ash pods*.

Basaltic to andesitic ash layers are usually dark, whereas dacitic/rhyolitic or phonolitic/trachytic ash layers are usually light gray to green. Fresh, translucent, brown, basaltic glass is called *sideromelane*, and opaque Fe/Ti-oxide microlites charged basaltic glass shards *tachylite*. Altered and recrystallized brownish basaltic glass is named *palagonite*. Secondary minerals can also color glassy ash particles. Alteration of ash to glauconitic minerals results in green colors, whereas coating by opaques results in dark or red colors due to pyrite or Fe-hydroxide incrustation.

3.4 Thickness and Areal Distribution

Tephra fallout layers can be widely distributed and have thicknesses in the centimeter range. Basaltic and rhyolitic ash layers drilled on the Vøring Plateau near the passive continental margin off Norway (ODP-Leg 104) amount to 0.73% of the total sediment and have thicknesses from < 1 to 14 cm (Fig. 1). Proximal North Atlantic ash layers can reach thicknesses up to 120 cm (Varet and Metrich 1979). Tephra layer thicknesses from three major Plinian and Phreato-Plinian eruptions at El Chichon-volcano/Mexico

Fallout Tephra Layers: Composition and Significance

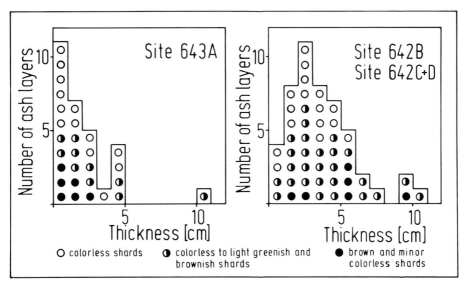

Fig. 1 Thickness and composition of ash layers from the Vøring Plateau, North Atlantic (Bitschene et al. 1989)

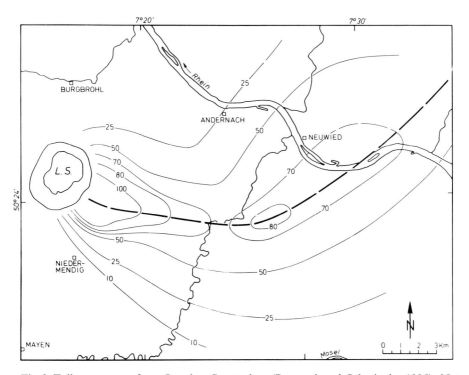

Fig. 2 Fallout pattern from Laacher See tephra (Bogaard and Schmincke 1985). Note curved axis of maximal thickness of tephra layer

in 1982 range from around 30 cm near the vent to 3 cm about 25 km away from the source (Carey and Sigurdsson 1986).

Tephra layer thicknesses may show secondary maxima several kilometers away from the source, which are most plausibly interpreted as due to aggregation of ash particles < 45 µm (Figs. 2, 3; Sorem 1982; Carey and Sigurdsson 1982; Schumacher and Schmincke 1990). Other processes leading to local changes in ash layer thicknesses are compaction (which can reduce primary thicknesses by about 50%), bioturbation, and sediment ponding (Watkins et al. 1978; Kennett 1981).

Fallout tephra sheets with circular dispersion patterns result from low eruption columns that are not distorted by strong winds (Walker and Croasdale 1971), whereas fan-shaped dispersion patterns form when high eruption columns encounter strong unidirectional winds. Sources of fan-shaped tephra layers may be inferred by projection of the axis, but must be done with care because fan axes commonly curve (Fig. 2). Some tephra deposits form lobate sheets resulting from ash flows or from subaerial tephra fallout, when rising ash clouds are dispersed by winds that blow in different directions at different altitudes.

The areal distribution of volcanic ash depends upon the energy of an eruption and height of the eruption column, the prevailing wind vectors, and the physical properties of ash particles. The eruption column of the May 18 Mt. St. Helens eruption, 1980, was 14 km high after 9 h of consecutive activity (Lipman and Mullineaux 1981). Monitoring of the ash cloud by radar showed that column height correlated directly to individual ash clouds (Harris et al. 1981). These ash clouds advanced initially with 250 km/h into high velocity wind layers at 10 to 13 km height, which then dispersed volcanic ash at velocities around 100 km/h (Sarna-Wojcicki et al. 1981). The varia-

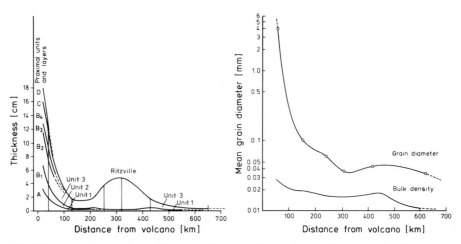

Fig. 3 Downwind variation of layer thickness, bulk density, and mean grain diameter of Mt. St. Helens ash (Sarna-Wojcicki et al. 1981)

tions in ash layer thickness, grain diameter, and bulk density of this May 18 eruption are shown in Fig. 3. The Mt. St. Helens ash layer displays the expected decrease in layer thickness (except for a distal thickness maximum about 320 km from the source), bulk density, and grain diameter downwind. The decrease in density is due to the fact that larger grains, heavier lithics, and crystal grains settled closer to the source.

3.5 Origin of Widespread Fallout Tephra Layers

Widespread ash layers that can be observed thousands of kilometers away from their volcanic source are the Toba ash in the Indian Ocean (e. g. Rose and Chesner 1987) and the Taupo tephra layers in the South Pacific Ocean (e. g. Walker 1981).

The prerequisites for wide ash dispersal are (1) fast ascent of eruption columns into regions of high shear wind velocity such as the tropopause (troposphere-stratosphere boundary) and (2) voluminous eruptions providing the heat and material necessary to feed the eruption column and hence the laterally dispersed ash cloud. Plinian to ultra-Plinian eruptions produce and disperse volcanic ash far enough to form widespread ash layers.

Most efficient use of magmatic heat, which propels the convective and major part of an eruption column, is made in purely magmatic eruptions, while a large portion of magmatic heat is used to convert external water into steam in phreatomagmatic eruptions. It is now recognized, following the pioneering work of GPL Walker and RSJ Sparks (e. g. Sparks et al. 1973; Sparks and Walker 1977; Sparks and Huang 1980), that many if not most widespread ash layers are the result of elutriation of fine ash from glowing avalanches (ignimbrite), especially when they enter the sea. Widespread ash layers are therefore often associated with large ignimbrites and calderas (e. g. Toba ash layers and caldera).

3.6 Physical Characteristics

Some physical properties and morphological features of ash particles affect settling velocities and provide valuable information on eruptive processes. Of importance are size, shape, vesicularity, and density.

The density of tephra particles is related to their structure (minerals or glass), chemical composition, and amount of vesicles. Most minerals are denser than glass and settle prior to glasses of similar size and shape. Within a specific sample, phenocryst fragments have smaller median diameters in ash layers than accompanying glass shards. Mafic or dark minerals have the highest densities, with $d > 2.8$ g/cm^3, whereas light colored, felsic minerals have densities around 2.5 to 2.8 g/cm^3. Similary, mafic glasses are denser ($2.5 < d < 2.9$) than intermediate ($2.3 < d < 2.55$), and felsic glasses ($2.3 < d < 2.4$).

Hole	Sample	Depth [mbsf]	Lithologic unit	Ash lithology	Relative abundance of colorless (c) and brown (b) shards (-100% c / -100% b)	Maximum grain size [μm] (250 / 750 / 1250)	Abundance [%] of crystal fragments	Mineral assemblages
642 B	6-1-131-132	41	I	L	●	X (250)	●	B, Kf, Cpx, Zr, Op, Q, Pl
642 B	9-3-95-96	71	IIA	L	● (mid)	X (250)	○	Q, Pl, Cpx, Cz, Zr, Op, Ca
642 B	9-3-106-107	71	IIA	E	● (mid)	X (250)	○	Q, Pl, Cpx, Op, Ca, Ru?
642 B	13 CC-10-12	114	IIC	L	●	X (750)	○	Q, Pl, Cpx, Op, Zr, Kf?
642 C	10-6-5-7	71	IIA	L	●	X (250)	○	Q, Pl, Cpx, Op, B?
642 C	15-2-86-87	114	IIC	L	●	X (750)	○	Q, Pl, Zr, Op, Cpx, Kf?
642 C	15-2-111-112	114	IIC	D	●	X (750)	○	Q, Pl, Zr, Op, Cpx?
642 C	15-4-88-93	117	IIC	L	●	X (250)	○	Q, Pl, Cpx, Zr, Op, Kf, Am?
642 C	19-3-71-74	153	IID	L	●	X (250)	○	Q, Pl, Cpx
642 D	7-6-70-74	247	III	P	●	X (250)	○	Pl, Cpx, Q, Op; Gl crackled + palagonit. glass
642 D	11-1-83-84	278	III	L	●	X (250)	○	Pl, Cpx, Q, Kf?, Am?
643 A	2-3-74-75	8	I	L	●	X (250)	○	Q, Pl, Cpx, Zr, Op
643 A	7-3-66-67	57	IIA	P	●	X (250)	◐	Q, Pl, Cpx
643 A	11-1-104-105	93	IIC	L	●	X (750)	○	Q, Pl, Cpx, Zr, Op
643 A	15-4-147-149	136	III	L	●	X (250)	◐	Q, Pl, Cpx, Zr, Cz, Op
643 A	19-4-116-117	173	III	L	●	X (250)	◐	Q, Pl, Cpx, Am?, Op, Zr

Note:
L = Discrete ash layer
E = Ash-rich layer
D = Disseminated ash
P = Ash pods

B = Biotite
Pl = Plagioclase
Kf = K-Feldspar
Q = Quartz
Cpx = Clinopyroxene
Am = Amphibole
Zr = Zircon
Ca = Calcite
Cz = Clinozoisite
Op = Opaques and tachylite
Gl = Glauconite
Ru = Rutile

○ rare, < 2%
◐ common, 2-5%
● abundant, > 5%

Fig. 4 Composition and maximum grain size of North Atlantic ash layers (Bitschene et al. 1989)

Fallout Tephra Layers: Composition and Significance

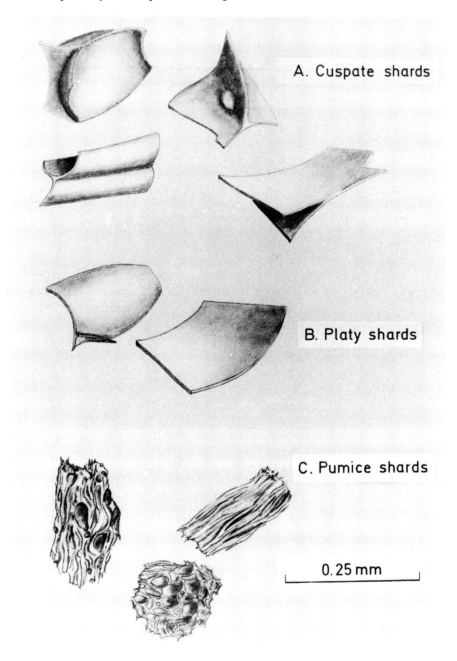

Fig. 5 Morphology of silicic glass shards (Fisher and Schmincke 1984)

Maximum grain sizes of individual ash layers are therefore associated with silicic shards, which have lower density and higher vesicularity (Fig. 4).

The morphology of glass shards (Fig. 5) is related to the viscosity and degree of vesicularity of the magma, which in turn depends on its primary composition, temperature, and concentration of volatiles. In general, basaltic ash particles are blocky or droplet-shaped and have low vesicularity, whereas silicic shards with higher initial volatile contents and higher viscosity are vesicular pumiceous shards or cuspate to platy bubble wall shards. Absence of vesicles and internal cracks especially of blocky basaltic glass shards indicate formation and instant cooling of the glassy particles under hydrostatic pressures high enough to prevent degassing of the magma and vesicle forming during cooling.

3.7 Chemical Characteristics

Refractive index, color, vesicularity, and shape of fresh shards allow to roughly estimate SiO_2^- concentration and thus the bulk chemical compositions of glass (Fig. 6; see Fisher and Schmincke 1984, pp 163–165). Silicic shards tend to be clear, highly vesicular, and have low refractive indices (1.48 < n < 1.57), whereas basaltic shards are brown, less vesicular, and have higher refractive indices (1.57 < n < 1.67). Brown shards, however, found in silicic ash layers from the Vøring Plateau, are not basaltic as anticipated from their color, but merely represent altered (Fe-oxidation) rhyolitic shards (Bitschene et al. 1989).

More precise methods to determine the chemical compositions of ash layers are analyses of bulk ash samples. These data have to be evaluated care-

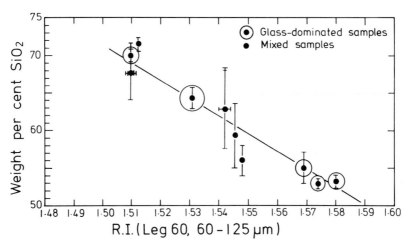

Fig. 6 Refractive indices and inferred SiO_2 concentrations of volcanic glass from ash layers in the North Pacific Ocean (DSDP Leg 60)

Table 1. Comparison of bulk chemistry and microprobe data of single shards from Toba distal ash layers (data from Ninkovich et al. 1978; Rose and Chesner 1987; and Dehn, ODP-Leg 121, unpublished)

	Ninkovich et al. 1978)			Dehn (unpublished)			Rose and Chesner (1987)
SiO_2	70.00	72.20	71.80	74.43	73.60	75.80	78.28
Al_2O_3	13.90	12.90	13.10	13.22	13.73	12.96	12.70
Fe_2O_3	2.00	0.30	0.15	1.53	1.95	1.26	
FeO	0.90	0.70	0.85				1.16
MnO	0.06	0.06	0.06	0.08	0.08	0.05	
MgO	0.80	0.35	0.25	0.34	0.65	0.13	0.08
CaO	1.80	2.20	1.65	0.93	0.19	0.58	0.74
Na_2O	3.40	2.80	3.70	3.80	4.03	3.40	2.71
K_2O	5.20	5.10	5.10	4.98	4.79	5.48	4.26
H_2O	1.40	3.10	3.00				
P_2O_5	0.06	0.03	0.05	0.05	0.06	0.02	
TiO_2	0.37	0.11	0.14	0.12	0.18	0.07	0.06
Sum	99.89	99.85	99.85	99.48	99.26	99.75	99.99

fully because of crystal/glass sorting and contamination processes prior and during deposition and subsequent alteration. Comparison of XRF and AA data on bulk ash samples and microprobe data on single fresh glass shards can show discordant Na, Fe, Mg, and Ti values for basaltic samples (Vallier et al. 1977). The silicic Toba ashes, when compared with bulk XRF data and microprobe data from single shards (Ninkovich et al. 1978; Rose and Chesner 1987), pose the same ambiguity (Table 1). The dilemma of alteration and contamination-dependent changes in concentration, can however, be partly overcome if element ratios (Na/K; K/Rb; Zr/Nb; Ti/Mg; and others) are considered. The chemical and mineralogical changes of volcanic ashes due to alteration have been reviewed in detail by Fisher and Schmincke (1984; pp 312–345).

3.8 Grain Size and Sorting

Ejected pyroclastic fragments are transported ballistically or in turbulent suspension resulting in distinct structural and textural patterns with increasing distance for fallout tephra deposition. Fallout tephra virtually blankets all terrains with slopes not steeper than about 25° to 30° (e.g. Wentworth 1938) and normally displays either density- or size-graded bedding. Reverse grading, however, can be caused by (1) progressive increase in gas velocities; (2) changes in vent morphology and particle ejection angles; (3) increase in density of the eruption column; (4) erratic changes in wind velocities; and (5)

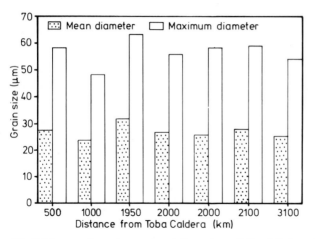

Fig. 7 Mean and maximum shard diameters of distal Toba ash, plotted as function of distance from the Toba eruptive center (Rose and Chesner 1987)

sorting due to settling through the water column (Self 1976). Bedding structures in near-source fallout deposits consist of alternating graded layers, whereas thin beds of distal fallout tephra lack internal structures.

Textural parameters provide information about physical conditions during eruption and deposition of tephra layers. Useful granulometric parameters are (1) maximum grain sizes; (2) median diameters; and (3) sorting values. Most studies (for references, see Fisher 1964; Fisher and Schmincke 1984; Carey and Sigurdsson 1986) showed a near exponential decrease of maximum particle diameters of pumiceous and lithic components away from the source. Denser lithics normally have about half the diameter of pumices. The same exponential decrease with distance is observed for the median diameter of fallout tephra. Graded bedding, multicomponent composition, and sampling bias, however, may complicate proper determination of median diameters. Median grain sizes from distal (> 300 km) fallout tephra do not vary significantly, and are generally in the range of 60 to 20 μm (6 to 4 phi; Figs. 3, 7). For more proximal layers (1 to 10 km), median grain sizes are in the range of 300 to 50 μm (2 to 4 phi) for weak eruptions (high fall velocity fragments, low velocity winds), and 64 to 1 mm (−4 to 0 phi) for powerful eruptions (low fall velocity fragments, high velocity winds; Fisher 1964).

Good sorting can be an important criterion to distinguish fallout ash from pyroclastic flow deposits. Sorting of fallout ash generally becomes better with increasing distance from the vent (Fisher 1964), but difficulties are posed by ash aggregation to form accretionary lapilli (Schumacher and Schmincke 1990), which lead to coarse aggregate tails on dry sieved grain distributions of distal samples (Varekamp et al. 1984).

4 Significance of Tephra Layers

4.1 Regional Distribution and Plate Tectonic Setting

Explosive volcanism occurs in all plate tectonic settings, but is most common above subduction zones. These zones include both active continental margins, such as the western margin of the Americas, and island arcs, such as those of the Carribbean and western Pacific. Magmas erupted in these settings have calc-alkaline compositions and range dominantly from andesitic through dacitic to rhyolitic but can also contain basalts.

Explosive volcanism also occurs in both continental and oceanic intraplate settings and at mid-ocean ridges (hot spot and rift volcanism). Magmas erupted in these environments are chiefly bimodal, tholeiite-basaltic and rhyolitic but may comprise alkali-basaltic, trachytic to phonolitic, and alkali-rhyolitic compositions. Constructive plate margins (mid-ocean ridges) mainly supply basaltic magma, and volcanic eruptions occur under hydrostatic pressure too high to allow explosive magma degassing and formation of widespread ash layers. Iceland, however, a volcanic landmass astride the Mid-Atlantic Ridge, has produced widespread tephra layers from the Miocene to the present, chiefly of bimodal, basaltic, and rhyolitic composition, as discussed in more detail below.

4.2 Correlation and Stratigraphy

Ash layers help in stratigraphic correlation when interstratified in marine sediments (e.g. cores from ODP Leg 104; Eldholm et al. 1987; Bitschene et al. 1989), or loesses and fluvioglacial sediments (Westgate and Gorton 1981; Bogaard and Schmincke 1988). Bentonites are valuable tools in intrabasin correlation or correlation of different basins (e.g. Central European coal basins; Burger 1982; Lippolt et al. 1986a).

Ash layer correlation is not straightforward (1) because different ash layers often have similar microscopic, and geochemical features, and (2) widespread ash layers may change in composition due to grain size and density sorting. During core handling on board the drill ship Joides Resolution, routine measurements of magnetic susceptibility proved to be indicative of admixture of volcanic ash. Similar information can be obtained from logging results, especially from the natural gamma-ray tool, which responds to the decay of ^{40}K and other radioactive elements (U and Th nuclides), which are concentrated in rhyolitic and alkaline magmas.

Consistent correlations require a multiple criterion approach. Tephra characterization and equivalence of samples should only be considered firmly established if their stratigraphic, paleontologic, paleomagnetic, and radiometric age relations are compatible, and the compositions of glass shards and phenocrysts agree after using grain-discrete methods of analysis.

Correlation based on major and trace element concentrations from glass shards and crystals is commonly done by comparison of data in divariant (e.g. $Na_2O + K_2O$ vs SiO_2 or TiO_2) or multivariant plots (e.g. Fe-Ca-Ti or Th-Sc-Hf triangles).

4.2.1 Intramarine Stratigraphic Correlation

Many prominent marine acoustic reflectors turned out to be ash layers (Worzel 1959) and can be used for correlation of piston and drill cores (e.g. Kennett 1981; Keller 1981; Rabek et al. 1985). Ash layers from piston cores in the Gulf of Mexico can be accurately distinguished by TiO_2, FeO and CaO concentrations of the glass shards. These elements are not affected by element gain or loss through reaction with seawater (for Ca this is only true for virtually pristine essential pyroclasts) or during microprobe analysis (cf. Sect. 2) and are the most constant within one layer but are distinctly different between layers (Fig. 8; Rabek et al. 1985). The major element chemistry was used for correlation of single ash layers between different sites in the Gulf of Mexico and corroborated, for instance, the correlation of the Y8 tephra layer in the Gulf of Mexico with ash layer D from the equatorial Pacific Ocean (Rabek et al. 1985).

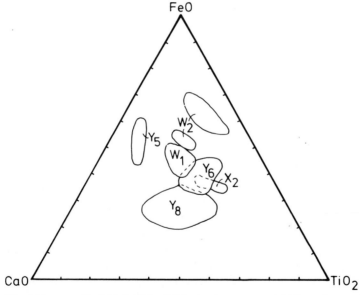

Fig. 8 Ternary CaO-FeO-TiO_2 (TiO_2 is the measured TiO_2 value multiplied by 5) plot showing chemical differences of six tephra layers in the Gulf of Mexico (Rabek et al. 1985). Ash layer Y8 is 84 000 a old and can be correlated with the tephra layer D in the equatorial Pacific Ocean. The ages of the other ash layers, obtained through oxygen isotope stratigraphy of the sedimentary sequences, are as follows; Y5 (30 000 a), Y6 (65 000 a), X2 (110 000 a), W1 (136 000 a), W2 (185 000 a). Four (Y5, Y6, X2, W2) of the latter five ash layers have limited distributions and probably come from sources in Mexico

A series of bentonites from Devonian marine deposits of the Eifel and Ardenne area, for instance, were correlated by zircon morphologies (Winter 1981). Other correlations performed by major element analysis of ash layers have been carried out with tephra layers from the Mediterranean (e. g. Keller 1981; Paterne et al. 1986, 1988) and from the North Atlantic (Bitschene and Schmincke 1990).

4.2.2 Land-Sea Correlation

For the North Atlantic region, many marine ash layers can be correlated to Icelandic volcanic activity. Distinct alkalic rhyolitic ash layers of Miocene to Oligocene age, however, do not fit into the well-known magmatic record from Iceland and thus herald explosive volcanism from other regions, for instance Greenland or the Jan Mayen magmatic province (Sigurdsson and Loebner 1981; Bitschene et al. 1989).

Sigurdsson and Carey (1981) compared microprobe data from single shards from marine tephra collected near the Lesser Antilles volcanic arc with data from the tephra deposits on the islands of the Antilles arc. They reached excellent agreement based on major element data and could correlate marine tephra to their terrestrial counterpart on the islands. Paterne et al. (1986, 1988) used K_2O/Na_2O ratios to divide marine tephra layers in the Mediterranean into two groups and to correlate them to probable sources in Ischia and Campi Flegrei.

Correlation of Quaternary marine tephra from the ocean basins surrounding Central America with terrestrial counterparts on the mainlands (Lake Atitlan caldera in Guatemala) was possible using geochemical fingerprints such as $FeO-CaO-TiO_2$ ternary diagrams (Ledbetter 1985). These elements are the most constant within one layer but are distinctively different between layers (Fig. 8; Rabek et al. 1985).

4.2.3 Ash Layers in Special Environments

During (cold) glacial periods, sedimentation patterns change and the deposition of tephra layers is restricted to regions where undisturbed sediment accumulation is possible. Ash layers interbedded with loess and river terraces are useful for correlation to the magnetostratigraphic and oxygen-isotope (marine) record, and for dating of the glacial periods (Lippolt et al. 1986b; Bogaard et al. 1987; Bogaard and Schmincke 1988). Ice cores with intercalated ash layers define another special environment where volcanic ash has been deposited and now serves as a stratigraphic marker in Quaternary stratigraphy and climate history.

4.3 Physical Dating and Calibration of Time Scale

The time span between eruption of hot magma and deposition as fallout tephra is very short (< 1 year) and insignificant when compared with the analytical errors of physical dating methods. The closing temperatures of minerals for loss of radioactive mother nuclides and/or radiogenic daughter nuclides, which are ultimately counted and are a measure for the time elapsed since the onset of the radiometric clock, are generally in the range < 200° to 500° C (Jäger and Hunziker 1979). These closing temperatures are instantly surpassed in retentive minerals from ash layers, which then behave as closed systems and serve as radiometric clocks without delay. No prolonged cooling history occurs and the ages obtained report the time of eruption and deposition. For this reason, all radionuclide retentive mineral phases from ash layers are especially useful for isotope dating, whereas temperature-sensitive methods (e.g. fission track or thermoluminescene) have to be treated with care.

Ash layers can be dated by several methods most of which are summarized by Naeser et al. (1981). Most useful is direct numerical dating by K-Ar, and, much less important, Rb-Sr, U, Th-Pb, and fission track methods. The analytical errors of the numerical ages depend on the physical properties of the sample and on the method applied. Advances in K-Ar dating techniques (e.g. single-grain laser dating and $^{40}Ar-^{39}Ar$ stepwise heating; York et al. 1981) and the reliability of ages obtained down to a few Ka clearly indicate the superiority of this method compared to others. The ash layers to be dated should (1) be stratigraphically well-controlled with additional paleomagnetic and paleontologic data; (2) contain sufficient representative and radiogenic nuclide-retentive mineral phases for precise dating (e.g. high sanidines for K-Ar methods, zircon for fission track or U-Pb method, compositionally zoned ash layers or phenocrysts with a wide range in Rb/Sr-ratios for Rb-Sr isochron dating); and (3) be mineralogically checked for diagenetic and/or metamorphic (thermal, burial) processes.

Tephra layers can provide numerical dates and therefore serve as orthostratigraphic frames for sedimentary sequences (e.g. Odin 1982; Odin et al. 1986). Tephra layers are especially important in dating of Quaternary sediments and bracketing of ice ages. High resolution calibration of the Quaternary time scale is a recent and most promising research field.

4.3.1 Potassium Argon Method and Calibration of Time Scale

The most precise method in numerical dating of ash layers is the K-Ar method, especially its $^{40}Ar-^{39}Ar$ variant (Merrihue and Turner 1966), when applied to K-bearing and Ar-retentive phenocrysts (sanidine, amphibole, phlogopite). Ages from low-K biotite (< 6% K!) can be unreliable as shown by Naeser et al. (1981; see also Hess et al. 1983, 1987). Dating of volcanic glass is especially problematic. Rad and Kreuzer (1987), for instance, obtained discordant results for the < 32 μm fraction (39.9 ± 0.7 Ma) and the > 32 μm fraction (44.8 ± 0.9 Ma) from K-Ar age determinations performed on glass shards from marine silicic ash from the Bermuda Plateau (DSDP Legs 93 and 95). Uncontrolled ^{40}Ar$_{rad}$ loss or gain and grain-size effects may be the reason for these discrepancies.

Reliable chronometers successfully used in tephra layer dating are especially sanidine and furthermore high-temperature anorthoclase and plagioclase as well as high-K biotite (> 6% K!), amphibole and leucite. Odin et al. (1986) dated high-K biotites from centimeter thick bentonites intercalated in Silurian limestones near Grötlingbo/Sweden to between 430.5 ± 3.0 and 425.8 ± 3.0 Ma (Wenlock). Damon and Teichmüller (1971) first dated Carboniferous bentonites (coal tonsteins) from the Ruhr area/West Germany to 298 ± 10 Ma applying the K-Ar method on sanidine separates. Their results were confirmed by Lippolt et al. (1986a), who employed the ^{40}Ar-^{39}Ar technique on high-sanidines from coal tonsteins and tuffs from the Carboniferous Ruhr and Bohemian basins. Together with age data obtained from sanidine from Central Europe bentonites, Hess and Lippolt (1986) then proposed new tie points for the Upper Carboniferous time scale (Fig. 9).

Calibration of the Middle Triassic time scale was attempted by Hellmann and Lippolt (1981) using bentonites from the Middle Triassic "Grenzbitumenzone". Sanidines from this stratigraphic marker horizon have

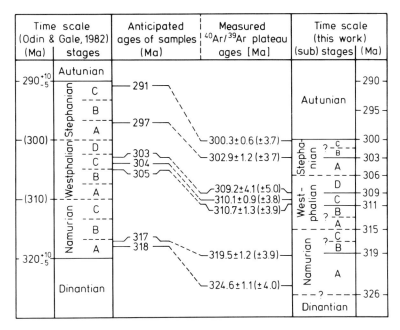

Fig. 9 Calibration of Upper Carboniferous time scale with new tie-points obtained from dating of sanidines from Central-European bentonites (Hess and Lippolt 1986)

been dated to 233 ± 9 Ma, and this age has been used to infer an age of 232 Ma for the Anisian-Ladinian boundary. Montanari et al. (1983) calibrated the Late Paleogene biostratigraphic and magnetostratigraphic time scale in pelagic sediments from Gubbio/Italy by dating biotite from bentonites with K-Ar and Rb-Sr methods. Their results from fresh biotites confined the Eocene-Oligocene boundary at 35.7 ± 0.4 Ma.

The use of bentonite dating for numerical calibration of the geologic time scale is evident from suggested age assignments for the biostratigraphically and lithostratigraphically derived subdivision of the geologic record, especially for the Carboniferous and Tertiary (Baadsgaard and Lerbekmo 1982; Odin 1982; Hess and Lippolt 1986). The reliability of theses absolute time markers increases when combined K-Ar and Rb-Sr age determinations are applied to fresh phenocrysts like biotite or sanidine from stratigraphically well-constrained bentonites. Harris and Fullagar (1989), for instance, dated biotite from a Middle Eocene smectitic bentonite with the conventional K-Ar method to 46.2 ± 1.8 Ma, and with a two-point Rb-Sr isochron to 45.7 ± 0.7 Ma.

Single-grain dating, with laser-induced mineral desintegration, coupled with the ^{40}Ar–^{39}Ar technique (e.g. York et al. 1981), is the most accurate method for dating of ash layers. This method requires small sample volumes due to low argon detection limits, and has the smallest analytical errors of all available methods. Stepwise heating experiments reveal distortions in the argon budget of minerals and can explain erratic K-Ar ages. Lobello et al. (1987), for instance, obtained saddle-shaped age spectra indicating severe distortions or contaminations, when dating individual K-feldspar samples from Quaternary tephra. Ultraclean sample preparation showed that two populations of feldspar were present. Single K-feldspar crystals dated by laser fusion then showed an accurate 0.58 ± 0.02 Ma age. Another possibility to date very young or low-K samples is the setup of the Ar-spectrometer to very discrete but then highly precise analytical conditions (Cassignol and Gillot 1982; Gillot and Cornette 1986). The improved analytical conditions, commonly circumscribed as the Cassignol technique, improve precision and accuracy of K-Ar dating, especially of Quaternary tephra layers. K-Ar dates as young as 2000 a can be obtained and precison for samples older than 100 000 a is better than 1.5%. The Quaternary (Panza and Ciglio-Serrara) tuffs from Italy were dated with the "Cassignol technique" to 29±3 and 33±1.8 Ka (Gillot and Cornetto 1986).

Several Pleistocene tephra layers interbedded with loess and fluvial sediments from Rhine valley terraces were dated using ^{40}Ar–^{39}Ar laser dating of single sanidine crystals and correlated to eruptive phases of the East Eifel volcanic field (Bogaard et al. 1987; Bogaard and Schmincke 1988). Early human occupation of the area, as indicated by tools found in loess interbedded with tephra layers at these sites, can be shown to be at least as old as 0.5 Ma in the Middle Rhine area (Fig. 10).

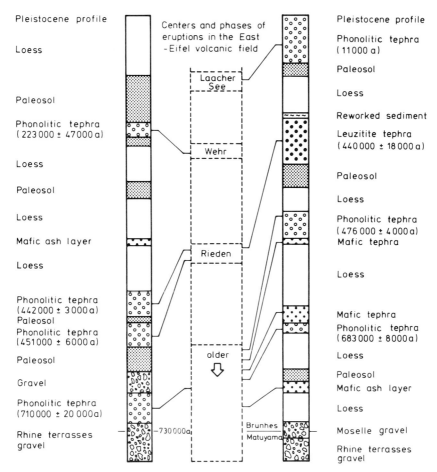

Fig. 10 Numerical dating and correlation of young tephra to eruptive cycles in the Eifel (Fed. Rep. Germany) volcanic area (Bogaard and Schmincke 1988)

4.3.2 Fission Track

Fission track (FT) dating is carried out on U-bearing glasses and heavy minerals such as zircon, apatite, and sometimes sphene. Silicic glasses usually have sufficiently high concentrations in U to produce enough fission tracks from the ^{238}U-decay to be quantitatively measured. Basaltic glasses are typically low in U and are thus not suitable for FT dating.

Zircon and glass from the Spooner tuff in New Zealand were dated successfully by Seward (1979) to 2.61 ± 0.28 and 2.46 ± 0.36 Ma. Izett and Naeser (1976) dated the Bishop Tuff (California) using zircon to 0.74 ± 0.05 Ma, and Naeser et al. (1973) and Seward (1979) reported concordant zircon and glass ages of 1.94 ± 0.12 and 1.93 ± 0.32 Ma for the Borchers ash from the Pearlette ash beds in the western United States. FT dating of glass is greatly hindered by track fading due to hydration or heating (Naeser et al. 1981; Seward 1979). Another disadvantage of FT dating are large analytical errors (commonly > 5 to 10% at a 2-sigma confidence level). Reproducible results can be achieved with the FT method, but are limited by grain size (at least about 75 µm diameter) and chemical composition (Naeser et al. 1981). But highly accurate FT ages with low analytical uncertainties were obtained from zircons from the tuffs of the Middle Miocene Bakate Formation (northern Kenya). This pantelleritic tuff overlies terrigeneous sediments with hominide(?) and other vertebrate fossils and was dated by the FT method applied to juvenile zircons with very homogeneous U-distributions to 16.2 ± 0.6 Ma (Hurford and Watkins 1987). Reproducibility of FT ages was tested by Carpena and Mailhe (1987), who obtained concordant ages on apatite and zircon from the Fish Canyon Tuff (26.8 ± 4.2 and 27.9 ± 2.2 Ma) using three different reactors for radiation.

4.3.3 Other Direct Methods

Dating by thermoluminiscence (TL) can be helpful for young volcanic ash (40 to 500 Ka) when other physical dating methods fail within this time window. Berger (1985) applied the TL method to the 2 to 11 µm glass component of both distal and proximal subaerial fallout ash. Accurate apparent TL ages were measured for young tephra deposits (Mazama ash: 7.8 ± 0.5 Ka). For ashes older than about 400 Ka only lower limits could be given (e.g. Bishop tuff: TL age of $> 380 \pm 40$ Ka). Fading of the TL signal is a major problem. Nevertheless, TL dating can give valuable lower limits for tephra layers not datable by other methods. Dating by electron spin resonance is a method related to TL.

Bouroz et al. (1972) used Rb-Sr isochrons from mineral separates and whole rock analyses to date Upper Carboniferous bentonites. The age obtained (311 ± 10 Ma) for a Stephanian bentonite contradicts previous time scales (Odin 1982) but fits within analytical error of more recent time scales (Hess and Lippolt 1986). Concordant Rb-Sr and K-Ar ages for biotites from Eocene bentonites were obtained by Harris and Fullagar (1989). The limits of Rb-Sr isochron dating are shown by Rundle (1986). He dated lava clasts from a Caradocian tuff to 454 ± 14 Ma, which serves as a rough time calibration but is certainly not accurate enough for bracketing epochs or stages. More successful approaches come from Rb-Sr isochrons obtained from plagioclase and different biotite separates. Baadsgaard et al. (1988), for instance, report ages of 64.5 ± 0.4 Ma using different biotite separates and plagioclase defining a Rb-Sr isochron.

Zircons from geologically old material (generally > ca. 100 Ma) can be dated using conventional U–Pb isotope evolution diagrams (see summary in

Faure 1986), U–Pb and ^{207}Pb–^{206}Pb age determinations on single zircons by whole grain evaporation (e.g. Kober 1986) or ion microprobe techniques. Geologically old, silicic tephra layers (Paleozoic or older) contain juvenile zircon populations (e.g. Winter 1981) which may have sufficient U and little common lead to be dated by single-grain techniques. Attempts in zircon dating by single-grain techniques until now are hindered by high technical requirements and still high analytical errors ($>$ 10 Ma at a 2-sigma level), which considerably limit the demand for zircon dating of Phanerozoic tephra layers. Successful U-Pb dating of phenocrystic zircon populations from bentonites, however, have been carried out by Baadsgaard et al. (1988). They separated zircon, biotite, and sanidine from a biostratigraphically and lithostratigraphically well-defined bentonite in North American coal seams. The bentonite occurs in three different sites between 30 and 80 cm above the K/T-boundary, which is well constrained by palynology and a sharp iridium anomaly. Bentonite samples were taken at three different sites and different zircon size fractions were separated. The age data obtained from U-Pb concordia diagrams (lower intercept) are 64.4 ($+$ 0.6, $-$ 0.8) and 64.3 ($+$ 0.6, $-$ 0.8) Ma. The U-Pb ages are well matched by K-Ar and Rb-Sr age data obtained from cogenetic biotites and sanidines. The weighed mean age of the bentonite derived from all age data is 64.4 $\pm$ 1.2 Ma, notwithstanding, however, possible systematic errors due to sample impurities (Baadsgaard et al. 1988). A similar study was carried out by Yanagi et al. (1988), who employed U-Pb, Rb-Sr, and K-Ar dating methods to estimate the age of a Middle Albian tuff in the Hulcross Formation (Canada).

Dating in the time span $<$ 250 Ka could also be attempted by the U-Th disequilibrium method. Minerals and glasses with different U/Th ratios from one volcanic eruption are analyzed for their respective U and Th isotopes and ages are calculated from isochrone slopes (e.g. Faure 1986).

4.3.4 Indirect Dating Methods

Most relative age assignments for ash layers as well as for all other sedimentary units are based on biostratigraphy and the paleomagnetic record. This parastratigraphic dating method is useful when undisturbed, complete, and geologically young (Cenozoic) sediment sequences with rather constant sedimentation rates are studied. Attributing numerical ages obtained from fossil and paleomagnetic correlations to single tephra layers becomes very doubtful when hiatuses or very condensed sections are analyzed. Tephra layers often contain organic particles (tree branches, shell fragments, bones), which can be dated by the ^{14}C method in the time span 0.5 to 50 Ka. The ^{14}C ages, therefore, give maximum ages of the host tephra layers. Other indirect methods include the warve method, dendrochronology, and lychenology, which eventually cover a time span of a few 1000 years.

4.4 Petrogenetic and Geodynamic Implications

Chemical and physical data from ashes not only help in stratigraphic correlation, but also provide the parameters neccessary for tracing the source region, estimating pT-conditions within the pre-eruptive magma chamber, and finally explaining apparent compositional heterogeneity of tephra through eruption of zoned magma bodies or magma mixing.

4.4.1 Source Regions and Petrogenesis

Chemical data on single glass shards from ash layers reflect differences in magma compositions. For any given magma type, variation of some major elements (Ti, Fe, P, K) and especially of immobile incompatible trace elements (REE, Zr, Nb) can be used to attribute different regional origins and evolution to widespread ash layers. Such source regions have been attributed to tephra layers in the Mediterranean (e. g. Keller 1981; Vinci 1985; Paterne et al. 1988), in the Pacific Ocean (Vallier et al. 1977), and in the North Atlantic Ocean (Varet and Metrich 1979; Sigurdsson and Loebner 1981; Morton and Keene 1984; Bitschene et al. 1989).

New data from the North Atlantic (Bitschene and Schmincke 1990) show that basaltic ash layers can be characterized by their Ti, Ca, Al, and K concentrations (Fig. 11) at equal MgO and SiO_2 concentrations. Basaltic ash layers with medium Ti-concentrations (N/T-MORB compositions) are stratigraphically older (Eocene to Lower Miocene) and may have been derived by partial melting of an upper mantle source only slightly enriched. Basaltic ash layers with higher Ti-concentrations (P-/E-/T-MORB compositions) are stratigraphically younger (Middle Miocene to Recent) and may have come from more enriched sources. The same holds true for rhyolitic ash layers, which can be well-characterized by some major elements (Fig. 11).

Apart from the age differences (old and depleted vs young and enriched), a spatial relationship, although not yet well constrained, can be recognized. The enriched sources and their resulting ash layers (OIB to T-MORB compositions) are best constrained near or at hot spots, (ocean islands or seamonts such as Kerguelen and Heard Island in the Indian Ocean, Jan Mayen in the North Atlantic Ocean indicate these hot spots). Depleted to slightly enriched sources and their resulting ash layers (N-MORB compositions) are confined to magmatic activity at mid-ocean ridges.

4.4.2 Compositional Heterogeneity of Magma Columns

Federman and Scheidegger (1984; see previous literature therein on the same issue!) used physical and chemical data from glasses and phenocrysts of distal

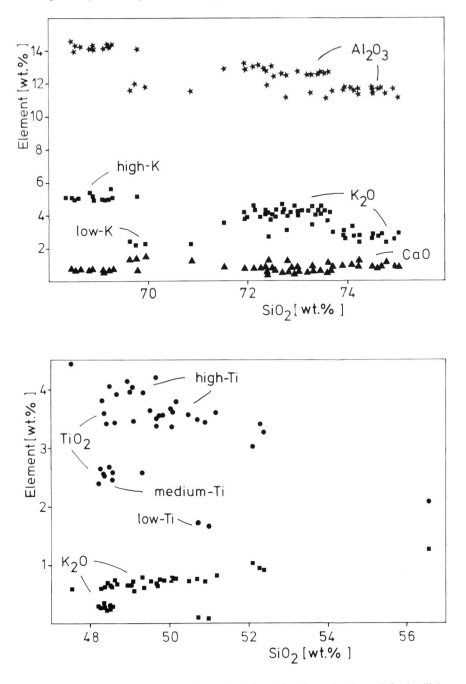

Fig. 11 Variation of major elements from fresh basaltic glass shards and fresh silicic (rhyolitic to rhyodacitic) glass shards from Vøring Plateau (North Atlantic, ODP-Leg 104) ash layers, data for low-Ti basalts from Sigurdsson (1981). The plots demonstrate the subdivision of basaltic shards into high-, medium-, and low-Ti types, and subdivision of rhyolitic shards into high- and low-K types

silicic ashes from the historic (1912) Novarupta eruption in Alaska to support previous findings that three fall units can be recognized and can be related to emptying of a zoned magma chamber. The lowermost unit containing quartz phenocrysts is silica-rich, and the overlying ashes are more mafic with up to 60% SiO_2. Temperature estimates show an increase from 860 °C for the silicic ash to 1000 °C for the basaltic ash during the course of the eruption. These results are consistent with eruption from a compositionally zoned (or mixed magma) reservoir starting with cool, highly silicic magma. Curved non-linear trends were observed in single variation diagrams implying crystal fractionation rather than magma mixing. Mineral and glass chemistry and derived PT estimates and oxygen fugacities allowed the evaluation of mixing and zonation in magma chambers. The volatile concentrations of glass inclusions in phenocrysts from different stratigraphic levels of the Quaternary Bishop Tuff (California) were found to have an inverse correlation of H_2O and CO_2 implying zonation of the magma chamber (Anderson et al. 1989).

4.4.3 Volcanic Episodicity

Deep-sea ash layers provide useful information on the compositional and eruptive evolution in time and space of explosive volcanism. DSDP/ODP ash layer frequencies and compositions from Late Cretaceous to Recent sediment records from the different geotectonic settings (island arc, convergent continental margin, intra plate, and ocean ridge) shed light on the old debate of episodic volcanism, especially during the Cenozoic (e. g. Donnelly 1975; Kennett and Thunell 1975; Scheidegger and Kulm 1975; Ninkovich and Donn 1976; Kennett et al. 1977; Vogt 1979; Donn and Ninkovich 1980; Sigurdsson and Loebner 1981; Bitschene et al. 1989).

To answer the question on episodic volcanism, geologic settings with high ash layer frequencies or ash accumulation rates, and undisturbed and continuous sediment records have to be encountered. Other important factors to be taken into account are age resolution of the interval investigated, hiatuses, alteration of ash, wind patterns, and plate motions, the latter accounting for significant changes in ash layer frequencies during the Quaternary. An oceanic plate moving towards the sites of volcanic activity (convergent margins) increases its tephra accumulation, whereas plates moving away from the foci escape tephra accumulation (Ninkovich and Donn 1976; Kennett 1981).

On a global scale, Vogt (1979) reported on synchronous hot spot (Hawaii-Emperor chain, Iceland) and subduction (circum-Pacific-belts)-related volcanism that is also matched by global ash layer frequencies in the oceans reported by Kennett et al. (1977). Vogt's peaks in magma discharge occur in the Middle Miocene, Plio-Pleistocene and Late Quaternary. These apparent episodes of enhanced magmatic and volcanic activity are mirrored in other provinces active during the Tertiary period (Fig. 12). In general,

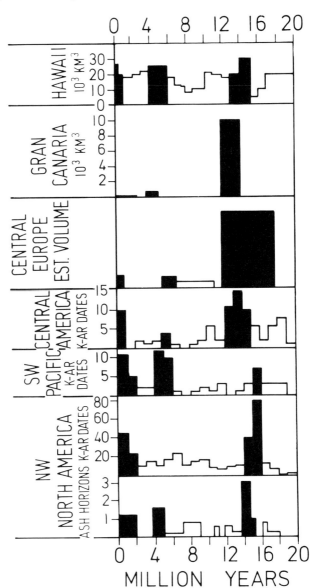

Fig. 12 Global synchronous volcanic episodes inferred from ash layer frequencies, K-Ar data, and magma discharge rates (Schmincke 1982)

magmatic activity, K-Ar dates and ash layer frequencies peak between 12 and 17 Ma (Middle Miocene), 2 and 5 Ma (Plio-Pleistocene), and < 2 Ma (Quaternary). A test of this hypothesis with the Canaries data set shows significant magmatic activity in the Middle Miocene and Plio-Pleistocene as evidenced by Gran Canaria magma supply rates (Fig. 12), ash layers, and pyroclastics from the adjacent ocean floor (Schmincke 1982).

On a regional scale episodic volcanic activity with increased ash layer frequencies is also reported from the Pacific ocean (Fig. 13). Fujioka (1986)

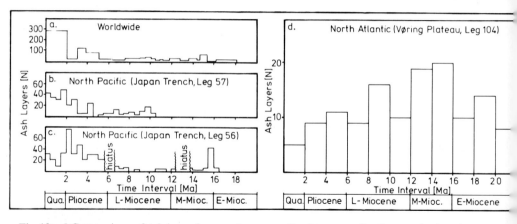

Fig. 13a–d Comparison of ash layer frequencies, normalized to nannofossil zones (**a-c**) and 2 Ma age intervals (**d**). North Atlantic ash layer frequencies have been exrapolated for Neogene hiatuses. Note the apparent ash layer frequency increase in **a** (global, from Kennett and Thunell 1975) and **b** (North Pacific, seaward, converging plate, from Fujioka 1986), and decrease in **d** (North Atlantic, divergent plate, after Bitschene and Schmincke 1990) towards the Recent

reported peaks in ash layer frequencies from the Japan trench region at 15 to 16 Ma (Middle Miocene) and 2 to 5 Ma (Plio-Pleistocene). The apparent increase of ash layer frequencies during Late Neogene to Recent times was governed by the movement of the Pacific Plate towards the source area, the Japanese Islands (Fujioka 1986). Donnelly (1975) found increased volcanic activity in the western Pacific at the Oligocene/Miocene boundary, during the Late Miocene (12 to 7 Ma) and since the Pliocene (5 Ma to Recent), and offshore SW Japan he found peaks in volcanic activity at the Miocene/Pliocene boundary (about 9 Ma) and from 5 Ma to the Recent.

Ash layer frequencies from the Gulf of Alaska appear to indicate a peak in volcanic activity at the Plio-Pleistocene boundary (5 to 2 Ma) and during the Late Quaternary (< 1 Ma); also cyclic variation of magma chemistry is inferred to occur every 2.5 Ma, when silicic ashes are erupted (Scheidegger and Kulm 1975). Regarding the time span between 7 and 8 Ma, after adopting the silicic cycles at 5, 2.5, and 0 Ma, the idea fails because only much less evolved volcanic activity is reported from this area. High age resolutions and continuous ash layer sedimentation or ash accumulation during entire geologic epochs have to be encountered to test episodicity in magma composition and thus magma chamber recharge and differentiation. The northern Atlantic Ocean (Iceland realm) and central Atlantic Ocean (Canaries; Azores) are suitable future sites to investigate short- and long-term eruptive cycles of volcanically active centers.

All the above-mentioned studies confirm global episodic volcanism with peaks in the Middle Miocene, Plio-Pleistocene, and Late Quaternary.

Episodicity of Cenozoic volcanism, however, is questioned by Ninkovich and Donn (1976), who claim continuous volcanic activity within the Pacific and Indonesian oceans. They suspect poor DSDP core recoveries and, most notably, plate motions towards the volcanic centers to be the reasons for apparent peaks in ash layer frequencies. The similarity in increasing ash layer frequency towards the Quaternary can be seen in Fig. 13a (ash layer frequencies worldwide) and Fig. 13b (subducting plate moving towards the volcanic foci) and supports the idea of increased ash layer frequencies due to plate movements. The opposite effect can be seen from the eastern margin of the North Atlantic (Fig. 13d), where ash layer frequencies gradually diminish due to moving the depot center away from the eruptive center (Iceland). This would indicate rather continuous volcanic activity during Neogene times.

To evaluate volcanic and magmatic episodicity (1) source regions should be clearly determined; (2) wind, water, and ice conditions taken into account; (3) plate motions and vertical uplift roughly estimated; and (4) most importantly of all, complete sediment records should be available and hiatuses, changes in sedimentation rates, and diagenesis and reworking properly considered. A case study from the North Atlantic, especially from the Norwegian-Greenland Sea and around Iceland (North Atlantic Ocean), reveals the magmatic evolution of the ocean floor during successful continental rifting and seafloor spreading, and illustrates in particular the hotly debated hypothesis on Cenozoic episodic explosive volcanism (Bitschene and Schmincke 1990).

Most successful in recovering ash layers from the Norwegian-Greenland Sea were Leg 38 (Talwani et al. 1976), which, however, suffered from poor core recovery down to only 4%, and especially Leg 104 (Eldholm et al. 1987), which had exceptionally good core recovery and provided a nearly complete record of North Atlantic ash layers since Eocene times. Contradictory interpretations on volcanic episodicity have been drawn (Sylvester 1978; Sigurdsson and Loebner 1981; Bitschene et al. 1989). Based on the ash layers from DSDP Leg 38 cores, Donn and Ninkovich (1980) claim high explosive activity in the North Atlantic Ocean in the Middle Eocene and Pliocene, whereas Sigurdsson and Loebner (1981) report apparent peaks in explosive volcanic activity in the Middle Eocene, Middle Oligocene, Early to Middle Miocene, and Pliocene to Pleistocene, using the same data base (Fig. 14). The outstanding good core recovery of undisturbed and nearly continuously deposited Neogene sediments from ODP Leg 104 showed that no significant high in explosive volcanism in the North Atlantic for Miocene to Recent epochs (Fig. 14) is evident. Low sedimentation rates (Smalley et al. 1986) are responsible for apparent high ash layer frequencies during the Miocene. Spreading of the North Atlantic and drifting of the Vøring Plateau away from the volcanic centers Iceland, Jan Mayen, and Greenland continental margin, and ice cover of the continent-near sea during Quaternary cold periods are natural reasons which led to the decline in ash layer frequency during the Pliocene and Quaternary (Figs. 13d, 14).

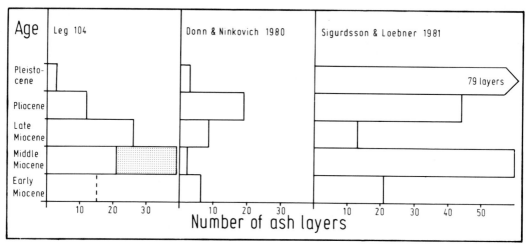

Fig. 14 Ash layer frequencies and related stratigraphic ages from the Vøring Plateau (North Atlantic, ODP-Leg 104) compared to North Atlantic ash layer frequencies from DSDP-Leg 38 (Bitschene et al. 1989)

From present knowledge of North Atlantic tephrostratigraphy, no episodicity of volcanic activity, as inferred previously, can be established. Average ash layer frequencies amount to between five and ten layers per Ma during Neogene epochs (Fig. 13d). An insignificant high ash layer frequency in the 14 to 16 Ma interval heralds the emergence of Iceland and proximity of the source region (this would not contradict the global high in magma discharge rates during the Middle Miocene as proposed by Vogt 1979). Constant explosive volcanic output on Iceland and drift of the Vøring Plateau away from Iceland can easily explain the slight decrease in ash layer frequencies towards younger ages. However, as time resolutions 0.1 Ma can be achieved by numerical dating of the ash layers, volcanic and magmatic espisodicity can be expected due to the lifetimes of single eruptive centers.

When comparing global magma discharge rates, ash layer frequencies and some age data (Figs. 12 and 13 a, b) from long-lived volcanic regions support episodic volcanic activity during certain Cenozoic epochs. This episodicity seems to be valid for the Pacific and partly for the Indian Ocean, which experienced episodic tephra input from highly explosive volcanic activities at convergent plate settings (island arc and continental margin). Support also comes from hot spot activated intraplate and ocean ridge settings, where magma discharge rates apparently match periods of enhanced volcanic activity from the circum-Pacific. The inferred episodes (Middle Miocene, Plio-Pleistocene, Late Quaternary) of magmatic and volcanic activity is also regarded to be synchronous with major geologic events such as subduction rates and orogenic movements and might have triggered volcanic ash dust veils responsible for cold periods ("volcanic winters") and glaciations (Lamb 1971; Kennett et al. 1977; Ninkovich and Donn 1976; Vogt 1979; Kennett 1981; Stothers et al. 1989).

On the other hand, North Atlantic ash layer frequencies and Neogene plate tectonics favor the concept of continuous explosive volcanism (Fig. 13 b, d), where apparent highs and lows in volcanism are governed by plate movements, ice covers, and prevailing wind patterns (Ninkovich and Donn 1976; Kennett 1981; Fujioka 1986; Bitschene et al. 1989).

Acknowledgements. Our work on marine tephra layers and dating was supported by DFG grants Schm 250/37-1 to 250/37-3. We thank B. Behncke and K. Mehl for helping in compilation of the literature, and H. Ferriz for thorough critical reading of the manuscript. We dedicate this contribution to German Müller on the occasion of his 60th birthday. Marine tephra was one of his early loves (Müller 1961) and our understanding of tephra diagenesis would, no doubt, have been greatly enhanced if he had continued to explore this subject.

References

Anderson AT, Newman S, Williams SN, Druitt TH, Skirius C, Stolper E (1989) H_2O, CO_2, Cl, and gas in Plinian and ash-flow Bishop rhyolite. Geology 17:221–226

Baadsgaard H, Lerbekmo JF (1982) The dating of bentonite beds. In: Odin GS (ed) Numerical dating in stratigraphy. John Wiley and Sons, New York, pp 423–440

Baadsgaard H, Lerbekmo JF, McDougall I (1988) A radiometric age for the Cretaceous-Tertiary boundary based upon K-Ar, Rb-Sr, and U-Pb ages of bentonites from Alberta, Saskatchewan, and Montana. Can J Earth Sci 25:1088–1097

Berger GW (1985) Thermoluminiscence dating of volcanic ash. J Volcanol Geotherm Res 25:333–347

Bitschene PR, Schmincke HU (1990) Composition and evolution of North Atlantic ash layers and the question of volcanic episodicity (in preparation)

Bitschene PR, Schmincke HU, Viereck L (1989) Cenozoic ash layers on the Vøring Plateau (ODP Leg 104). Proc Ocean Drill Prog Initial Rep 104B. US Gov Print Off, Washington DC (in press)

Bogaard Pvd, Schmincke HU (1985) Laacher See tephra: A widespread isochronous late Quaternary tephra layer in central and northern Europe. Geol Soc Am Bull 96:1554–1571

Bogaard Pvd, Schmincke HU (1988) Aschenlagen als quartäre Zeitmarken in Mitteleuropa. Geowissenschaft 6:75–84

Bogaard Pvd, Hall CM, Schmincke HU, York D (1987) $^{49}Ar/^{39}Ar$ Laser dating of single grains: ages of Quaternary tephra from the East Eifel volcanic field, FRG. Geophys Res Lett 14:1211–1214

Bouroz A, Tourette M, Vialette Y (1972) Signification de mesures d'ages de cinérites, tonsteins et rhyolites de bassins houillers francais. Mem BRGM 77:951–956

Bramlette MN, Bradley WH (1942) Geology and biology of North Atlantic deep-sea cores between Newfoundland and Ireland, pt. 1, Lithology and geologic interpretations. US Geol Surv Prof Pap 196A:1–55

Burger K (1982) Kohlentonsteine als Zeitmarken, ihre Verbreitung und ihre Bedeutung für die Exploration und Exploitation von Kohlenlagerstätten. Z Dtsch geol Ges 133:201–255

Carey S, Sigurdsson H (1982) Influence of particle aggregation on deposition of distal tephra from the May 18, 1980 eruption of Mount St. Helens volcano, J Geophys Res 87:7061–7072

Carey S, Sigurdsson H (1986) The 1982 eruptions of El Chichon volcano, Mexico (2): Observations and numerical modelling of tephra-fall distribution. Bull Volcanol 48:127–141

Carpena J, Mailhe D (1987) Fission-track dating calibration of the Fish Canyon tuff standard in French reactors. Chem Geol (Isot Geosci Sect) 66:53–59

Cassignol C, Gillot PY (1982) Range and effectiveness of unspiked potassium-argon dating: experimental groundwork and applications. In: Odin GS (ed) Numerical dating in stratigraphy. John Wiley & Sons, New York, pp 159–179

Damon PE, Teichmüller R (1971) Das absolute Alter des sanidinführenden kaolinitischen Tonsteins im Flöz Hagen 2 des Westfal C im Ruhrrevier. Fortschr Geol Rheinl Westf 18:53–56

Donn WL, Ninkovich D (1980) Rate of Cenozoic explosive volcanism in the North Atlantic Ocean inferred from deep sea cores. J Geophys Res 85:5455–5460

Donnelly TW (1975) Neogene explosive volcanic activity of the Western Pacific: Sites 292 and 296. In: Karig DE, Ingle JC, et al. (eds) Initial Rep Deep Sea Drill Proj 31. US Gov Print Off, Washington DC, pp 577–597

Eldholm O, Thiede J, Taylor E et al. (1987) Proc Ocean Drill Prog Initial Rep 104. US Gov Print Off, Washington DC, pp 1–783

Faure G (1986) Principles of isotope geology, 2nd edn. John Wiley & Sons, New York, pp 1–464

Federman AN, Scheidegger KF (1984) Compositional heterogeneity of distal tephra deposits from the 1912 eruption of Novarupta, Alaska. J Volcanol Geotherm Res 21:233–254

Fisher RV (1964) Maximum size, median diameter, and sorting of tephra. J Geophys Res 69:341–355

Fisher RV, Schmincke HU (1984) Pyroclastic rocks. Springer, Berlin Heidelberg New York, pp 1–472

Fujioka K (1986) Synthesis of Neogene explosive volcanism of the Tohoku Arc, deduced from the marine tephra drilled around the Japan Trench Region, DSDP Legs 56, 57 and 87B. In: Kagami H, Karig DE, Coulbourn WT, et al. (eds) Initial Rep Deep Sea Drill Proj 87. US Gov Print Off, Washington DC, pp 703–723

Gillot PY, Cornette Y (1986) The Cassignol technique for potassium-argon dating, precision and accuracy: examples from the Late Pleistocene to recent volcanics from southern Italy. Chem Geol (Isot Geosci Sect) 59:205–222

Harris DM, Rose WI Jr, Roe R, Thompson MR (1981) Radar observations of ash eruptions. In: Lipman PW, Mullineaux DR (eds) The 1980 eruptions of Mount St. Helens, Washington. US Geol Surv Prof Pap 1250:323–333

Harris WB, Fullagar PD (1989) Comparison of Rb-Sr and K-Ar dates of Middle Eocene bentonite and glauconite, southeastern Atlantic Coastal Plain. Geol Soc Am Bull 101:573–577

Heiken G, Wohletz K (1985) Volcanic ash. Univ Cal Press, Berkely, pp 1–246

Hellmann KN, Lippolt HJ (1981) Calibration of the Middle Triassic time scale by conventional K-Ar and $^{40}Ar/^{39}Ar$ dating of alkali feldspars. J Geophys 50:73–88

Hess JC, Lippolt HJ (1986) $^{40}Ar/^{39}Ar$ ages of tonstein and tuff sanidines: new calibration points for the improvement for the upper Carboniferous time scale. Chem Geol (Isot Geosci Sect) 59:143–154

Hess JC, Backfisch S, Lippolt HJ (1983) Konkordantes Sanidin- und diskordante Biotitalter eines Karbontuffs der Baden-Badener Senke, Nordschwarzwald. N Jahrb Geol Paläontol Mh 1983:277–292

Hess JC, Lippolt HJ, Wirth R (1987) Interpretation of $^{40}Ar/^{39}Ar$ spectra of biotites: evidence from hydrothermal degassing experiments and TEM studies. Chem Geol 66:137–149

Huff WD, Türkmenoglu AG (1981) Chemical characteristics and origin of Ordovician K-bentonites along the Cincinati arch. Clays Clay Minerals 29:113–123

Hurford AJ, Watkins RT (1987) Fission-track age of the tuffs of the Buluk Member, Balkate Formation, northern Kenya: a suitable fission-track age standard. Chem Geol (Isot Geosci Sect) 66:209–216

Izett GA, Naeser CW (1976) Age of the Bishop Tuff of eastern California as determined by the fission-track method. Geology 4:587–590

Jäger E, Hunziker JC (eds) (1979) Lectures in istope geology. Springer, Berlin Heidelberg New York, pp 1–328

Jezek PA, Noble DC (1978) Natural hydration and ion exchange of obsidian: an electron microprobe study. Am Mineral 63:266–273

Keller J (1981) Quaternary tephrochronology in the Mediterranean region. In: Self S, Sparks RSJ (eds) Tephra studies. Reidel, Dordrecht, pp 227–244

Kennett JP (1981) Marine tephrochronology. In: Emiliani C (ed) The oceanic lithosphere. The Sea. John Wiley and Sons, New York: 1373–1436

Kennett JP (1983) Marine geology. Prentice Hall, Englewood Cliffs, NJ, pp 1–813

Kennett JP, Thunell RC (1975) Global increase in Quaternary volcanism. Science 187:497–503

Kennett JP, McBirney AR, Thunnell (1977) Episodes of Cenozoic volcanism in the circum-Pacific Region. J Volcanol Geotherm Res 2:145–163

Kober (1986) Whole-grain evaporation for $^{207}Pb/^{206}Pb$-age-investigations on single zircons using a double-filament thermal ion source. Contrib Mineral Petrol 93:482–490

Lamb HH (1971) Volcanic dust in the atmosphere, with a chronology and assessment of its meteorological significance. Philos Trans R Soc London Ser A 266:425–553

Ledbetter MT (1985) Tephrochronology of marine tephra adjacent to Central America. Geol Soc Am Bull 96:77–82

Lipman PW, Mullineaux DR (eds) (1981) The 1980 eruptions of Mount St. Helens, Washington. US Geol Surv Prof Pap 1250:1–844

Lippolt HJ, Hess JC, Holub VM, Pesek J (1986a) Correlation of Upper Carboniferous deposits in the Bohemian Massif (Czechoslovakia) and in the Ruhr district (FR Germany). Evidence from $^{40}Ar/^{39}Ar$ ages of tuff layers. Z Dtsch Geol Ges 137:447–464

Lippolt HJ, Fuhrmann U, Hradetzky H (1986b) $^{40}Ar/^{39}Ar$ age determinations on sanidines of the Eifel volcanic field (Fed. Rep. of Germany): constraints on age and duration of a Middle Pleistocene cold period. Chem Geol (Isot Geosci Sect) 59:187–204

Lo Bello P, Feraud G, Hall CM, York D, Lavina P, Bernat M (1987) $^{40}Ar/^{39}Ar$ Stepheating and laser fusion dating of a Quaternary pumice from Neschers, Massif Central, France: the defeat of xenocrystic contamination. Chem Geol (Isot Geosci Sect) 66:61–71

McCoy FW jr (1981) Areal distribution, redeposition and mixing of tephra within deep-sea sediments of the eastern Mediterranean Sea. In: Self S, Sparks RSJ (eds) Tephra studies. Reidel, Dordrecht, pp 245–254

Merrihue C, Turner G (1966) Potassium-argon dating by activation with fast neutrons. J Geophys Res 71:2852–2857

Montanari A, Drake R, Bice DM, Alvarez W, Curtis GH, Turrin BD, Depaolo DJ (1983) Radiometric dating of the Eocene-Oligocene boundary at Gubbio, Italy. In: Pomerol C, Premolisilva I (eds) Terminal Eocene events. Dev Paleontol Stratigr 9:41–47

Morton AC, Keene JB (1984) Paleogene pyroclastic volcanism in the southwest Rockall Plateau. In: Roberts DG, Schnitger D, et al. (eds) Initial Rep Deep Sea Drill Proj 81. US Gov Print Off, Washington DC, pp 633–641

Müller G (1961) Die rezenten Sedimente im Golf von Neapel. Mineral-Neu- und Umwandlungen in den rezenten Sedimenten des Golfes von Neapel. Ein Beitrag zur Umwandlung vulkanischer Gläser durch Halmyrolyse. Beitr Mineral Petrol 8:1–20

Naeser CW, Izett GA, Wilcox RE (1973) Zircon fission-track ages of Perlette family ash beds in Meade County, Kansas. Geology 1:187–189

Naeser CW, Briggs ND, Obradovich JD, Izett GA (1981) Geochronology of Quaternary tephra deposits. In: Self S, Sparks RSJ (eds) Tephra studies. Reidel, Dordrecht, pp 13–47

Ninkovich D, Donn WL (1976) Explosive Cenozoic volcanism and climatic implications. Science 194:899–906

Ninkovich D, Sparks RSJ, Ledbetter MT (1978) The exceptional magnitude and intensity of the Toba eruption, Sumatra: an example of the use of deep-sea tephra layers as a geological tool. Bull Volcanol 41:286–298

Odin GS (1982) Numerical dating in stratigraphy. John Wiley & Sons New York, pp 1–1040

Odin GS, Hunzicker JC, Jeppsson L, Spjeldnaes N (1986) Ages radiometriques K-Ar de biotites pyroclastiques sedimentees dans le Wenlock de Gotland (Suede). Chem Geol (Isot Geosci Sect) 59:117–125

Paterne M, Guichard F, Labeyrie J, Gillot PY, Duplessy JC (1986) Tyrrhenian sea tephrochronology of the oxygen isotope record for the past 60 000 years. Mar Geol 72:259–285

Paterne M, Guichard F, Labeyrie J (1988) Explosive activity of the South Italian volcanoes during the past 80 000 years as determined by marine tephrochronology. J Volcanol Geotherm Res 34:153–172

Rabek K, Ledbetter MW, Williams DF (1985) Tephrochronology of the western Gulf of Mexico. Quat Res 23:403–416

Rad Uv, Kreuzer H (1987) Composition, K-Ar dates and origin of a Mid-Eocene rhyolitic ash layer at Deep sea Drilling Project Sites 605 and 613, New Jersey transect, Legs 93 and 95. In: Hinte JEv, Wise SW, et a. (eds) Initial Rep Deep Sea Drill Proj 93. US Grov Print Off, Washington DC, pp 977–981

Rose WI, Chesner CA (1987) Dispersal of ash in the great Toba eruption, 75 Ka. Geology 15:913–917

Ruddiman WF, Glover LK (1972) Vertical mixing of ice-rafted volcanic ash in the North Atlantic sediments. Geol Soc Am Bull 83:2817–2836

Rundle CC (1986) Radiometric dating of a Caradocian tuff horizon. Chem Geol (Isot Geosci Sect) 59:111–115

Sarna-Wojcicki AM, Meyer CE, Woodward MJ, Lamothe PJ (1981) Composition of airfall ash erupted on May 18, May 25, June 12, July 22, and August 7. In: Lipman PW, Mullineaux DR (eds) The 1980 eruptions of Mount St Helens, Washington. US Geol Surv Prof Pap 1250:667–681

Scheidegger KF, Kulm LD (1975) Late Cenozoic volcanism in the Aleutian arc: information from ash layers in the northeastern Gulf of Alaska. Geol Soc Am Bull 86:1407–1412

Schmincke HU (1982) Volcanic and chemical evolution of the Canary Islands. In: Rad Uv, Hinz K, Sarnthein M, Seibold E (eds) Geology of the Northwest African continental margin. Springer, Berlin Heidelberg New York, pp 273–306

Schmincke HU, Sunkel G (1987) Carboniferous submarine volcanism at Herbornseelbach (Lahn-Dill area, Germany). Geol Rundsch 76:709–734

Schmincke HU, Bogaard Pud (1990) Tephra layers and tephra events. In Event stratigraphy. Einsele S, Ricken W, Seilacher A (ed) 2nd ed Springer Verlag (in press)

Schumacher R, Schmincke HU (1990) Internal structure and occurences of accretionary lapilli. Bull Volcanol (in preparation)

Self S (1976) The recent volcanology of Terceira, Azores. J Geol Soc London 132:645–666

Self S, Sparks RSJ (eds) (1981) Tephra studies. Reidel, Dordrecht, pp 1–481

Seward D (1979) Composition of zircon and glass fission-track ages from tephra horizons. Geology 7:479–482

Sigurdsson H (1981) First order major element variation in basalt glasses from the Mid-Atlantic Ridge: 29° N to 73° N. J Geophys Res 86:9483–9502

Sigurdsson H, Carey SN (1981) Marine tephrochronology and Quaternary explosive volcanism in the Lesser Antilles arc. In: Self S, Sparks RSJ (eds) Tephra studies. Reidel, Dordrecht, pp 255–280

Sigurdsson H, Loebner B (1981) Deep-sea record of Cenozoic explosive volcanism in the North Atlantic. In: Self S, Sparks RSJ (eds) Tephra studies. Reidel, Dordrecht, pp 289–316

Smalley PC, Nordaa A, Raheim A (1986) Geochronology and paleothermometry of Neogene sediments from the Vøring Plateau using Sr, C and O isotopes. Earth Planet Sci Lett 78:368–378

Sorem RK (1982) Volcanic ash clusters: tephra rafts and scavengers. J Volcanol Geotherm Res 13:63–71

Sparks RSJ, Huang TC (1980) The volcanological significance of deep-sea ash layers associated with ignimbrites. Geol Mag 117:425–436

Sparks RSJ, Walker GPL (1977) The significance of vitric-enriched air-fall ashes associated with crystal-enriched ignimbrites. J Volcanol Geotherm Res 2:329–341

Sparks RSJ, Self S, Walker GPL (1973) Products of ignimbrite eruption. Geology 1:115–118

Stothers RB, Rampino MW, Self S, Wolff JA (1989) Volcanic winter? Climatic effects of the largest volcanic eruptions. In: Latter JH (ed) Volcanic hazards. IAVCEI Proc Volcanol 1:3–9

Sylvester AG (1978) Petrography of volcanic ashes in deep sea cores near Jan Mayen island: Sites 338, 345–350 DSDP Leg 38. In: Talwani H, Udintsev G, et al. (eds) Initial Rep Deep Sea Drill Proj Suppl 38, 39, 40, 41. US Gov Print Off, Washington DC, pp 101–109

Talwani H, Udintzev G, et al. (1976) Initial Rep Deep Sea Drill Proj 38. US Gov Print Off, Washington DC, pp 1–1256

Thorarinsson S (1944) Tefrokronologiska studier pa Island. Geogr Ann 26:1–217

Thorarinsson S (1981) Tephra studies and tephrochronology: a historical review with special references to Iceland. In: Self S, Sparks RSJ (eds) Tephra studies. Reidel, Dordrecht, pp 1–12

Vallier TL, Bohrer D, Moreland G, McKee EH (1977) Origin of basalt microlapilli in lower Miocene pelagic sediment, northeastern Pacific Ocean. Geol Soc Am Bull 88:787–796

Varekamp J, Luhr J, Prestegaard K (1984) The 1982 eruptions of El Chichon volcano (Chiapas, Mexico): character of the eruptions, ash fall deposits, and gas phase. J Volcanol Geotherm Res 23:39–68

Varet J, Metrich N (1979) Ash layers interlayered with the sediments of holes 407 and 408, IPOD Leg 49. Initial Rep Deep Sea Drill Proj 49. US Gov Print Off, Washington DC, pp 437–441

Vinci A (1985) Distribution and chemical composition of tephra layers from eastern Mediterranean abyssal sediments Mar Geol 64:143–155

Vogt PR (1979) Global magnetic episodes: new evidence and implication for the steady-state mid-oceanic ridges. Geology 7:93–98

Walker GPL (1981) Plinian eruptions and their products. Bull Volcanol 44:223–240

Walker GPL, Croasdale RC (1971) Two Plinian-type eruptions in the Azores. J Geol Soc London 127:17–55

Watkins ND, Sparks RSJ, Sigurdsson H, Huang TC, Federman A, Carey S, Ninkovich D (1978) Volume and extent of the Minoan tephra from Santorini: new evidence from deep-sea sediment cores. Nature (London) 271:122–126

Wentworth CK (1938) Ash formations on the island of Hawaii. 3rd Spec Rep Hawaiian Volcano Observatory, Honolulu, Hawaii, pp 1–183

Westgate JA, Gorton MP (1981) Correlation techniques in tephra studies. In: Self S, Sparks RSJ (eds) Tephra studies. Reidel, Dordrecht, pp 73–94

Winter J (1981) Exakte tephrostratigraphische Korrelation mit morphologisch differenzierten Zirkonpopulationen (Grenzbereich Unter-/Mitteldevon, Eifel-Ardennen). N Jahrb Geol Paläontol Abh 162:97–136

Worzel LJ (1959) Extensive deep-sea sub-bottom reflections identified as white ash. Proc Natl Acad Sci USA 45:349–355

Yanagi T, Baadsgaard H, Stelck CR, McDougall I (1988) Radiometric dating of a tuff bed in the Middle Albian Hulcross Formation at Hudson's Hope, British Columbia. Can J Earth Sci 25:1123–1127

York D, Hall CM, Yanase Y, Hanes JA, Kenyon WJ (1981) $^{40}Ar/^{39}Ar$ dating of terrestrial minerals with a continuous laser. Geophys Res Lett 8:1136–1138

Isotope Geochemistry of Primary and Secondary Carbonate Minerals in the Shaban-Deep (Red Sea)

P. Stoffers, R. Botz, and J. Scholten[1]

CONTENTS

Abstract	83
1 Introduction	83
2 Tectonic Setting	84
3 Material and Methods	84
4 Results	88
5 Discussion	88
5.1 Core 379 KL	88
5.2 Core 378 KL	90
6 Conclusions	92
References	92

Abstract

The isotopic composition of carbonate sediments from the Shaban Deep indicates that minerals of low- and high-Mg calcite and aragonite formed in near-isotopic equilibrium with Red-Sea-water bicarbonate at normal sedimentary temperatures.

However, within organic-rich layers, diagenetically formed dolomite and rhodochrosite are found. The isotopic composition of the poorly ordered dolomite is characterized by negative $\delta^{13}C$ values ($\delta^{13}C_{min} = -5.7‰$). This indicates that ^{12}C-rich biogenic CO_2 contributed to the carbonate formation. It is believed that this ^{12}C-rich biogenic Co_2 was formed during the anaerobic oxidation of organic matter during sulphate reduction.

1 Introduction

The Red Sea sediments, in particular the metalliferous deposits in the various deeps within the axial valley, have attracted the attention of sedimentologists

---

[1] Geologisch-Paläontologisches Institut u. Museum, Universität Kiel, Olshausenstr. 40–60, D-2300 Kiel, FRG

and mineralogists since their first description by Degens and Ross (1969) and Bischoff (1969). In addition, many studies have focused on the sedimentary environment of the Red Sea. These studies generally indicate considerable climatic fluctuations within the Holocene and upper Pleistocene, reflected by changes in fauna and sediment characteristics (Milliman et al. 1969; Deuser and Degens 1969; Rossignol-Strick 1987; Thunell et al. 1988). Carbonates are major constituents of Red Sea sediments (Herman 1965; Gevirtz and Friedman 1966; Milliman et al. 1969; Stoffers and Botz 1989). The minerals are generally precipitated from primary seawater (Milliman et al. 1969; Ellis and Milliman 1985), although early diagenetic cementation by aragonite, Mg calcite and/or dolomite is known (Gevirtz and Friedman 1966; Stoffers and Botz 1989).

2 Tectonic Setting

The northern Red Sea is dominated by sedimentation and salt tectonics. The fault-dominated graben structure, typical of the axial trough in the central Red Sea, is replaced by a shallow depression which is offset twice by fracture zones striking in the "Aquaba" direction. Within this zone, a few deeps occur (Fig. 1; e.g., Bannock, Vema, Kebrit, Shaban, Oceanographer), some of which contain brines and are characterized by high heat-flow values and volcanic injections.

The Shaban Deep, discovered in 1981 during the Arabian month of Shaban by F. S. Valdivia, lies near 26°15'N, 35°22'E, and has a maximum depth of 1490 m (uncorrected). According to detailed seabeam surveys (Pautot et al. 1984; Puchelt and Laschek 1984), the deep measures about 10 × 6 km. It is subdivided into two large subbasins by a prominent ridge; a less pronounced sill, perpendicular to this ridge, separates two other smaller subbasins. Layered brines, with a salinity of 190‰, and a temperature of 24.8° C, have been observed in the two main subbasins. The acoustic surface of the brines is at 1325 m (corrected).

3 Material and Methods

Two 6-m-long box cores were taken in the two basins. Superpenetration was achieved at Station 378 located in the SW basin.

The bulk sediments were analyzed by X-ray diffraction analysis (Cu kα radiation). The carbonate content was measured by applying the method of Müller and Gastner (1971). The amount of organic carbon was determined using a Carlo-Erba element analyzer.

Carbon and oxygen isotopic analyses of 70 bulk carbonate samples were performed by applying standard techniques (McCrea 1950) The results are

Isotope Geochemistry of Carbonate Minerals in the Shaban-Deep 85

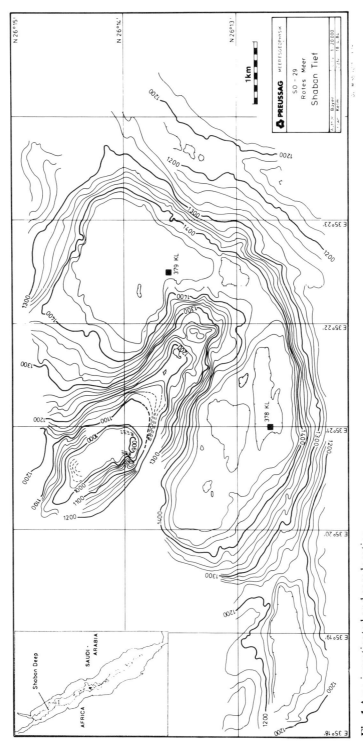

Fig. 1 Area investigated and core locations

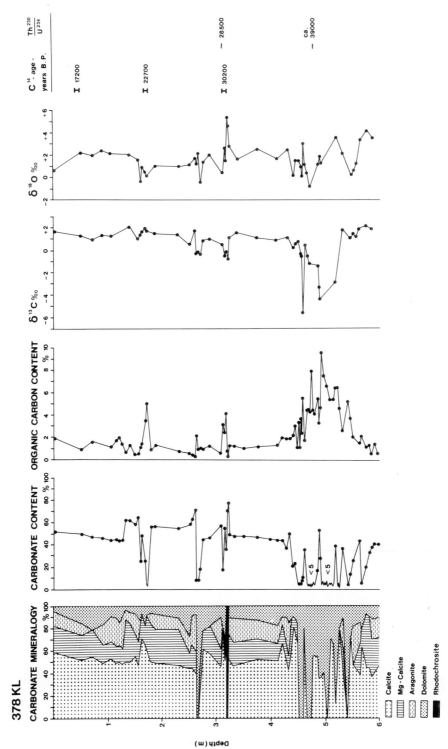

Fig. 2 The carbonate mineralogy, the inorganic and organic carbon content, the δ values of the carbonates and the ^{14}C and ^{230}Th ages of core 378 KL

Isotope Geochemistry of Carbonate Minerals in the Shaban-Deep 87

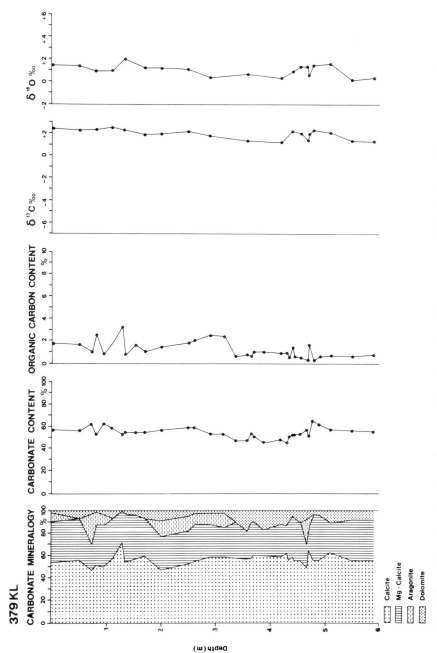

Fig. 3 The carbonate mineralogy, the inorganic and organic carbon content and the δ values of the carbonates of core 379 KL

given in per mil deviations from the PDB standard using the standard δ notation:

$$\delta = \frac{R_{sample} - R_{standard}}{R_{standard}} \times 1000\ (‰);$$

$$R = {}^{13}C/{}^{12}C;\ {}^{18}O/{}^{16}O.$$

The $\delta^{18}O$ values of dolomite samples (> 50% dolomite) were corrected by $-0.8‰$ (Sharma and Clayton 1965).

4 Results

The carbonate content of the sediments from the Shaban deep varies between < 5 and 87%. Figures 2 and 3 show that the carbonate content of core 379 KL is less variable (49 to 71%) than that of core 378 KL. In the latter core, well-defined carbonate-rich (up to 87%) and carbonate-poor (< 5%) layers are recognizable. The predominant carbonate minerals are low- and high-Mg calcite. Aragonite is a minor constituent. However, poorly ordered dolomite and rhodochrosite may be the major carbonate minerals within organic-rich layers (Fig. 2). The mean organic carbon content of the sediments is 1 to 2%. However, organic-rich horizons of core 378 KL (lower core half) contain up to 9.8% organic carbon (Figs. 2 and 3). The $\delta^{13}C$ values of all carbonate samples investigated vary between 2.5 and $-5.7‰$. Their $\delta^{18}O$ values range from 5.5 to $-1.0‰$. Core 379 KL reveals more homogeneous δ values than core 378 KL (Fig. 3). Negative $\delta^{13}C$ values are characteristic for dolomite-rich samples in 454–519 cm depth of core 378 KL. Pure dolomites in core 378 KL have $\delta^{13}C$ values between -0.2 and $-5.7‰$. Their $\delta^{18}O$ values range from -1 to $3‰$.

The most positive $\delta^{18}O$ values were measured for rhodochrosite in core 378 KL ($\delta^{18}O = 4.6$ and $5.5‰$, respectively). A rough time scale for the sediments (core 378 KL) is given in Fig. 2.

5 Discussion

The mineralogy of the sediments from the Shaban Deep is predominated by primary carbonate minerals. These are low- and high-Mg calcite and aragonite. The origin of the primary minerals in the Red Sea is discussed by Gevirtz and Friedman (1966) and Milliman et al. (1969). However, diagenetic mineral formation is indicated by stable isotope analyses.

5.1 Core 379 KL

The $\delta^{13}C$ values of the carbonate sediments from core 379 KL indicate that the carbonates precipitated are close to equilibrium with atmospheric CO_2, which has a $\delta^{13}C$ value of $-7‰$ (Keeling 1958). The fractionation between solid calcium carbonate and gaseous carbon dioxide is 10.2‰ at 20° C (Emrich et al. 1970), which gives a $\delta^{13}C$ value of 3.2‰ for solid calcium carbonate in isotopic equilibrium with the atmospheric CO_2. However, carbonate formation by organisms such as molluscs, foraminifera, or coccoliths is a nonequilibrium process in shell building as they rely exclusively on organic template phenomena. These shells have $\delta^{13}C$ values identical or close to that of seawater bicarbonate (Deuser and Degens 1969; Degens 1976). Schoell and Stahl (1972) reported the isotopic composition of the dissolved inorganic carbon in Red Sea water as $-1.5‰$ PDB.

The carbonates of the Red Sea are of both biogenic and nonbiogenic origin. Low- and high-Mg calcite are the predominant carbonate minerals present in core 379 KL. Low-Mg calcite is derived from shells (mainly coccoliths), whereas high-Mg calcite is believed to be mainly of inorganic origin (Gevirtz and Friedman 1966; Milliman et al. 1969). Thus, the $\delta^{13}C$ values in the range of 1.1 and 2.5‰ for low- and high-Mg calcite mixtures reflect the mineral formation near isotopic equilibrium with the dissolved seawater bicarbonate.

Primary variations in the oxygen-isotope composition of carbonates are due to changes in water salinity and/or temperature during carbonate formation. Present-day Red Sea water has a $\delta^{18}O$ value of 1.8‰ (Schoell and Faber 1978). The ^{18}O values measured for bulk carbonates from the top of core 379 KL are around 1.5‰, which is characteristic for carbonate formation from this water under sedimentary temperatures (18°C using the equation of Shackleton and Kennett 1975).

However, since the bulk carbonate, rather than a single specific component, was analyzed, the $\delta^{18}O$ values can be interpreted only tentatively in terms of temperature and/or salinity fluctuations during carbonate formation. The reason is that the $\delta^{18}O$ (and $\delta^{13}C$) values also depend on the type of carbonate. For instance, the ^{18}O content of aragonite at 25°C is 0.6‰ (and the ^{13}C content 1.8‰) higher than in the coexisting calcite (Rubinson and Clayton 1969). Furthermore, ^{18}O is concentrated in Mg-calcite relative to pure calcite. For each mol % $MgCO_3$ in calcite, the ^{18}O content increases by 0.06‰ (Tarutani et al. 1969).

The carbonate mineralogy of core 379 KL is uniform (Fig. 3), consisting predominantly of low-Mg calcite (47–63%) and high-Mg calcite (21–43%). The high-Mg calcite of Red Sea sediments usually contains 12 ± 1.4 mol % Mg (Milliman et al. 1969). The present study revealed Mg contents most common around 11 mol %. A Mg content of 11 mol % would result in approximately 0.7‰ more positive $\delta^{18}O$ values for pure Mg-calcites relative to Mg-free calcites.

In discussing the oxygen isotope composition of carbonates, one must consider postdepositional changes of the orginal isotope composition. For example, Deuser (1968) found negative $\delta^{18}O$ values (-1.3 to $-0.6‰$) with only small fluctuations in foraminifera tests from the Discovery Deep in the

Red Sea. One core, taken 6 km away from the deep, revealed significantly larger (primary) fluctuations in $\delta^{18}O$ (-0.9 to $+0.7$‰). Deuser (1968) explained the more uniform and lower $\delta^{18}O$ values in foraminifera from the Discovery Deep as the result of postdepositional changes in the oxygen isotope ratios of foraminifera towards equilibrium with hot brines. At present, the temperature of the brine in the Discovery Deep is 44.7°C. This is a considerably higher temperature than that of normal Red Sea deep water, which is about 22°C below 500 m. However, the present temperature of the brine in the Shaban Deep is 24.8°C (Pautot et al. 1984). This is a temperature only 2° to 3°C higher than that of normal bottom water. Thus, postdepositional changes in $\delta^{18}O$ of carbonates due to the present brine are unlikely.

In summary, the more or less uniform δ values of carbonate sediments from core 379 KL reflect the primary formation of the carbonate minerals in seawater at normal sedimentary temperatures. Probable fluctuations in $\delta^{18}O$ (and/or temperature) of Red Sea water caused by glacially induced eustatic changes in sea level may explain the $\delta^{18}O$ range of 1.9‰ for bulk carbonate in core 379 KL.

5.2 Core 378 KL

A more complex situation was found for the sediments of core 378 KL. Here, rapid changes in both the mineralogical composition and the carbon and oxygen isotope geochemistry can be seen. One important feature of core 378 KL is the occurrence of organic-rich horizons with up to 9.8% organic carbon. These sediments are generally characterized by very low carbonate contents. Furthermore, the carbonate minerals are dolomite (or mixtures of dolomite and low-Mg calcite) and, at a depth of 315 to 323 cm, rhodochrosite.

The carbon isotope composition of carbonate minerals is determined by the ^{13}C content of dissolved HCO_3^-, which depends on the source of the HCO_3^-. The $\delta^{13}C$ values of primary carbonates from core 379 KL indicate that the minerals formed from HCO_3^- in, or close to, isotopic equilibrium with atmospheric CO_2. The same is true for carbonate sediments from the top of core 378 KL, which have similar $\delta^{13}C$ values (1 to 2‰). However, with increasing depth, the $\delta^{13}C$ values of the carbonates within the organic-rich layers become distinctly more negative (Fig. 2). For example, the most negative $\delta^{13}C$ values (-5.7 to -2.9‰) were found for the dolomite-rich samples between 458 and 519 cm depth, in core 378 KL. The negative $\delta^{13}C$ values of the carbonates suggest that organically derived, ^{12}C-rich HCO_3^- contributed to the carbonate formation. Marine organic matter has negative $\delta^{13}C$ values in the approximate range of -20 to -30‰. Thus, contributions of CO_2, derived from the degradation of organic matter during very early stages of diagenesis may result in negative $\delta^{13}C$ values of diagenetically formed carbonate minerals.

After Claypool and Kaplan (1974), Irwin (1980), Irwin et al. (1977), Pisciotto and Mahoney (1981) and Kelts and McKenzie (1980), four diagenetic zones in anoxic sediments can be differentiated from top to base: (1) a bacterial oxidation zone; (2) a bacterial sulphate reduction zone; (3) a carbonate reduction zone; (4) a deep zone where thermal processes cause the degradation of organic matter. Microbial oxidation and anaerobic sulphate reduction can result in negative $\delta^{13}C$ carbonate values as low as $-21‰$. However, positive $\delta^{13}C$ carbonate values (to 15‰) can be found below this zone, where CH_4 generation by CO_2 reduction results in residual ^{13}C-rich bicarbonate (the process of CH_4 generation by CO_2 reduction is related to the disappearance of dissolved sulphate). Thermocatalytic decarboxylation at greater depth can result in isotopically light CO_2 and, therefore, with increasing depth, the $\delta^{13}C$ values of authigenic carbonates may decrease again.

In core 378 KL, the negative $\delta^{13}C$ values were found only in sediments with high organic contents where anoxic conditions prevail. These sediments are also characterized by a relatively high pyrite content. Bacterial sulphate reduction processes are therefore believed to be responsible for the negative $\delta^{13}C$ values of carbonates within those horizons. The actual carbon isotope composition of the diagenetically formed carbonate minerals is determined by the relative amounts of HCO_3^- with $\delta^{13}C$ values around 0‰ (possibly derived from the dissolution of primary minerals such as high-Mg calcite and/or aragonite) and the organically derived ^{12}C-rich HCO_3^-, formed during sulphate reduction processes.

The $\delta^{18}O$ values of all carbonate samples from core 378 KL range from -0.9 to 5.5‰. This is a considerably larger range than that of core 379 KL. It is probable that the sediments of core 378 KL were deposited in waters which had undergone large fluctuations in $^{18}O/^{16}O$ ratio, salinity, and temperature, possibly as a result of glacially induced eustatic changes in sea level in the Pleistocene. This is indicated by the sediment sequence of core 378 KL, e.g., the alternation of "normal marine" sediments consisting predominantly of low- and high-Mg calcite and the carbonate-poor, organic-rich sediments. The latter were deposited when the water flow in the basin was restricted, and stagnant conditions prevailed. Deuser and Degens (1969) suggest the repeated occurrence of periods of evaporation in the Red Sea during the last 80000 to 100000 years. They believed that these periods coincided with those of lowered sea level, which severely restricted the exchange of water between the Red Sea and the Indian Ocean.

However, the carbonates of core 378 KL are of mixed primary and secondary origin. The $\delta^{13}C$ values showed that dolomite formed during the anaerobic degradation of organic matter. The oxygen isotope fractionation of sedimentary dolomite is still unclear. Theoretically, dolomite, formed in isotopic equilibrium with calcite under sedimentary temperatures, should be enriched in ^{18}O by 4 to 7‰ relative to the calcite (Epstein et al. 1964; O'Neil and Epstein 1966; Northrop and Clayton 1966). Land (1980) and McKenzie (1981) suggested equilibrium values of $\Delta^{18}O$ dolomite/calcite near 3‰ at

25 °C. However, the oxygen isotope composition of sedimentary dolomite and calcite was often found to be similar (Degens and Epstein 1964; Northrop and Clayton 1966; O'Neil and Epstein 1966; Fritz and Smith 1970; Matthews and Katz 1977). Dolomites from core 378 KL have a wide (4‰) range of $\delta^{18}O$ values. If these dolomites precipitated in isotopic equilibrium with pore waters at a certain temperature, the $\delta^{18}O$ values of the pore waters must have been variable during dolomite formation. Even more positive $\delta^{18}O$ values (4.6 and 5.5‰) were found for rhodochrosite at a depth of 320 cm in core 378 KL. Wada et al. (1982) also found rhodochrosite enriched in ^{18}O ($\delta^{18}O = 7.4‰$) in deep-sea sediments. The authors conceived that ^{18}O enrichment was produced by squeezing water depleted in ^{18}O through clay layers (Kharaka et al. 1973). Thus, it is not clear to what extent primary changes in the seawater $\delta^{18}O$, or secondary processes during diagenenis, determined the isotope composition of the carbonate minerals dolomite and rhododrosite.

6 Conclusions

The sediments in the Shaban Deep are dominated by the primary low- and high-Mg calcite and little aragonite. The isotope composition of these minerals reflects their formation near isotopic equilibrium with Red Sea water bicarbonate at normal sedimentary temperatures. However, in the lower half of core 378 KL organic-rich layers occur. These layers, although carbonate-poor, contain poorly ordered dolomite and occasional rhodochrosite. The isotope composition of the dolomite (rhodochrosite) indicates its formation during diagenesis, when ^{12}C-rich bicarbonate from the anaerobic degradation of organic matter (sulphate reduction) was available for carbonate formation.

References

Bischoff JL (1969) Red Sea geothermal brine deposits: Their mineralogy, chemistry and genesis. In: Degens ET, Ross DA (eds) Hot brines and recent heavy metal deposits in the Red Sea. Springer, Berlin Heidelberg New York, pp 368–401
Claypool GE, Kaplan IR (1974) The origin and distribution of methane in marine sediments. In: Kaplan IR (ed) Natural gases in marine sediments: marine science, vol 3. Plenum, New York, pp 99–139
Degens ET (1976) Molecular mechanisms of carbonate, phosphate and silica deposition in the living cell. Springer, Berlin Heidelberg New York, 112 pp
Degens ET, Epstein S (1964) Oxygen and carbon isotope ratios in coexisting calcites and dolomites from recent and ancient sediments. Geochim Cosmochim Acta 28:23–44
Degens ET, Ross DA (1969) Hot brines and recent heavy metal deposits in the Red Sea. Springer, Berlin Heidelberg New York, 600 pp
Deuser WG (1968) Postdepositional changes in the oxygen isotope ratios of Pleistocene foraminifera tests in the Red Sea. J Geophys Res 73:3311

Deuser WG, Degens ET (1969) O-18/O-16 and C-13/C12 ratios of fossils from the hot brine deep area of the central Red Sea. In: Degens ET, Ross D (eds) Hot brines and recent heavy metal deposits. Springer, Berlin Heidelberg New York, pp 336–347

Ellis JP, Milliman JD (1985) Calcium carbonate suspended in Arabian Gulf and Red Sea waters: Biogenic and detrital, not "chemogenic". J Sediment Petrol 55:805–808

Emrich K, Ehhalt DH, Vogel JC (1970) Carbon fractionation during the precipitation of calcium carbonate. Earth Planet Sci Lett 8:363–371

Epstein S, Graf DL, Degens ET (1964) Oxygen isotope studies on the origin of dolomites. In: Craig H, Miller SL, Wasserburg GJ (eds) Isotopic and cosmic chemistry. Elsevier/North-Holland Biomedical Press, Amsterdam New York, pp 169–180

Fritz P, Smith DGW (1970) The isotopic composition of secondary dolomites. Geochim Cosmochim Acta 43:1161–1173

Gevirtz JL, Friedman GM (1966) Deep sea carbonate sediments of the Red Sea and their implications on marine lithification. J Sediment Petrol 36:143–152

Herman YR (1965) Etudes des sédiments quaternaires de la Mer Rouge. PhD Theesis, Univ Paris. Masson, Paris, pp 341–415

Irwin H (1980) Early diagenetic carbonate precipitation and pore fluid migration in the Kimmeridge clay of Dorset, England. Sedimentology 27:577–591

Irwin H, Curtis C, Coleman M (1977) Isotopic evidence for source of diagenetic carbonates formed during burial of organic-rich sediments. Nature (London) 269:209–213

Kharaka YK, Berry AF, Friedman I (1973) Isotopic composition of oil-field brines from Kettleman North Dome, California and their geological implications. Geochim Cosmochim Acta 37:1899–1908

Keeling CD (1958) The concentration and isotopic abundance of carbon dioxide in rural areas. Geochim Cosmochim Acta 13:322–334

Kelts KR, McKenzie JA (1980) Formation of deep sea dolomites in anoxic diatomaceous oozes. 26th Int Geolical Congr, Paris. Fr (Abstr)

Land LS (1980) The isotope and trace element geochemistry of dolomite: the state of the art. Spec Publ Soc Econ Paleontol Mineral 28:87–110

Matthews A, Katz A (1977) Oxygen isotope fractionations during the dolomitization of calcium carbonate. Geochim Cosmochim Acta 41:1431–1438

McCrea JM (1950) On the isotopic chemistry of carbonates and a paleotemperature scale. J Chem Phys 18:849–857

McKenzie JA (1981) Holocene dolomitization of calcium carbonate sediments from the coastal sabkas of Abu Dhabi. U. A. E.: a stable tope study. J Geol 89:185–198

Milliman JD, Ross DA, Ku T (1969) Precipitation and lithification of deep sea carbonates in the Red Sea. J Sediment Petrol 39, 2:724–736

Müller G, Gastner M (1971) The carbonate bombe, a simple device for the determination of the carbonate content in sediments, soils and other materials. N Jahrb Mineral 10:466–469

Northrop DA, Clayton RN (1966) Oxygen-isotope fractionation in systems containing dolomite. J Geol 74:174–196

O'Neil, JR, Epstein S (1966) Oxygen isotope fractionation in the system dolomite-calcite-carbon dioxide. Science 152:198–201

Pautot G, Guennoc P, Contelle A, Lyberis N (1984) Discovery of a large brine deep in the northern Red Sea. Nature (London) 310:133–136

Pisciotto KA, Mahoney JJ (1981) Isotopic survey of diagenetic carbonates. Deep Sea Drilling Project Leg 23. In: Yeats RS, Haq BU et al. (eds) Init Rep Deep Sea Drill Proj 63. US Gov Print Off, Washington DC, pp 595–609

Puchelt H, Laschek D (1984) Marine Erzvorkommen im Roten Meer. Fridericiana Z Univ Karlsruhe, vol 34

Rossignol-Strick M (1987) Rainy periods and bottom water stagnation initiating brine accumulation and metal concentrations: 1. The late quaternary. Paleoceanography 2:379–394

Rubinson M, Clayton RN (1969) Carbon C-13 fractionation between aragonite and calcite. Geochim Cosmochim Acta 33:997–1002

Schoell M, Faber E (1978) New isotopic evidence for the origin of Red Sea brines. Nature 275:436–438

Schoell M, Stahl W (1972) The carbon isotopic composition and the concentration of dissolved anorganic carbon in the Atlantis II Deep brines, Red Sea. Earth Planet Sci Lett 15:206–211

Shackleton NJ, Kennett JP (1975) Paleotemperature history of the Cenozoic and the initiation of Antarctic glaciation: oxygen and carbon isotope analyses in DSDP sites 277, 279 and 282. In: Kennett JP, Houtz RE (eds) Init Rep Deep Sea Drill Proj 29. US Gov Print Off, Washington DC, pp 743–755

Sharma T, Clayton RN (1965) Measurement of O-18/O-16 ratios of total oxygen of carbonates. Geochim Cosmochim Acta 29:1347–1353

Stoffers P, Botz R (1989) Carbonate Crusts in the Red Sea: Their Composition and Isotope Geochemistry In: Ittekott V, Kempe S, Michaelis W, Spitzy A, (Eds.) Facets of modern biogeochemistry, Festschrift for E T Degens Springer, Berlin Heidelberg New York (in press)

Tarutani I, Clayton RN, Mayeda TK (1969) The effect of polymorphism and magnesium substitution on oxygen isotope fractionation between calcium carbonate and water. Geochim Cosmochim Acta 33:987–996

Thunell RC, Locke SM, Williams DF (1988) Glacioeustatic sea-level control on the Red Sea salinity. Nature (London) 334:601–604

Wada H, Niitsuma N, Nagasawa K, Okada H (1982) Deep sea carbonate nodules from the Middle American Trench area off Mexico, Deep Sea Drilling Project Leg 66. In: Watkins JS, Moore JC et al: (eds) Init Rep Deep Sea Drill Proj 66. US Gov Print Off, Washington DC, pp 453–474

Biogenic Constituents, Cement Types, and Sedimentary Fabrics

Their Interrelations in Lower Cretaceous (Valanginian to Hauterivian) Peritidal Carbonate Sediments (Trnovo, NW Slovenia)

R. KOCH[1] and B. OGORELEC[2]

CONTENTS

	Abstract	96
1	Introduction	96
2	Geologic Setting and Location of the Trnovo Profile	97
3	The Valanginian-Hauterivian Interval within the Trnovo Profile	101
3.1	Biogenic Constituents	103
4	Cement Types and Their Facial Interpretation	107
4.1	Marine Phreatic Cements	107
4.1.1	Isopachous Rims with Bladed Relict Textures	107
4.1.2	Isopachous Rims with Needle Relict Textures	111
4.1.3	Isopachous Radiaxial Fibrous Rims	114
4.2	Marine Vadose Cements	115
4.2.1	Gravitational Cements with Bladed Relict Textures	115
4.2.1.1	Gravitational Cements with Beginning Growth of Dogtooth Crystals	116
4.2.2	Gravitational Radiaxial Fibrous Cements	116
4.2.3	Gravitational Cements with Algal Relict Textures	116
4.2.4	Other "Gravitational" Textures	116
4.3	Meteoric Vadose Cements	116
4.4	Meteoric Phreatic Cements	117
5	Other Diagenetic Characteristics	117
5.1	Vadose Crystal Silt	118
5.2	Dolomite	118
5.3	Twisted Calcite	118
6	Results	119
	References	121

---

[1] Institut für Paläontologie, Loewenichstr. 28, 8520 Erlangen, FRG.
[2] Geoloski Zavod Ljubljana, Dimičera 14, 61000 Ljubljana, Yugoslavia.

Abstract

Valanginian to Lower Hauterivian carbonates from the Dinaric carbonate platform are analyzed in regard to the interrelation of biogenic constituents, cement types and sedimentary fabrics. Biogenics are predominantly *benthic* foraminifera, *green algae (Bankia striata, Cayeuxia* sp.), ostracods, gastropods, and stromatolitic algal-laminations. *Cements observed in the limestones are calcitic throughout. Nevertheless, from the relic textures observed, their primary mineralogy can be deduced. Evidence is given that they originally consisted of either aragonite, Mg-calcite or calcite.* Cements are marine phreatic and vadose and meteoric vadose cements (isopachous rims with bladed and needle relict textures and isopachous radiaxial fibrous rims; gravitational cements with bladed, needle and algal relict textures and gravitational radiaxial fibrous cements; granular gravitational cements). Fine- to coarse-granular and poikilotopic cements occur commonly but cannot be used for this analysis. Sedimentary fabrics are birds-eyes, solution molds, algal laminations, geopetal micrite and vadose crystal silt.

Sediments formed in the subtidal marine phreatic environment show both high faunal diversity and numbers. Ostracods do not occur. Isopachous rims are the most common cements. Birds-eyes and vadose silts are missing.

Sediments from the intertidal, predominantly marine vadose environment, show small bentic foraminifera with thin walls and ostracods. Isopachous rims reveal neomorphic overprinting by meteoric waters. Marine vadose gravitational cements are common. Locally, birds-eyes and stromatolitic fabrics occur.

In the meteoric vadose zone, sediments are present without any fauna. Only stromatolitic layers are found. Granular gravitational cements and birds-eyes are common.

These interrelating features can be traced in the Lower Cretaceous over the Dinaric carbonate platform, due to a homogeneous facies development with very low relief on the platform. After the tectonic and facial differentiation of the platform in the Middle Cretaceous, only local emergence, for example, can be demonstrated due to high facial diversity and morphology (reefs and grabens) present on the platform.

1 Introduction

Having studied Cretaceous outcrops in Yugoslavia since 1975, we noticed, especially in the Lower Cretaceous, that a close relationship exists between faunal composition, porosity types, cement types, and sedimentary textures. This is generally expected, but is rarely emphasized in the literature documented by single discipline papers. Equally important is the question why there is no such relationship for the Upper Cretaceous of the same area. What is the time span in which organisms can survive extreme conditions, compared to the time needed to form cement types usable as keys to environmental and diagenetic history?

No general answer can be given, but an example from the Trnovo profile, which stresses these problematic points, can be offered. This is especially important in the determination of sedimentary cycles where knowledge about the density of sampling used is necessary. Additionally, we try to give explanations for the distribution of fossils and cement types in relation to both the tectonic position and development within the Dinaric carbonate platform.

2 Geologic Setting and Location of the Trnovo Profile

Yugoslavia is generally subdivided into nine tectonic units (Fig. 1). The subdivision was established by Petković (1958), Sikošek and Medwenitsch (1965), Sikošek and Vukašinović (1975), Andjelković (1976, 1978), Buser (1976, 1987) and Miljuš (1976).

The strata studied are situated in the High Karst Area of the Outer Dinarides, which are subdivided from East to West into the High Karst Area, the Adriatic-Ionian Zone and the autochthonous region of Istria. Whereas flysch sediments occur in the Adriatic-Ionian Zone, carbonate platform sediments are found in the High Karst Area and in the autochthonous region of Istria.

These areas represent the major parts of the Dinaric carbonate platform which has an extension of more than 600 km (NW-SE, from the Venetian Alps to Montenegro) and a width of up to 250 km. *In the Middle Cretaceous the platform was broken into smaller tectonic units.* In NW Slovenia, where the studied area is located, the important small tectonic units form overthrusts with a lateral transport of 30 km or less. From NE to SW in a zone with a width of about 100 km these units are the (1) Trnovo overthrust; (2) Hrušica overthrust; (3) Snežnik slice; (4) Komen slice; (5) Čičarija slices; (6) Koper slice; and (7) Istrian autochthonous region. From 1975 to 1982 characteristic profiles were studied from some of these smaller tectonic units (Koch 1977, 1978, 1988). The Trnovo profile represents a standard section for the sedimentological development of the Cretaceous in the area of the Trnovo overthrust. As shown in Fig. 2 other Mesozoic rocks occur in this area, too.

Early biostratigraphic studies were carried out by numerous scientists (Polšak 1965; Turňsek and Buser 1966; Radoičić 1966; Šribar 1979). Subsequently, Orehek and Ogorelec (1979) conducted a sedimentological, biostratigraphic study of the area, and Koch (1988) made more detailed sedimentological, geochemical studies.

The Trnovo profile has a thickness of about 1400 m. It is located in the Trnovo forest near Nova Gorica on the Italian border (Fig. 3).

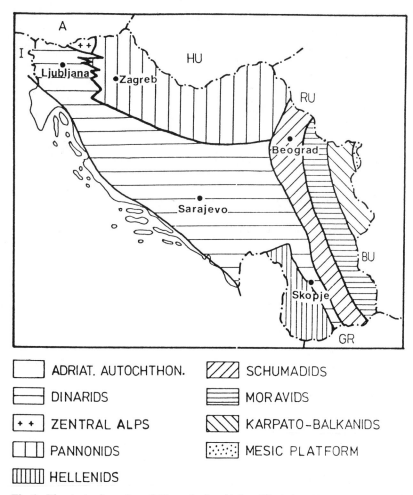

Fig. 1. The tectonic units of Yugoslavia. (After Sikošek and Medwenitsch 1965; Andjelković 1976, 1978; Buser 1976, 1987)

A detailed description of the microfaunal and diagentic development of the carbonate rocks is given by Koch et al. (1989). The biostratigraphic zonal subdivision with the most important microfaunal elements is documented in Fig. 4. Generally, the facies development can be briefly summarized as follows:

The *Berriasian to Barremian* shows a 950-m-thick sequence of creamy brown, well-bedded limestones and minor dolomites with beds of centimeter to decimeter thickness. Therein, platy dolomitic limestones are locally intercalated. In general, this is a zone of low diversity and numbers of specimens.

In the *Barremian-Aptian*, small patch reefs (up to 3 m in thickness) occur embedded in normal brownish, lagoonal, well-bedded limestones. The patch reefs show predominantly corals, hydrozoans, *algae (Bacinella irregularis)*,

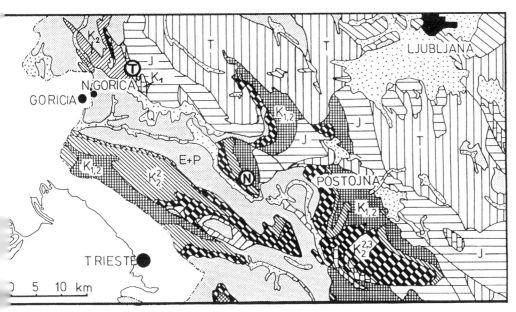

Fig. 2. Detail of the geologic map of NW Slovenia with location of the Trnovo profile (*T*) and the Nanos profile (*N*; Koch 1977, 1978). *T* Triassic; *J* Jurassic; K_1 Lower Cretaceous; $K_{1,2}$ Lower-Upper Cretaceous; K_2^2 Turonian; $K_2^{2,3}$ Turonian-Senonian; E + P Eocene and Paleocene

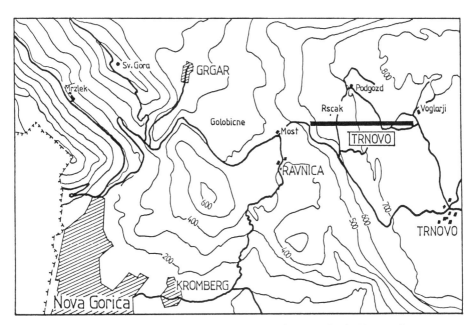

Fig. 3. Location of the Trnovo profile near the village of Trnovo in the Trnovo forest

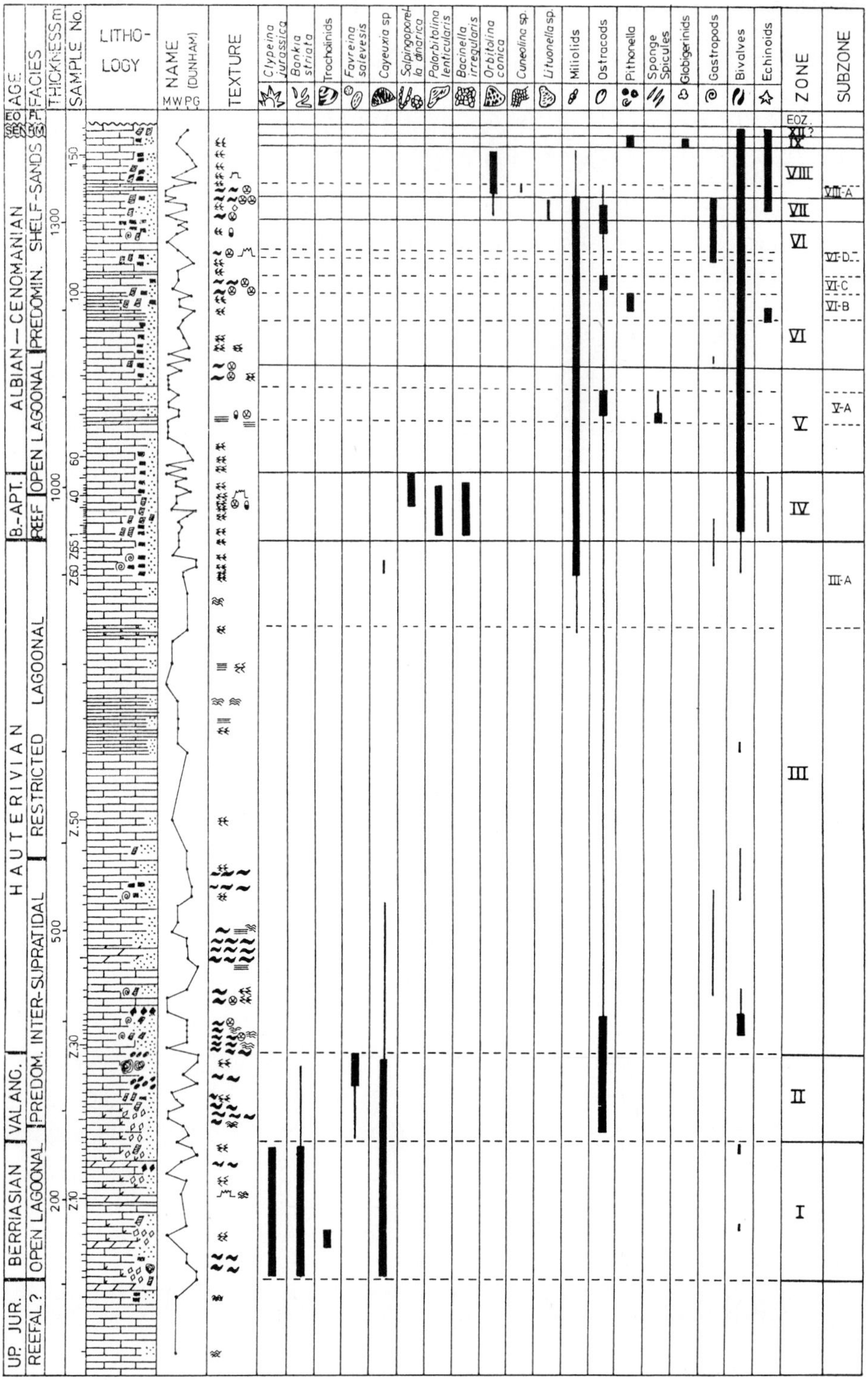
AGE
FACIES
THICKNESS m
SAMPLE No.
LITHO-LOGY
NAME (DUNHAM)
MWPG
TEXTURE
Clypeina jurassica
Bankia striata
Trochoiinids
Favreina salevesis
Cayeuxia sp
Salpingoporella dinarica
Palorbitolina lenticularis
Bacinella irregularis
Orbitolina conica
Cuneolina sp.
Lituonella sp.
Miliolids
Ostracods
Pithonella
Sponge Spicules
Globigerinids
Gastropods
Bivalves
Echinoids
ZONE
SUBZONE
UP. JUR.
BERRIASIAN
VALANG.
HAUTERIVIAN
B.-APT.
ALBIAN—CENOMANIAN
REEFAL ?
OPEN LAGOONAL
PREDOM. INTER-SUPRATIDAL
RESTRICTED LAGOONAL
REEF
OPEN LAGOONAL
PREDOMIN. SHELF-SANDS
I
II
III
IV
V
VI
VII
VIII
IX
III-A
V-A
VI-B
VI-C
VI-D
VIII-A

molluscs and echinoids. Foraminifera are abundant. *Dasycladaceans* (*Salpingoporella dinarica*) are most characteristic for this stratigraphic interval.

The *Albian* (about 100 m thick) shows brown micritic limestones with predominant miliolids. Locally, platy limestones with chert nodules and layers, and dolomitic limestones are intercalated.

The overlying *Albian-Cenomanian* interval has a thickness of about 200 m and consists of cream to whitish (in the upper part) limestones with miliolids, bivalves, gastropods and echinoids. In the Upper Cenomanian, large Foraminifera (*Orbitolina conica*) and others occur in arenitic limestones.

The Turonian is characterized by abundant rudists (mounds), although these are not present everywhere, due to faulting. Therefore, the *Senonian* often occurs immediately above the Cenomanian. White limestones contain abundant silicified mollusc shells (*Sabinia* sp.).

Above another fault the sequence is overlain by *Eocene flysch*.

3 The Valanginian-Hauterivian Interval within the Trnovo Profile

The interval of interest in the present study has a thickness of about 200 m (Fig. 4). It comprises nearly the complete Valanginian and the lowermost Hauterivian. Characteristic *biogenic constituents* for the Valanginian *are algae* (*Bankia striata* and *Cayeuxia* sp.) *and crustacean pellets* (*Favreina salevensis*). Additionally, some miliolids and molluscs (bivalves and gastropods) occur. The lowermost Hauterivian contains only traces of *Cayeuxia* sp., mollusc fragments and some miliolids. Ostracods occur throughout the analyzed section. The studied interval is characterized by abundant parallel bedding, birds-eyes, algal laminations and fractures. Following the environmental determination of birds-eyes (Shinn 1968), it can therefore be generally described as an intertidal sequence. However, the intertidal sequence includes conditions ranging from shallow subtidal to supratidal, revealing no information about the time at which a sediment remained in the different environments which were probably created due to sea-level fluctuations. Therefore, the main purpose of this chapter is to reveal the details of the different stages, included in the term "intertidal", for the interval analyzed. The most important biogenic constituents and cement types are documented in Plates 1–6. The schematic distribution of the main parameters is given in Fig. 5.

Fig. 4. The Trnovo profile and the distribution of the most important microfossils for stratigraphic subdivision (After Koch et al. 1989).

Legend: pellets; intraclasts; coquina; gastropods; onkoids; ooids; single dolomite rhombohedrons; dedolomite; parallel laminations; stromatolitic structures; birds-eyes; vadose silt; stylolites; tectonic fractures; geopetal fillings. *M*, Mudstone; *W*, wackestone; *P*, packstone; *G*, grainstone

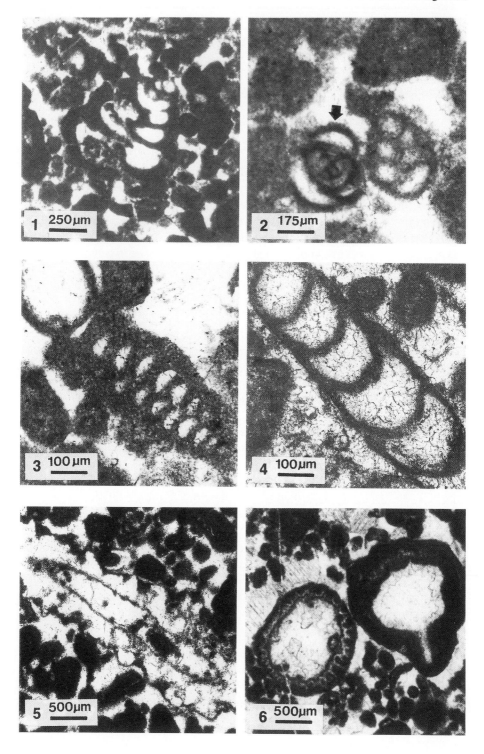

3.1 Biogenic Constituents

Only microfossils occur in the interval analyzed.

Benthic foraminifera of differing sizes and appearance are most common. In addition to textulariid forms (Plate 1,*1*) other bi- or triserial (Plate 1,*3*) and uniserial specimens (Plate 1,*4*) occur. Miliolids (Plate 1,*2*) are common locally. It is obvious that miliolids show an especially strong variation of their wall thickness in relation to rock type and diversity, reflecting environmental conditions. In samples with higher diversity, miliolids have relative massive, strong chamber walls, as do the other *benthic* foraminifera present in the same samples. On the other hand, very thin and fragile chamber walls are found in biogenic mudstones in which only miliolids (Plate 2,*3*) and some ostracods (Plate 2,*4*) are found. This is also true for algally laminated rocks. Therefore, miliolids appear to be very sensitive to extreme environmental conditions and environmental changes and respond by varying the thickness of their chamber walls. Other *benthic* foraminifera (described above) are not able to withstand such extreme conditions, which range from slightly elevated salinity to brackish waters.

Finally, both packstones and grainstones, accumulated in higher energy (wave activity) setting, show *benthic* foraminifera, primarily deposited in a shallow marine environment of probably normal marine conditions. Due to the local abundance and fossil associations, it is assumed that the sites of deposition were not far from the original habitats. Therefore, this fauna may always reflect primary marine conditions. Of course, meteoric conditions can

---

Plate 1. Biogenic constituents.

1 Sample TZ 39. Lowermost Hauterivian.
   Large textulariid and other small foraminifera occur in a poorly washed intrabiopelsparite (packstone).

2 Sample TZ 39. Lowermost Hauterivian.
   Abundant small foraminifera as miliolids (*arrow*), etc. occur in addition to the above shown larger ones.

3 Sample TZ 39. Lowermost Hauterivian.
   Biserial benthic foraminifera (*center*) are common. Some **problematica** (*upper left*) are found, too.

4 Sample TZ 38. Lowermost Hauterivian.
   Uniserial benthic foraminifera are common locally.

5 Sample TZ 25. Valanginian.
   Not further determined dasycladacean fragment (oblique section) in a Favreina grainstone.

6 Sample TZ 27. Valanginian.
   Bankia striata in various sections occurs commonly in onkoid-packstones-grainstones. In these samples also *Cayeuxia* sp. is found.

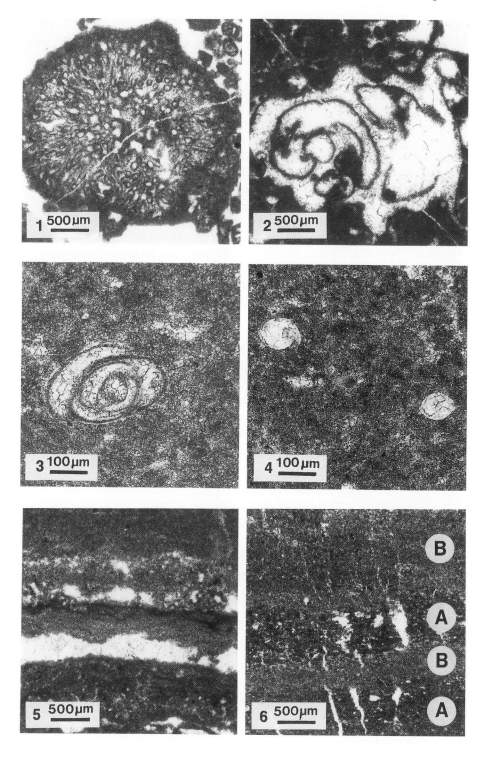

influence the sediments at a very early stage due to *sea-level* fluctuations. *Dasycladaceans (Plate 1,5) and other green algae (Bankia striata, Plate 1,6 and Cayeuxia sp., Plate 2,1) show similar behavior to the benthic foraminifera.* They occur in favreina- and/or onkoid-rich packstones and grainstones, probably indicating conditions of slightly elevated salinity.

Ostracods (Plate 2,4) were present in most samples analyzed. Locally, they occur abundantly. Layerwise, smaller specimens are predominant.

Gastropods (Plate 2,2) can be found locally and have a size ranging from a few millimeters to centimeters. They are therefore interpreted as being embedded within the sediment (pellet packstones) on which they originally grazed. Their primary aragonitic shells are generally dissolved due to an early freshwater influx. The former fossil molds are now filled with clear, granular calcite cement. Only the shape of the molds (Plate 4,2) and micritic envelopes (Plate 2,2) and (Plate 4,5), sometimes developed, indicate their former presence.

Stromatolitic algal laminations are abundant (Plate 2,5 and 6). They are often found together with birds-eyes and/or stromatactis-like elongated cavities parallel to bedding planes. Synsedimentary fractures were formed locally due to stratified water bodies of different salinities (Plate 2,6). These "synaeresis cracks" only occur in pelletoidal layers which alternate with layers stabilized more by algal laminations. In these more coherent sediments, only traces of thin cracks are formed.

Plate 2. Biogenic constituents.

1 Sample TZ 27. Valanginian.
Cayeuxia sp. with characteristic sections of single small tubes, circular in the center and more elliptical towards outer oblique sections, occur commonly in this onkoid-packstone-grainstone.

2 Sample TZ 33. Lowermost Hauterivian.
A gastropod, preserved as fossil mold and micritic envelopes of the former shell. The solution of the primary aragonitic shell is ascribed to an early diagenetic freshwater influx.

3 Sample TZ 20. Valanginian.
In biogenic mudstones locally thin-walled miliolids are found, reflecting extreme living conditions in a very restricted environment.

4 Sample TZ 20. Valanginian.
In addition to the above described miliolids, only some ostracods occur.

5 Sample 31. Lowermost Hauterivian.
Stromatolitic layers are abundant in this part of the profile. The limestones are generally barren of other biogenics constituents, indicating an extreme environment with probably elevated salinity.

6 Sample TZ 20. Valanginian.
Alternating layers of pelletoidal and stromatolitic mudstones are affected by the early diagentic formation of cracks due to stratification of waters with different salinity. These synaeresis cracks are only well developed in the pelletoidal layers (*A*), whereas the stromatolitic binding (*B*) hinders the fracturing of the sediment

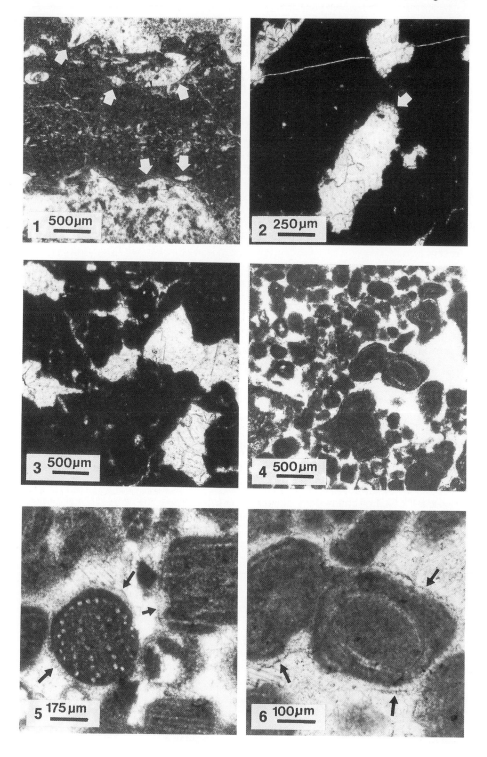

4 Cement Types and Their Facial Interpretation

All cements observed in the limestones studied consist of calcite. Nevertheless, due to the characteristic relict textures found in the cement crystals, the primary mineralogy (aragonite, Mg-calcite and calcite) can be interpreted in comparison with recent structures of identical morphology.

Following the general subdivision of environments, most important for the formation of recent carbonate cements (Longman 1980; Harris et al. 1985; Prezbindowski 1985), the present chapter describes cements in relation to the (1) marine phreatic; (2) marine vadose; (3) meteoric phreatic; and (4) meteoric vadose zones. Subsequently, descriptions of the cement types found are given. Their general relation to environments is briefly discussed on the basis of literature data, and the conclusions drawn for the cement types described are given.

4.1 Marine Phreatic Cements

4.1.1 Isopachous Rims with Bladed Relict Textures

Rims of this cement type generally have a thickness of 25–40 μm (Plate 3,*5* and *6*; Plate 4,*1*). They reveal a relatively smooth homogeneous surface.

Plate 3. Fabrics and cement types.

1 Sample TZ 20. Valanginian.
 In limestones with abundant algal laminations (Plate 2,*6*) also gympsum pseudomorphs (*arrows*) occur in layers. The crystals are now replaced by calcite. They reveal the typical lozenge shape described from sabkha facies, and gypsum crystals grown on the surface or within the first upper centimeters of muddy sediment.

2 Sample TZ 23. Valanginian.
 Fenestrae are commonly enlarged to vugs due to further freshwater influx. These vugs are generally filled with coarse granular, clear calcite. At the roofs of some vugs, small rims of very fine crystalline pendular cements are also found (*arrow*), interpreted as meteoric vadose cements.

3 Sample TZ 21. Valanginian.
 Abundant birds-eyes are filled by coarse granular, clear calcite.

4 Sample TZ 27. Valanginian.
 Onkoid packstones to grainstones are intercalated in the muddy sequence. In these rocks, marine phreatic cements are developed. Biogenic particles commonly show intensive micritization.

5 Sample TZ 28. Valanginian.
 Favreina salevensis is locally very abundant in packstones and grainstones consisting nearly exclusively of crustacean pellets. Here isopachous rim cements of primary Mg-calcite mineralogy are developed (*arrows*).

6 Sample TZ 27. Valanginian.
 Detail of *4*. Isopachous rim cements around a micritized miloilid. The relict textures in the cement, and especially the relatively smooth surface of the cement rim, indicate a primary Mg-calcite mineralogy

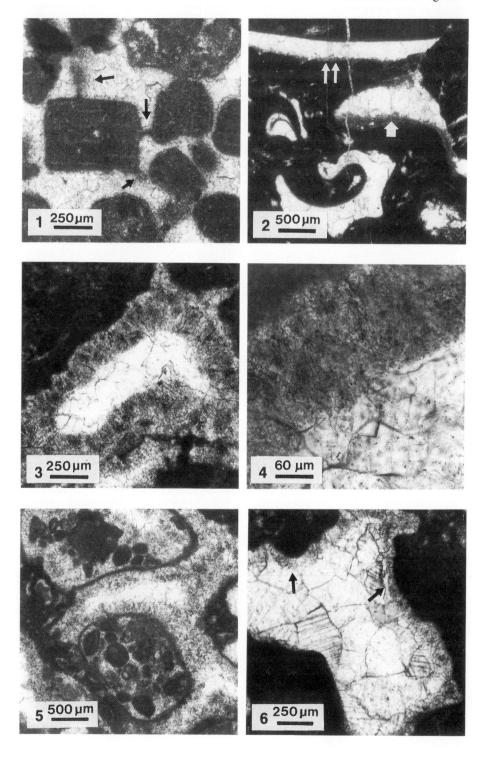

Commonly, laterally enlarged bladed crystals of 10–30 µm width occur, oriented perpendicularly to the substrate, and containing fine relict textures. Each crystal is composed of two to five smaller, single "blades", combined to form one larger crystal with scalenohedral terminations. The crystals are clear and without inclusions.

Similar rims of a "dental Mg-calcite" have been described by Macintyre (1984) from the Belize Barrier reef platform. Meyers (1985) describes "clear fringes of phreatic microcrystalline Mg-calcite cement subsequent to meniscus cryptocrystalline cement" from Maui, Hawaii. Prezbindowski (1985) using microprobes analyzed much higher Mg-contents in similar cements from the Stuart City Trend (Aptian/Albian) than in the surrounding allochems and later cements. Pierson and Shinn (1985) document textural differences of Mg-calcite and aragonite cements from Pleistocene limestones of the Hogsty Reef, Bahamas.

On the basis of these data, the above described isopachous rims are interpeted as being of primary Mg-calcite mineralogy and of having been formed under marine phreatic conditions. The aggrading neomorphism, to coarser bladed crystals, is caused by a freshwater influx, although the same features can occur in the deep burial environment due to the different ionic composition of pore waters. Locally, these cement rims reveal different degrees of diagenetic overprinting due to the influence of waters depleted in Mg. These features are enriched in layers. An early diagenetic meteoric influx is thus as-

Plate 4. Cement types.

1 Sample TZ 25. Valanginian.

In a *Favreina* grainstone locally micrite bridges between particles can be observed, probably indicating short-time vadose soil conditions. Generally, abundant isopachous rim cements of a primary Mg-calcite can be found.

2 Sample TZ 35. Lowermost Hautervian.

Due to an early freshwater influx, all aragonitic particles were dissolved. Locally, pelletoidal micritic geopetal infillings occur (*arrow*) which also form thin dusty layers (*double arrow*). The fossil molds are cemented by clear, medium granular meteoric phreatic calcite.

3 Sample TZ 30. Lowermost Hauterivian.

In solution enlarged cavities, thick radiaxial fibrous isopachous rims grow. The dirty appearance is caused by abundant inclusions.

4 Sample TZ 30. Lowermost Hauterivian.

Detail of *3*. Characteristic relict textures within the radiaxial fibrous cement mimicking a former needle cement.

5 Sample TZ 33. Lowermost Hauterivian.

An earlier dissolved gastropod is only recognizable due to the outlines of the fossil mold, and to micritic envelopes. The mold is filled with radiaxial fibrous cement.

6 Sample TZ 30. Valanginian.

Solution mold filled with gravitational radiaxial fibrous cement at the roof (*arrows*). At the *bottom*, only coarse granular calcite crystals occur filling the complete mold

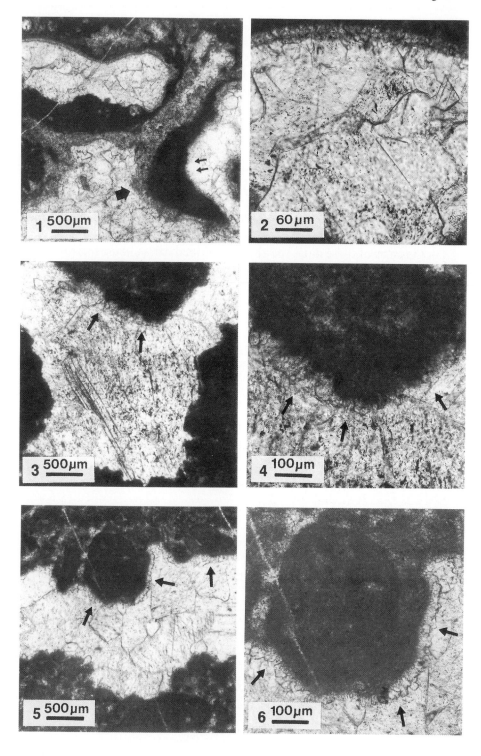

sumed. According to Folk (1973), a predominant sideward growth of the recrystallization products is the result.

4.1.2 Isopachous Rims with Needle Relict Textures

These rims have a thickness of 120–160 μm and occur very sporadically. The relict textures reveal primary needles growing perpendicularly to sub-perpendicularly to the surface of the substrate and form a very indistinct surface due to single crystals of varying length and orientation. The "dirty" lines between the original needles are caused by relicts of primary organic matter separating the single needles as described by Hall and Kennedy (1967). This is characteristic of primary aragonite formed in recent marine phreatic environments.

Alexandersson (1969, 1972) described examples from the Mediterranean where he studied Mg-calcite and aragonite in biogenic borings. Schroeder (1972) *found* these cements in Holocene cup reefs on the Bahamas. Moore (1973) reported similar cements from intertidal sediments of Grand Cayman in the West Indies. Other publications documenting the occurrence of

◀───

Plate 5 Cement types.

1 Sample TZ 33. Lowermost Hauterivian.
    Varying gravitational cements can be found in directly neighboring enlarged fossil molds. In the upper mold, the roof is covered by a gravitational cement, with a relatively smooth surface, and consisting of bladed crystals indicating a primary Mg-calcite mineralogy. In the lower mold, asymmetric radiaxial fibrous cement occurs. The cements grow locally only on one side of a wall (*arrow*), whereas they are missing at the other side (*double arrow*).

2 Sample TZ 33. Lowermost Hauterivian.
    Detail of *1*. Gravitational cement from the roof of the upper mold with primary Mg-calcite mineralogy.

3 Sample TZ 26. Valanginian.
    Solution enlarged mold with local gravitational cement at the roof (*arrows*), revealing characteristic texture of microstalactitic development. Note the absence of the cement at all other walls of the mold. There, only direct contacts of the later formed coarse granular cement can be observed.

4 Sample TZ 26. Valanginian.
    Detail of *3* revealing the relict texture of the cement consisting of single bladed crystals with internal relicts (*arrows*).

5 Sample TZ 40. Lowermost Hauterivian.
    Microstalactitic cements (*arrows*) are found locally at roofs of enlarged molds, oriented parallel to the bedding in pelletoidal limestone. Note that the bottom of the cavity only contacts medium-coarse-granular cements.

6 Sample TZ 40. Lowermost Hauterivian.
    Detail of *5* showing texture of the gravitational cement, made up of single bladed crystals

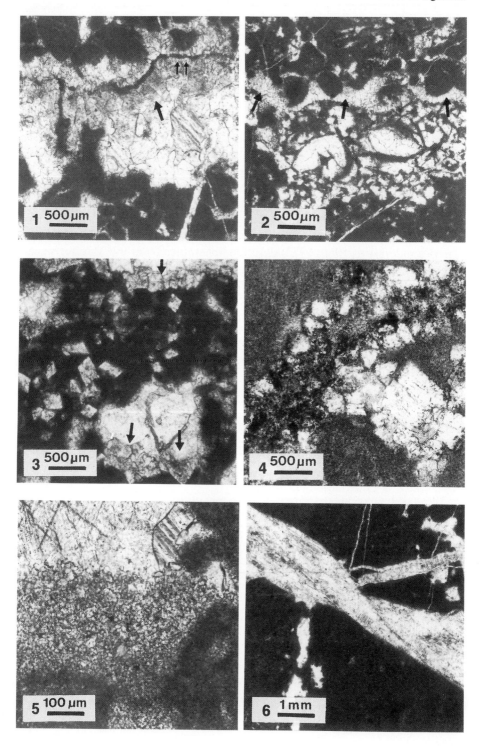

aragonite needle cements are Friedman et al. (1974) from the Red Sea, Macintyre (1984) from the Belize reef platform; Beier (1985) from Quaternary beach rocks of the Bahamas; and Lighty (1985) from Holocene barrier reefs of the Bahamas. Sandberg (1985) discussed abundant recent and fossil examples and described aragonite crystals in aragonite fringe cements. The crystals are usually of quite variable length and diverge perpendicularly to the substrate.

Isopachous rim cements with needle relict textures documented in the present study are interpreted as being of primary aragonite mineralogy. They occur subordinately and therefore, probably indicate special conditions necessary for their preservation. In a normal open system with large transport of pore fluids, aragonite is generally expected to be dissolved along with gastropod fragments or other aragonitic biogenic allochems. The preservation of the relict textures is ascribed to stagnant conditions sensu Longman (1980) in the pore water system. This results in only minor undersaturation with respect to aragonite, and therefore establishes a solution

◄───

Plate 6. Cement types, dolomite and others.

1 Sample TZ 38. Lowermost Hauterivian.

Gravitational cement (*arrow*) filling the upper part of a solution cavity in a pelletoidal packstone. This cement shows a dirty seam (*double arrow*) running parallel to the roof from where the inclusion-rich crystals grow. This is ascribed to the activity of microorganisms which contributed to the formation of the gravitational cement, as described by Monty and Mas, from the Lower Cretaceous of eastern Spain.

2 Sample TZ 38. Lowermost Hauterivian.

Asymmetric position of a botryoidal cement (*arrows*) at the roof of a solution cavity. The rest of the cavity is filled with alveolar structures indicating soil formation. The location of the cement points to a vadose origin.

3 Sample TZ 22. Valanginian.

Dolomite rhombohedrons grow in the micrite as well as at the bottom of solution cavities (*arrows*). The roofs of cavities show granular cements revealing dogtooth characteristics. The geopetal formation of dolomite in the cavities is ascribed to the early filling with Mg-calcite mud, which later resulted in growth of idiomorphic dolomite crystals. The granular cements at the roof are ascribed to short-time meteoric vadose influences.

4 Sample TZ 21. Valanginian.

Dolomite crystals growing on both sides of cracks indicate their late formation probably under deeper burial conditions. These crystals are very rich in iron and also show outer iron-rich seams.

5 Sample TZ 37. Lowermost Hauterivian.

Some molds are geopetal, filled with poorly sorted vadose silt, which consists of fine crystal fragments of up to 40 µm size. This corresponds to the characteristics of vadose silt as originally described by Dunham (1969).

6 Sample TZ 29. Lowermost Hauterivian.

The growth of twisted, long, fibrous calcite in tectonic fractures is ascribed to their syngenetic slow growth during the opening of the fractures and dislocation of parts of the limestone reprecipitation system on an ultra-scale. In this system, insoluble organic residues were not removed from their original sites.

4.1.3 Isopachous Radiaxial Fibrous Rims

Limestones with this cement type are predominantly found within the Trnovo profile, in the section described here. They occur in intercalations (decimeter to meter thickness) of intertidal sediments. They are characterized by their "dirty" appearance, due to numerous inclusions, and by a surface seemingly homogeneous at first glance (Plate 4,*3–6*; Plate 5,*1*). Higher magnification reveals bundles of needles with different orientations of their axes, and curved cleavage planes, forming an irregular surface. The thickness of the rims is 160–400 µm, and can reach even more than 1 mm. Locally, the cements are cleaner, (Plate 4,*5*, and *6*) and contain only minor inclusions. Clear, radiaxial, fibrous cements without inclusions, as described by Saller (1986), cannot be found.

Radiaxial fibrous cements (Krebs 1969; Bathurst 1971; Kendall and Tucker 1973; Kendall 1985) are generally composed of crystals with undulatory extinction, very rich in inclusions. Three types can be differentiated according to Kendall (1985), who gives detailed descriptions of the optical differences. In general, the interpretations of the sites of formation of radiaxial fibrous cements differ in a wide range. Krebs (1969) and Mountjoy and Krebs (1983) considered it a marine phreatic formation. Kendall and Tucker (1973) explained the optical appearance, and deduced radiaxial fibrous calcite to a complex replacement of primary aragonite. For a long time this was the major thesis, regardless of the fact that Bechstädt (1974) had already described radiaxial fibrous calcite from the supratidal environment, as had Schneider (1977) from meteoric pore waters under a very thin sedimentary cover. Kendall (1985) disproved in Devonian material from Australia, the transformation hypothesis of Kendall and Tucker (1973), as well as the formation under meteoric influence. He interpreted radiaxial fibrous calcite to be a primary Mg-calcite, formed in the marine phreatic environment. Prezbindowski (1985) analyzed elevated Mg-contents in radiaxial fibrous calcites from the Lower Cretaceous Stuart City Reef Trend and supported this interpretation. Moreover, he distinguished another type as radial-radiaxial calcite, which differs from the radiaxial fibrous calcite of Bathurst (1971) and Kendall and Tucker (1973) by its great diversity in the orientation of optical axes. Although Saller (1986) discussed some *possibilities* of the formation of radiaxial fibrous calcite, he preferred the idea that it originally formed as marine phreatic Mg-calcite. Sandberg (1985) gave a similar interpretation. Städter and Koch (1987) pointed to different generations of radiaxial fibrous calcite, one formed under marine phreatic conditions, and another of probably meteoric origin.

In the present study, radiaxial fibrous calcite is found in intercalated intertidal sediments formed mainly in the marine vadose zone. This is proven by the fossil associations as well as by other cement types. As described above, radiaxial fibrous cements are found, on the one hand, as isopachous rims within molds, vugs and cavities and surrounding allochems. On the other hand, they often show an asymmetric distribution within solution vugs (Plate 4,*6*; Plate 5,*1*). *Probably all other cements found in the same rocks were primarly Mg-calcite.* All primary aragonitic allochems are dissolved, creating molds and vugs which were later partly filled by radiaxial fibrous cement and partly by granular blocky calcite. Additionally, radiaxial fibrous calcite reveals well-preserved textures.

All of these observations point to the primary Mg-calcite mineralogy of the radiaxial fibrous calcite in recently studied material. Only primary Mg-calcite cement can be transformed with good preservation of the primary structures by processes running in an ultra-scale, as described by Oti and Müller (1985). The limestones with radiaxial fibrous calcite generally have a large minus cement porosity (molds, vugs, and interparticle pores). Due to abundant solution enlarged pores parallel to the bedding, a fairly good permeability was present. Therefore, an active diagenetic environment sensu Longman (1980) can be assumend. This results in dissolution of aragonitic cements, as proven by primary aragonitic particles. Thus, the interpretation of radiaxial fibrous calcite of primary Mg-calcite mineralogy is most apparent.

4.2 Marine Vadose Cements

These cements are found in the roofs of particle molds, in molds enlarged by solution, in shelter pores, and in pores parallel to bedding planes (Plate 4,*6*; Plate 5,*1–6*; Plate 6,*1*). They generally show the characteristic position of vadose cements, as described by Dunham (1971) and Müller (1971), including a sidewards thinning of the seams towards the walls. Within the group of marine vadose cements, different types can be distinguished.

Dunham (1971) and Müller (1971) first described vadose cements and their distribution in sediments from the Bahamas (also including meniscus cements, not present in the recently studied material). The reasons for the formation of fibrous and bladed morphologies of the crystals are discussed by Folk (1973). Prezbindowski (1985) describes "pendulous cements of 80–180 µm thickness, which are composed of bladed to fibrous crystals", and found in carbonate rocks which underwent a strong neomorphism. Additionally, dissolution of biogenic particles and parts of the rock can be observed.

4.2.1 Gravitational Cements with Bladed Relict Textures

This type is most common (Plate 5,*1–4*). It shows similar relict textures and crystal size as described for the marine phreatic cement with bladed relict textures. Therefore, it is interpreted as being of primary Mg-calcite mineralogy formed from marine pore waters.

4.2.1.1 Gravitational Cements with Beginning Growth of Dogtooth Crystals. Some of the cement rims described above show the beginning growth of scalenohedral crystal terminations (Plate 5,*5* and *6*), growing on cements with bladed relict textures. They probably formed under short meteoric influx subsequent to their formation in the marine vadose environment. Due to the lowered Mg/Ca ratio in the meteoric water, the single blades recrystallized. Additionally, the growth of scalenohedrals occurred at crystal terminations (dog tooth).

4.2.2 Gravitational Radiaxial Fibrous Cements

Gravitational cements with radiaxial fibrous textures (Plate 4,*6*) are commonly found in molds enlarged by solution. They are "dirty" due to many inclusions, and correspond completely to the radiaxial fibrous cements described in Section 4.1.3. Due to their position in the pores, they must have formed under marine vadose conditions. A primary Mg-calcite mineralogy is assumed. Locally, asymmetric radiaxial fibrous cements are found. They were formed during the transition from the marine phreatic to the vadose zone. The first generation of isopachous radiaxial fibrous cement formed in the marine phreatic environment.

4.2.3 Gravitational Cements with Algal Relict Textures

Locally, cements are associated with stromatolitic textures and incorporate algal material in layers (Plate 6,*1*). The growth of cements is formed at sites where abundant algal layers occur. They form seams of up to 250 µm thickness. The relict textures reveal the contribution of algal filaments to the growths of cements. Comparable cements are described by Monty and Mas (1981), from the Lower Cretaceous of Spain, and by Herrmann and Koch (1985), from Tertiary sediments in the Mainz Basin in W. Germany.

4.2.4 Other "Gravitational" Textures

Very minor cements seams similar to microbotryoids occur at the roofs of solution enlarged cavities (Plate 6,*2*). The remaining pore space is filled by alveolar textures, indicating a probable soil environment for their formation.

4.3 Meteoric Vadose Cements

Meteoric vadose cements can only be observed very locally. They form gravitational cements at the roofs of molds and vugs (Plate 3,*2*) and consist of clear fine granular calcite crystals of 5–20 µm size, without any impurities.

Gravitational cements of similar appearance are described by Dunham (1971) and Müller (1971) from Pleistocene sediments from the Bahamas. Koch and Schorr (1986) and Wirsing and Koch (1986) document similar cements from Upper Jurassic sponge algal mud mounds in Southern Germany.

Crystal morphology and sites of formation point to a primary calcite mineralogy. They were formed in the meteoric vadose environment as microstalactitic, pendulous cements.

4.4 Meteoric Phreatic Cements

It is very difficult to prove the presence of meteoric phreatic cements by microscope observations alone; in most cases it is impossible. Generally abundant fine, medium, and coarse granular cements (10–150 μm crystal size) as well as equant spar occur. In the recently studied material, they are found as late formation in nearly all interparticle pores, molds, vugs, and caverns, as well as in fractures and channels, which sometimes show a relation to primary sedimentary fabrics.

Granular cements of different crystal size can be formed in all early diagenetic stages, in shallow burial (eogenetic stage) on in the telogenetic stage (according to Choquette and Pray 1970). In the deeper burial diagenesis (mesogenetic stage), generally coarse granular equant spar and/or poikilotopic calcite crystals are formed, due to only minor oversaturation and slow transport of pore fluids. These cements show no orientation of the crystals to the substrate. They can be formed from pore waters similar to meteoric waters depleted in Mg and Sr. Freeman-Lynde et al. (1986), Moldovanyi and Lohmann (1984) and Prezbindowski (1985) give good examples, and show that their formation sites occur under deep marine to meteoric to deep burial conditions.

Due to the difficulty of relating these cements to certain environments without additional analyses, we do not include them in our discussion. We cannot find any relation of their sites and different types of formation, to the biogenics and sedimentary fabrics studied in the recently discussed interval. Therefore, we view these types as fine to coarse granular cements.

5 Other Diagenetic Characteristics

Other diagenetic characteristics are internal sediments (micrite, pellets, pelletoidal micrites and crystal silts) and microsparite. Locally, alveolar fabrics are found. Additionally, dolomite and carbonate cement in tectonic fractures can be observed. Only the most important features are documented in the photoplates, and will be discussed briefly. Micritic and pelletoidal internal sediments are generally interpreted as being of marine origin, and therefore, are not of major interest for the recent study.

5.1 Vadose Crystal Silt

Locally, a poorly sorted internal sediment can be observed in molds and solution enlarged molds consisting of very angular to angular crystal fragments of 5–40 µm size (Plate 6,*5*).

Dunham (1969) described crystal silt of the same composition with supratidal origin, and called it "vadose silt". Lang (1964) described the process of destruction of limestones and dolomites resulting in crystal silt from the Upper Jurassic of SW Germany. Aissaoui and Purser (1983) reported this as having been formed in the supratidal to continental environment, and pointed to some additional problems; "although its presence implies emergence of the substratum, this continental detritus may filter down to occupy spaces some 1–2 m below the emerged paleosurface and thus become associated with host sediments which have never emerged into the vadose zone".

On the basis of these descriptions, the associated fabrics and the absence of fossils, crystal silts found in the recently studied rocks are considered to be characteristic vadose silt of supratidal origin.

5.2 Dolomite

Early diagenetic dolomite, in relation to algal mats, is not found in the sediments studied here. Three types of dolomite can be distinguished. (1) Dolomite in fractures (Plate 6,*4*), growing on both sides of the fractures towards the inner pore space, are late diagenetic formations of a deep burial environment formed after fracturing of the limestones. (2) Idiomorphic scattered dolomite crystals in a micritic matrix (Plate 6,*3*) are interpreted as having been formed under shallow burial conditions. Magnesium was made available by captured marine pore waters. (3) Idiomorphic dolomite crystals, occurring geopetally, at the bottom of vugs and cavities (Plate 6,*3*), reveal clear crystal terminations at the top, and are "dirty" in the lower part. Locally, a horizontal boundary can be recognized, separating the two areas. This boundary can often be traced through all neighboring crystals in the pores and is interpreted as marking the surface of a former internal sediment. Therefore, this type of dolomite is interpreted as having grown under elevated P/T-conditions from an internal sediment of primary Mg-calcite mineralogy.

5.3 Twisted Calcite

In tectonic fractures, large fibrous calcite crystals can be observed (Plate 6,*6*) which show orientation of their c-axes parallel to the fracture walls. Each crystal is somewhat twisted around the neighboring crystals. Finally, a generally twisted appearance of the complete bundle of crystals results. This growth form is ascribed to a general stress combined with slight rotation of rock fragments during the growth of the crystals. Therefore, the "twisted calcite" can be interpreted as a late diagentic calcite of syngenetic growth with tectonic activities (e. g. overthrusting in this area).

6 Results

Regarding faunal distribution, cement types, and sedimentary fabrics together, the interval studied can be subdivided into much greater detail than previously thought. It becomes obvious that the presence or absence of distinct biogenic allochems excludes the presence of certain cement types and of some sedimentary fabrics. The subsequently presented results are summarized in Fig. 5.

1. Sediments formed in the *subtidal marine phreatic environment* show the highest diversity and number of specimens. Characteristic is the general absence of ostracods.

The cements nearly exclusively consist of isopachous rims with bladed relict textures (Mg-calcite). Rims of primary aragonite (needle relict textures) and radiaxial fibrous rims occur only very rarely. Birds-eyes and vadose silt do not occur. Only micritic and pelletoidal subtidal internal sediments are found.

2. The *intertidal, predominantly marine vadose environment* is characterized by a very restricted fauna. Only small bethic foraminifera with very thin walls are found besides ostracodes of different size. Gatropods occur locally. Stromatolitic fabrics are locally enriched. This fauna indicates a freshwater influx, and probably brackish conditions for most of the time.

Cements found here are isopachous rims and mostly abundant gravitational cements. The isopachous rims consist of bladed crystals (Mg-calcite) with strong neomorphic overprinting, and of radiaxial fibrous rims. Gravitational cements formed in the marine vadose environment are cements with bladed relict textures (Mg-calcite) and cements with incorporated algal filaments. Additionally, asymmetric and gravitational radiaxial fibrous cement seams are found. Granular gravitational cements were formed in the meteoric vadose environment, and consist of clear calcite crystals.

Birds-eyes, vadose crystal silts and stromatolitic fabrics are most common.

3. In sediments formed in the *supratidal (metoric vadose) environment*, only stromatolitic fabrics are found. Other biogenic elements do not occur.

Granular gravitational cements are most abundant. Locally, gravitational cements with relict textures are found. They are interpreted as gravitational cements formed primarily in the marine vadose zone, and overprinted by meteoric waters.

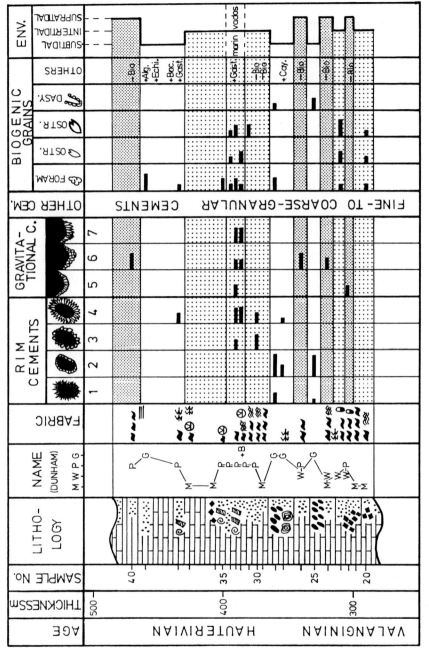

Fig. 5. Interrelation between biogenic constituents, cement types, and sedimentary fabrics. Cement types: Isopachous rims with *1* needle relict textures; *2* bladed relict textures; *3* with bladed relicts and neomorphic overprinting; *4* radiaxial fibrous. Gravitational cements: *5* with bladed relict textures, marine vadose; *6* granular, meteoric vadose; *7* radiaxial fibrous, marine vadose. *Alg.* Dasycladaceans; *Echin.* echinoids; *Bac. Bacinella irregularis*; *Gast.* gastropods; *Cay. Cayeuxia* sp.; *Bio* no biogenic allochems

Birds-eyes can be commonly observed.

The results demonstrate the close interrelation between biogenic allochems, cement types and sedimentary fabrics. Features formed in a special environment are overtaken into the subsequently etablished facies. They can be overprinted, destroyed or preserved, and will be associated with features just newly formed.

In the Lower Cretaceous rocks of W Slovenia and Istria, the above described results can be traced over long distances in many profiles. This is not valid for the Upper Cretaceous. We explain this with the uniform coherence of the Dinaric carbonate platform up to the Hauterivian, and only a moderate relief present on the platform. Therefore, sea-level fluctuations result in facies changes which can be traced over the complete platform. After the platform was broken into smaller tectonic units in the Barremian/Aptian, high differentiation of facies developed. Therefore, sea-level fluctuations now results in locally different features, depending on the relief formed by the formation of reefs and grabens. In these sediments, for example, features of emergence are only found locally, and cannot be traced over the Dinaric carbonate platform, as in Lower Cretaceous sediments.

Acknowledgements. We wish to thank P. Scholle (during a visit in Heidelberg) and A. Dunham (Erlangen) for reading most parts of the manuscript and for numerous improvements. We thank E. Flügel and L. Hottinger for supporting the basic idea of this paper. The study was partly funded by the German Research Foundation (DFG) and by the SCR (Slovenian Research Council).

References

Aissaoui DM, Purser BH (1983) Nature of origins of internal sediments in Jurassic limestones of Burgundy (France) and Fnoud (Algeria). Sedimentology 30:273–283

Alexandersson T (1969) Recent littoral and sublittoral high-Mg-calcite lithification in the Mediterranean. Sedimentology 12:47–61

Alexandersson T (1972) Intergranular growth of marine aragonite and Mg-calcite: evidence of precipitation from supersaturated sea water. J Sediment Petrol 42:441–446

Andjelković M (1976) Structural zoning of the Inner and Central Dinarids of Yugoslavia. Geol An 40:1–22

Andjelković M (1978) Tectonic rayonisation of Yugoslavia. Zb Radova, 9th Congr Geol Jug, Sarajevo, pp 7–13

Bathurst RGC (1971) Carbonate sediments and their diagenesis. Development in sedimentology, vol 12. Elsevier, Amsterdam, 620 pp

Bechstädt T (1974) Sind Stromatactis und radiaxial fibröser Calcit Faziesindikatoren? N Jahrb Geol Paläontol Mh 1974:643–663

Beier JA (1985) Diagenesis of Quaternary Behamian beachrock: petrographic and isotopic evidence. J Sediment Petrol 55:755–761

Buser S (1976) Tektonischer Aufbau Südwest-Sloweniens. 8th Congr Geol Jug Ljubljana, vol 3, pp 45–48

Buser S (1987) Development of the Dinaric and the Julian Carbonate Platforms and of the intermediate Slovenian Basin. (NW Yugoslavia). Hem Soc Geol Ital 40:313–320

Choquette, PW, Pray LC (1970) Geologic nomenclature and classification of porosity in sedimentary carbonates. Am Assoc Petrol Geol Bull 54:207–250

Dunham RJ (1969) Early vadose silt in Townsend Mound (reef), New Mexico. In: Friedman GM (ed) Depositional environments in carbonate rocks. SEPM Spec Publ 14:139–181

Dunham RJ (1971) Meniscus cements. In: Bricker OP (ed) Carbonate cements. Johns Hopkins Stud Geol 19:297–300

Folk RL (1973) The natural history of crystalline calcium carbonate: effect of magnesium and salinity. J Sediment Petrol 44:30–53

Freeman-Lynde RP, Whitley KF, Lohmann KC (1986) Deep marine origin of equant spar cements in Bahama escarpment limestones. J Sediment Petrol 56:799–811

Friedman GM, Amiel AJ, Schneidermann N (1974) Submarine cementation in reefs: example from the Red Sea. J Sediment Petrol 44:816–825

Hall A, Kennedy WJ (1967) Aragonite in fossils. Proc R Soc London Ser B 168:377–412

Harris PM, Kendall CGStC, Lerche J (1985) Carbonate cementation: a brief review. In: Schneidermann N, Harris PM (eds) Carbonate cements. Soc Econ Paleontol Mineral Spec Publ 36:79–95

Herrmann T, Koch R (1985) Auftauchphasen und Süßwasser-Zementation in den oberen Cerithien-Schichten (Aquitan) des Mainzer Beckens. Münchner Geowiss Abh A 6:51–74

Kendall AC (1985) Radiaxial fibrous calcite: a reappraisal. In: Schneidermenn N, Harris PM (eds) Carbonate cements. Soc Econ Peleontol Mineral Spec Publ 36:59–77

Kendall AC, Tucker ME (1973) Radiaxial fibrous calcite: as a replacement after acicular carbonate. Sedimentology 20:365–389

Koch R (1977) Mikrofazielle Untersuchungen in der Kreide Sloweniens. Diss Univ Heidelberg, 123 pp

Koch R (1978) Fazies und Diagenese eines Biostrom-Komplexes (Oberkreide, Jugoslawien). Ergänzungsband Erdöl und Kohle, Compendium 78/79. Echterdingen, V Hernhaussen, Echterdingen, pp 41–64

Koch R (1988) Mikrofazielle und diagenetische Entwicklung kretazischer Karbonatgesteine im jugoslawischen Raum. Habilschrift, Univ Heidelberg, vol 1:210 pp, 90 Figs, 5 Tabs, 8 Plates; vol 2: Photodokumentation der Mikrofaziestypen, 41 Tabs

Koch R, Schorr M (1986) Diagenesis of Upper Jurassic sponge algal reefs in SW Germany. In: Schroeder JH, Purser BH (eds) Reef diagenesis. Springer, Berlin, Heidelberg New York Tokyo, pp 224–244

Koch R, Ogorelec B, Orehek S (1989) Microfacies and diagenesis of Lower and Middle Cretaceous carbonate rocks of NW-Jugoslavia (Slovenia, Trnovo area). FACIES 21:135–170

Krebs W (1969) Early void filling cementation in Devonian fore-reef limestones (Germany). Sedimentology 12:279–299

Lang HB (1964) Dolomit und zuckerkörniger Kalk im weißen Jura der mittleren Schwäbischen Alb. N Jahrb Geol Paläontol Abh 120:253–299

Lighty RG (1985) Preservation of internal reef porosity and diagenesis sealing of submerged early Holocene barrier reef, southeast Florida shelf. In: Schneidermann N, Harris PM (eds) Carbonate cements Soc Econ Paleontol Mineral Spec Publ 36:123–151

Longman MW (1980) Carbonate diagenetic textures from near-surface diagenetic environments. Am Assoc Petrol Geol Bull 64:461–487

Macintyre IG (1984) Extensive submarine lithification in a cave in the Belize Reef Platform. J Sediment Petrol 54:221–235

Meyers JH (1985) Marine vadose beachrock cementation by cryptocrystalline magnesian calcite – Maui, Hawaii. J Sediment Petrol 57:558–570

Miljuš P (1976) Dinarids geosyncline – tectonic framework and the evolution of the eugeosyncline. 8th Congr Geol Jug Ljubljana, vol 3:139–156

Moldovanyi EP, Lohmann KC (1984) Isotopic and petrographic record of phreatic diagenesis: Lower Cretaceous Sligo and Cupido Formation. J Sediment Petrol 54:972–985

Monty CLV, Mas JR (1981) Lower Cretaceous (Wealdian) blue-green algal deposits of the Province of Valencia, eastern Spain. In: Monty C L V (ed) Phanerozoic stromatolites. Case histories. Springer, Berlin Heidelberg New York, pp 209–229

Moore CH (1973) Intertidal carbonate cementation, Grand Cayman, West Indies. J Sediment Petrol 43:591–602

Mountjoy EW, Krebs W (1983) Diagenesis of Devonian reefs and buildups, western Canada and Europe – a comparison. Z Dtsch Geol Ges 134:5–60

Müller G (1971) Gravitational cement: an indicator for the vadose zone of subaerial environments. In: Bricker OP (ed) Carbonate cements. Johns Hopkins Stud Geol 19:301–302

Orehek S, Ogorelec B (1979) Sedimentologic features of the Jurassic and Cretaceous carbonate rocks of Trnovski Gozd. Geol Vjesn 32:185–192

Oti M, Müller G (1985) Textural and mineralogical changes in coralline algae during meteoric diagenesis: an experimental approach. N Jahrb Mineral Abh 151:163–195

Petković K (1958) Neue Erkenntnisse über den Bau der Dinariden. Jahrb Geol Wien B A 101:1–24

Pierson BJ, Shinn EA (1985) Cement distribution and carbonate mineral stabilization in Pleistocene limestones of Hogsty Reef, Bahamas. In: Schneidermann N, Harris P M (eds) Carbonate cements. Soc Econ Paleontol Mineral Spec Publ 36:153–168

Polšak A (1965) Stratigraphie des couches jurassiques et cretacees de l'Istrie centrale. Geol Vjesn 18/1:176–184

Prezbindowski DR (1985) Burial cementation – is it important? A case study, Stuart City Trend, south central Texas. In: Schneidermenn N, Harris PM (eds) Carbonate cements. Soc Econ Paleontol Mineral Spec Publ 36:241–264

Radiočić R (1966) Microfacies du Jurassique des Dinarides externes de la Yougoslavie. Geologija 9:5–378

Saller AH (1986) Radiaxial calcite in Lower Miocene strata, subsurface Enewetak Atoll. J Sediment Petrol 56:743–762

Sandberg P (1985) Aragonite cements and their occurrence in ancient limestones. In: Schneidermenn N, Harris PM (eds) Carbonate cements. Soc Econ Paleontol Mineral Spec Publ 36:33–57

Schneider W (1977) Diagenese devonischer Karbonatkomplexe Mitteleuropas. Geol Jahrb D 21:107 pp

Schroeder JH (1972) Fabrics and sequences of submarine carbonate cements in Holocene Bermuda cup reef. Geol Rundsch 61:708–730

Shinn E (1968) Practical significance of birdseye structures in carbonate rocks. J Sediment Petrol 38:215–223

Sikošek B, Medwenitsch W (1965) Neue Daten zur Fazies und Tektonik der Dinariden. Verh Geol B A Wien Sonderh G:86–102

Sikošek B, Vukašinović M (1975) Geotectonical evolution of the Inner Dinarides. Radovi Znan Sav Naftu Jazu A 5:176–183

Šribar L (1979) Biostratigraphy of Lower Cretaceous Beds from the Logatec plain. Geologija 22:277–308

Städter T, Koch R (1987) Mikrofazielle und diagenetische Entwicklung einer devonischen Karbonatfolge (Givet) am SW-Rand des Briloner Sattels. FACIES 17:422–428

Turnšek D, Buser S (1966) The development of Lower Cretaceous beds in the western part of Trnovski Gozd. Geologija 9:527–548

Wirsing G, Koch R (1986) Algen-Schwamm-Bioherme des Flachwasserbereichs (Schwäbische Alb, Weißjura delta 3). FACIES 14:285–308

The Southern Permian Basin and its Paleogeography

E. Plein[1]

CONTENTS

Abstract 124
1 Introduction 125
2 The Volcanic Rotliegend 126
3 The Sedimentary Rotliegend 128
3.1 Schneverdingen Formation 129
3.2 Hannover Wechselfolge/Slochteren Hauptsandstein . . . 130
4 Results 132
References 132

Abstract

Based on three paleogeographic maps, newly published results of the southern Permian basin are integrated with existing data to show the development of the southern Permian basin for the larger lithostratigraphic units.

The Volcanic Rotliegend is controlled by fault tectonics and volcanism. New, fault-controlled basins of smaller extent, came into existence in the Variscan foldbelt, as well as in the foreland. A compilation of reliable isotopic age determination on various volcanic rocks in the basin gives some evidence that volcanic activity seems to occur in the Early Permian rather simultaneously, and within a relatively short interval of 15 million years. Magnetostratigraphic investigations (indicating the Illawarra Reversal) now give some evidence of a rather long-time gap between the main volcanic activity and the deeper part of the Sedimentary Rotliegend in the foreland area. Towards the end of the Schneverdingen formation and the beginning of the Hannover Wechselfolge, a tectonic structural modification of the entire southern Permian basin took place. The late Variscan wrench tectonic became inactive during this time, too. This process might be caused by a stabilization and thermal cooling of the lithosphere. It is a platform develop-

---

[1] Institut für Sedimentforschung der Universität Heidelberg, Postfach 103020, 6900 Heidelberg, FRG.

ment which then came into existence and continued to extend from the Upper Permian till today. It can also be said that the conditions which began then and were accompanied by the extension of the southern Permian basin to the west and east prepared, together with the northern Permian basin, the connection to the proto-Atlantik, which was created by the ingression of the Zechstein sea.

1 Introduction

In Central Europe, the Rotliegend represents the lower main part of the Permian. It is intercalated between the continental paralic Upper Carboniferous and the marine Zechstein. The Rotliegend sediments are of exclusively continental origin. Within the Variscan foldbelt, the Rotliegend occurs in numerous intramontane basins. In the northern foreland of the Variscan belt, Rotliegend occurs in broad basins, now deeply buried by a thick Meso-Cenozoic cover beneath the NW European basin (Ziegler 1982), also called "Mitteleuropäische Senke", or Central European lowland (Katzung 1972, 1988).

Whereas the intramontane basins are characterized by short-distant lateral variability of lithology, in the foreland basins the lithology of the sedimentary sequences changes over longer distances, and successions from distant locations can be compared more easily. The foreland basins can be divided into a northern and southern Permian basin, separated by the Mid-North Sea Ringkøbing-Fyn trend of highs, which came into existence during Late Carboniferous/Early Permian (Ziegler 1982).

The aim of this chapter is to extend a former publication (Plein 1978) by including and integrating data published since. With the help of three paleogeographic maps for the southern Permian basin, the present knowledge concerning the lateral distribution of the various types of continental sediments will be given, and some open litho- and time-stratigraphic problems discussed. This compilation should be considered a progress report. The maps presented are not final, and will require revision as new information becomes available from areas and regions for which access to only very limited data existed.

Before beginning it seems necessary to make some remarks concerning the use of litho- and biostratigraphic names in both the past and present. The early tripartition of the Rotliegend used in the past century, and based on the assumption that eruptive rocks were restricted to the middle part of the Rotliegend, was abandoned, as eruptive rocks occurred in the deepest part of some Rotliegend sequences. As the terrestrial fauna and flora in this type of continental sediments are only locally developed, the Rotliegend was then divided into Lower and Upper Rotliegend, on the basis of an unconformity caused by the "Saalian tectonic phase". In this bipartition, the Lower Rotliegend is characterized by coexistent grey and red sediments and in part by tight interfingering of sedimentary and volcanic sequences. The Upper Rotliegend consists of red beds and some salt intercalations with or without very rare volcanic intercalations. It usually begins with coarse conglomerates. At higher levels, the sediments turn to finer grades and tend to overlap the earlier outlines of sedimentation. Medium- and large-scale cycles in these sequences have a high value for lithostratigraphic correlations.

Twenty years ago, arguments urging the mistrust of the Saalian unconformity as a stratigraphic mark arose (Lorenz and Nicholls 1976). The observed unconformities are not the result of a widespread orogenic pulse, but are due rather to local differences in subsidence and sedimentation. Lützner (1988) has recently proven this concept in the Thuringian Forest, a part of the Saale trough. He found that fault movements and shifting of depocentres and elevations also gave rise to numerous disconformities. There was no evidence of a general regionally significant unconformity at the base of the Upper Rotliegend. The Saalian unconformity in the Thuringian Forest only represents the last in a series of intra-Rotliegend disconformities.

In the last 20 years, some authors have used the biostratigraphic terms Autunian and Saxonian as synonyms of Lower and Upper Rotliegend, without supporting evidence for identification. No international agreement has been achieved on this matter. Since the biostratigraphic definition is still under discussion, a great deal of confusion has arisen from the use of all the stratigraphic terms. Therefore, they will be avoided here. In accordance to the German subcommission for Permian stratigraphy, only the descriptive names "Volcanic Rotliegend" and "Sedimentary Rotliegend" will be used.

2 The Volcanic Rotliegend

The Variscan orogeny was caused by a subduction zone along the Variscan deformation front, stretching from South Ireland to Poland (Meissner et al. 1984). Oceanic crust was driven below the continental crust of the Variscan, leading to a thermal destabilization of the lithosphere, which, in turn induced a remobilization of the subducted crust. Late to postorogenic calc alkaline intrusives are very widespread in the Variscan foldbelt, and are followed by calc alkaline volcanic rocks. In connection with the crustal shortening, a period of uplifting, as well as the development of the Late Variscan wrench tectonic occurred (Ziegler 1982). The formation of a series of tectonically induced intramontane troughs was followed by block faulting and graben/horst structures in the foreland (Katzung 1982; Gast 1988), between the North Sea and Poland. The wrench movements were accompanied by intense eruptions, producing accumulations of thick volcanic sequences in different parts of the area (Fig. 1). Intercalated sediments indicate periods of volcanic quiet. The main portion of the volcanic rocks is of acidic and intermediate composition; basic volcanic rocks are subordinated (Eckhardt 1979; Benek 1983). The rocks are Ca-poor and alkali-rich; they belong to the calc

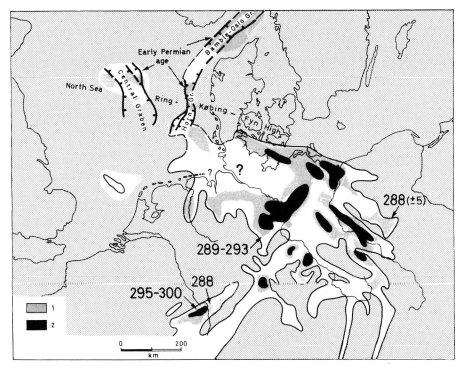

Fig. 1 Distribution and thickness of the Volcanic Rotliegend (After Benek et al. 1983; Katzung 1988; Ryka 1981; Siemaszko 1981; Ziegler 1982). *1* Volcanic rocks 50 to 1000 m thick; *2* volcanic rocks > 1000 m thick; *white areas* mostly clastic sedimentation. *Numbers* indicate isotopic age determinations in million years (After Lippolt 1982; Hess 1989 ps. comm.)

alkali rock suite. Rhyolitic, as well as most of the andesitic rocks, are thought to be of crusted origin at different levels. Only basaltic rocks are thought to be descended from subcrustal depths (Ryka 1981).

Figure 1 attempts to combine the data published by various Polish and East German authors on volcanites of the "Lower Rotliegend" with data of authors on the West. The areas with predominantly clastic deposits are shown in white. The wide distribution of the volcanites can be seen. It can also be noted that the greater thickness of volcanites is largely bound to lineaments in N-S and NW-SE directions.

To the west, the distribution of the volcanics ends. In the Elbe estuary, the development of Permian sequences remains unknown. They are thought to be located at depths of 6000 to 7000 m. Along the River Ems, as well as south of Bremen, thicker volcanics have been found. Early Permian volcanites are also described within the Oslo-Bamble-Horn graben, as well as from the margins of the North Sea Ringkøbing-Fyn High, which came into existence during this period (Ziegler 1982). Volcanic activity is mostly bound to faults of the later central graben. Glennie (1984) has concluded that the central graben formation began at this time.

Permian volcanism was apparently restricted to parts of the subaerial mid-European continent, and activity was confined to its depressions; graben or half-grabens. Of what age are these various volcanic rocks? Only a small number of reliable isotopic age determinations exist. Most of the basaltic to intermediate Rotliegend rocks in the Federal Republic of Germany and in Poland are frequently spilitized, and thus might be products of slight thermal metamorphism and metasomatism during a period of an increased geothermal gradient (Lorenz and Nicholls 1976).

"The Laboratorium für Geochronologie der Universität Heidelberg" has been working on the dating of Rotliegend volcanics for years. Some available results are shown on the map (Fig. 1). This work has been carried out using conventional K/Ar and ^{40}Ar/^{39}Ar dating methods on biotites and sanidines. For all of these samples, Early Permian ages between 288 and 300 million years were inferred. The results show that not only did the "Grenzlager" volcanism of the Saar-Nahe region (which covers an area of about 10^4 km^2) erupt almost simultaneously (Lippolt and Hess 1983), but also the Rotliegend volcanite in the Wrzesnia well 30 km east of Posen (Lippolt et al. 1982), and the volcanites in the Ilfeld basin south of the Harz mountains (Hess 1989, pers. comm.).

Thick Early Permian volcanics also occur in the Horn graben as well as at the margins of the mid-North Sea Ringkøbing-Fyn High, and are evident, too, in the Bamble trough (Ziegler 1982).

This limited age evidence is sufficient, however, to indicate that in the entire area, volcanic activity seems to occur in the Early Permian rather simultaneously, and within a relatively short time interval (15 million years). Nevertheless, much research and refining must still be done.

3 The Sedimentary Rotliegend

The Sedimentary Rotliegend, also often called Upper Rotliegend, is divided into two parts in the Federal Republic of Germany: the Schneverdingen Formation, and the Hannover Wechselfolge/Slochteren Hauptsandstein above. In East Germany the lower sequence is called the Havel-Folge; the upper sequence the Elbe-Folge. In Poland, the nonvolcanic sediment sequence is divided into an upper and lower megacycle.

Recently, the Upper Rotliegend was subdivided by means of magnetostratigraphic investigations (= Illawarra Reversal) into two sections, the lower Upper Rotliegend and upper Upper Rotliegend (Menning 1988). Menning dates the Illawarra Reversal at 259 million years. According to detailed investigations of cores of an E. German well, near the axis of the Rotliegend basin, the Illawarra Reversal was found at a level more than 1000 m below the Zechstein base. This position is within the lowermost part of the Havel-

Folge. As no magnetostratigraphic investigations exist in our part of the basin, we can only speculate that the Illawarra Reversal can be found within the lower part of the Schneverdingen Formation.

If the dating of the Illawarra Reversal is correct, and the main volcanic activity in the whole basin occured 300–285 million years ago, a time gap of about 25 million years remains (Menning 1988). Within this gap, relatively few sedimentary sequences are known in the foreland area to the present.

3.1 Schneverdingen Formation

The lithostratigraphic term Schneverdingen Sandstone and Fanglomerate war first proposed by Drong et al. (1982). The over 700-m-thick sediment sequence, found in a graben, was thought to be located in the Lower Rotliegend. In the conclusions of the revised lithostratigraphic subdivision of the Rotliegend in NW Germany, the unit was placed at the base of the Upper Rotliegend (Hedemann et al. 1984). Nevertheless, the stratigraphic definition of this formation is still under discussion.

The Schneverdingen Formation is mostly developed east of the Weser in grabens and half-grabens. According to Gralla (1988), the sediment sequences in the grabens can be subdivided into three units: the bottom unit consists predominantly of alluvial fans, with coarse, unsorted breccias and fanglomerates composed mainly of volcanics. They rest, with a gap, of carboniferous rocks. In other grabens, volcanics alternate with sediments below the first unit (Fig. 7 in Gast 1988). The second unit consists of fluvial fanglomerates and sandstones, followed by thick sands of aeolian deposition.

A hierarchy of aeolian sand bodies is often developed. The bulk of the sand is fine grained and beautifully cross-bedded on a meter scale, with very little interstitial clay or carbonate or andhydrite cements (Drong et al. 1982). The third unit began with some fluviatile conglomerates at the base of the Slochteren Hauptsandstein, belonging to the next subformation.

Basinwards, aeolian sediments have been found near the later axis of the basin, according to Gast (1988). The area of sedimentation could thus be defined as an intracontinental depression with a completely arid climate. Gast also points out that some North Sea wells possess a saline sequence down to the basis of the Sedimentary Rotliegend, and are underlain by volcanic conglomerates. Gralla (1988) also mentions the development of a playa lake and a sabkha sendimentation in the central part of the basin, where evaporites are also developed.

Fig. 2 attempts to combine the data published by various Polish and individual E. German authors on the distribution of the "lower megacycle" and "Havel-Folge", with data on the Schneverdingen Formation published in the Federal Republic of Germany. Areas with very limited data have been incorporated following a general tendency. The author is fully aware that this can only be a first attempt, which still retains a number of deficiencies. Nevertheless, this figure can be regarded as a basis for discussion, especially

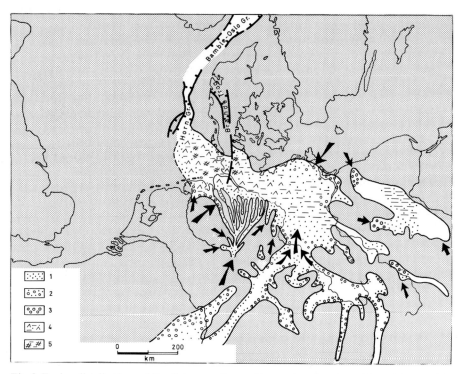

Fig. 2 Facies distribution of the Schneverdingen Formation (upper unit) and its possible equivalent clastic sequences (After Pokorski 1981; Katzung 1988; Gralla 1988; Gast 1988). *1* Mostly aeolian sands; *2* fluvial wadi flood with aeolian; *3* fluviatile deposits; *4* Playa, partly silty or anhydritic; *5* clay and evaporite. *Arrows* indicate areas of sediment input.

with regard to the possible regional extent of this formation. In order to show the Saxon I of Katzung (1988), the intramontane basins are also depicted. They indicate the possible paths of sediment transport.

A comparison of Fig. 2 to Fig. 1 indicates that the area of sedimentation did not increase during the Schneverdingen Formation; indeed, it became smaller, if one also takes the volcanic areas into account. This is also the case in Poland, where, however, the Polish sediment basin came into existence west of the Weichsel (Pokorski 1981).

3.2 Hannover Wechselfolge/Slochteren Hauptsandstein

The Subsidence of the grabens and half-grabens in the southern Permian basin slowly decreased during the Slochteren Hauptsandstein (Drong et al. 1982). Most of the sedimentation was still linked to the graben areas, where much greater thicknesses of the basal Hannover sequences have been observed (Gast 1988). There are also many indications that the sedimentation

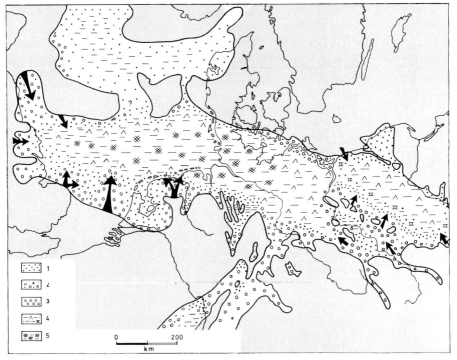

Fig. 3 Facies distribution of the Hannover Wechselfolge/Slochteren Hauptsandstein (After Pokorski 1981; Ziegler 1982; Katzung 1988; Gast 1988; Gralla 1988). *1* Aeolian/shoreline sands; *2* fluviatile and sheet sands + aeolian sands; *3* fluviatile deposits + alluvial fans; *4* playa, partly silty; anhydritic, carbonatic; *5* clay + evaporite. *Arrows* indicate alluvial fans.

area shifted from the intramontane basins of the inner Variscan belt to the foreland. According to Ziegler (1982), Variscan wrench tectonics became inactive during this time, and rapid erosion and isostatic uplift of the Variscan fordbelt went hand in hand with the subsidence of the foreland basin. This process might have been caused by a stabilization and a cooling of the lithosphere. Due to this large-scale (epirogenetic) subsidence, the local depression zones (such as former grabens) joined, and the basin expanded more and more; first along its axis to the west and east, and afterwards especially to the south, onlapping on its marginal parts (Fig. 3). In this way, the sediments occupied the entire basin with a quite uniform character. Complete successions with playa lake sediments and some halite intercalations can be correlated over larger areas (Gralla 1988). Locally (along temporary inland lakes), aeolian or beach sands were also deposited. Some of the beach deposits can be mapped over a distance of nearly 300 km (Gast 1988).

Sequences of aeolian sands locally reach thicknesses of 100 to 200 m. Analyses of orientation of the aeolian bedding, both in outcrops and wells, indicate that both transverse and seif dunes were formed by "E-W trade winds" (Glennie 1984). In the eastern part of the basin, aeolian sands are found only locally. By cyclicity, the sedimentary sequences in the basin are subdivided into several small units (see Gralla 1988). The small- and medium-scale cycles together form an asymmetric progressive megacycle, from coarse (at the bottom) to very fine clastics at the top of the Sedimentary Rotliegend.

It can be stated in summary that with the thermal cooling of the lithosphere in northern Central Europe, a tectonic structural regeneration tending towards a platform stage obviously took place. These platform conditions have continued through the Upper Permian and Mesozoic until the present. It can also be said that the conditions which came into being during the Hannover-Wechselfolge, and accompanied by the extension of the southern Permian basin towards the west and east, together with the northern Permian basin, prepared the connection to the proto-Atlantic. This came into existence at the end of the Rotliegend, with the ingression of the Zechstein sea.

4 Results

1. According to present knowledge, most of the Permian volcanic activity in the mid-European area seems to occur rather simultaneously and within a relatively short time interval (300 to 285 million years).

2. There is now growing evidence for a time gap (nearly 25 million years?) between the main volcanic activity and the Illawarra Reversal at the lower part of the Havel-Folge (=Schneverdingen Formation?).

3. With the thermal cooling of the lithosphere towards the end of the Schneverdingen Formation and the beginning of the Hannover-Wechselfolge, a tectonic structural modification of the entire basin tending towards a platform stage took place, and continued in the Upper Permian and Mesozoic.

4. With the extension of the southern Permian basin towards the west and east, and the formation of the northern Permian basin, the connection to the proto-Atlantic, which come into existence with the ingression of the Zechstein sea, was prepared.

References

Benek R (1983) Über Beziehungen des permosilesischen Vulkanismus zum Bruchmuster, speziell zum Elbe-Lineament. In: Contributions on geological development of molasses in some regions of Europe and USSR and on block tectonics of Elbe-Lineament. Veröff Zentralinst Phys Erde 77 pp 177–186

Drong HJ, Plein E, Sannemann D, Schuepbach MA, Zimdars J (1982) Der Schneverdingen-Sandstein des Rotliegenden – eine äolische Sedimentfüllung alter Grabenstrukturen. Z Dtsch Geol Ges 133:699–725

Eckhardt FJ (1979) Der permische Vulkanismus Mitteleuropas. Geol Jahrb D35:3–84

Falke H (1972) The continental permian in North and South Germany. Int Sediment Petrogr Ser 15:43–113
Gast RE (1988) Rifting im Rotliegenden Niedersachsens. Geowissenschaften 6:115–122
Gersemann J (1989) Vulkanotektonische Strukturen im Permokarbon Ost-Niedersachsens. Nachr Dtsch Geol Ges 41 (in press)
Glennie KW (1984) Early Permian-Rotliegend. In: Glennie KW (ed) Introduction to the petroleum geology of the North Sea. Blackwell, Oxford, pp 41–60
Gralla P (1988) Das Oberrotliegende in NW-Deutschland-Lithostratigraphie und Faziesanalyse. Geol Jahrb A106:3–59
Gralla P, Nieberding F, Sobott R (1988) Der Wustrow-Sedimentationszyklus des Oberrotliegenden im Bereich einer NW-deutschen Erdgaslagerstätte. Nachr Dtsch Geol Ges 39 pp 22–23
Hedberg HD (ed) (1976) International stratigraphic guide. John Wiley & Sons, New York, XVII + 200 pp, 14 Figs, 3 Tabs
Hedemann HA, Naschek W, Paulus B, Plein E (1984) Mitteilung zur lithostratigraphischen Gliederung des Oberrotliegenden im nordwestdeutschen Becken. Nachr Dtsch Geol Ges 30:100–107
Katzung G (1972) Stratigraphie und Paläogeographie des Unterperms in Mitteleuropa. Geologie 21:570–584
Katzung G (1982) Explanatory notes to the lithotectonic molasse profile of the Central European Depression (NE-German Depression) G.D.R. Veröff Zentralinst Phys Erde 66:209–244
Katzung F (1988) Tectonics and sedimentation of Variscan molasses in Central Europe. Z geol Wiss 16:823–843
Kelch HJ, Paulus B (1980) Die Tiefbohrung Velpke-Asse Devon 1. Geol Jahrb A 57:1–175
Lippolt HJ, Hess JC (1983) Isotopic evidence for the stratigraphic position of the Saar-Nahe Rotliegend volcanism. N Jahrb Geol Paläontol Mh 12:713–730
Lippolt HJ, Raczek J, Schleicher H (1982) Isotopenalter eines Unteren Rotliegend-Biotits aus der Bohrung Wrzesnia/Polen. Aufschluß 33:13–25
Lorenz V, Nicholls IA (1976) The Permocarboniferous Basin and Range Province of Europe: an application of plate tectonics. In: Falke H (ed) The continental permian in Central, West and South Europe. NATO Adv Stud Inst Ser C, Math Phys Sci Reidel, Dordrecht, pp 313–342
Ludwig AO (1984) Zur Bruchtektonik während der variszischen Morphogenetappe (Mittlerer Teil der DDR). Z Geol Win 12:215–234
Lützner H (1988) Sedimentology and basin development of intramontane rotliegend basins in Central Europe. Z Geol Win 16:845–863
Lützner H, Falk F, Ellenberg I, Grumbt E, Ludwig AO (1979) Übersicht über die variszische Molasseentwicklung in Mitteleuropa und am Ural. Z Geol Win 7:1157–1167
Meissner R, Springer M, Flüh E (1984) Tectonics of the Variscides in north-western Germany based on reflection seismic measurements. In: Hutton DHW, Sanderson DJ (eds) Variscan tectonics of the North Atlantic region. Blackwell, Oxford, pp 23–32
Mennig M, Katzung G, Lützner H (1988) Magnetostratigraphic investigations in the Rotliegendes (300–252 Ma) of Central Europe. Z Geol Wiss 16:1045–1063
Plein E (1978) Rotliegend-Ablagerungen im Norddeutschen Becken. Z Dtsch Geol Ges 129:71–97
Pokorski J (1981) Paleogeography of the Upper Rotliegendes in the Polish Lowland. In: Int Symp Central European permian proceedings, Warsaw
Ryka W (1981) Some problems of the Autunian volcanism in Poland. In: Int Symp Central European permian proceedings, Warsaw
Siemaszko E (1981) Autunian intrusives in the Fore-Sudetic Monocline. In: Int Symp Central European permian proceedings, Warsaw
Ziegler PA (1982) Geological atlas of western and central Europe. Shell Int Petrol Maatsch BV

Integrated Hydrocarbon Exploration Concepts in the Sedimentary Basins of West Germany

MICHAEL BLOHM[1]

CONTENTS

Abstract 134
1 Introduction 134
2 Classical Exploration Concept 136
3 Integrated Exploration Concept 137
3.1 Landsat Imagery Analysis 138
3.2 Non-Seismic Methods 140
3.3 Advanced Log Interpretation 140
3.4 Quantitative Basin Modelling 146
3.5 Probabilistic Methods 147
4 Conclusion 150
References 150

Abstract

Consequent application of a modern integrated hydrocarbon exploration concept demonstrated by selected case histories shows that even in a highly matured exploration area like West Germany combinations of less conventional exploration techniques can minimize the risk and improve the success ratios, while reducing the exploration expenditures.

1 Introduction

About one-half of West Germany is covered by prospective sedimentary basins (Ziegler 1982). Hydrocarbon exploration has been known since 1859. Since then more than 3000 exploration wells have been drilled and some 770 × 10^6 tons of oil and some 800 × 10^9 m^3 of gas have been found in this area (Fig. 1). During the last three decades gas has become the main economic exploration target in West Germany (Plein 1985). The main producing reservoirs are found in the Lower Triassic, Permian and Upper Carboniferous

---

[1] BEB Erdgas und Erdöl GmbH, Riethorst 12, D-3000 Hannover 51, FRG.

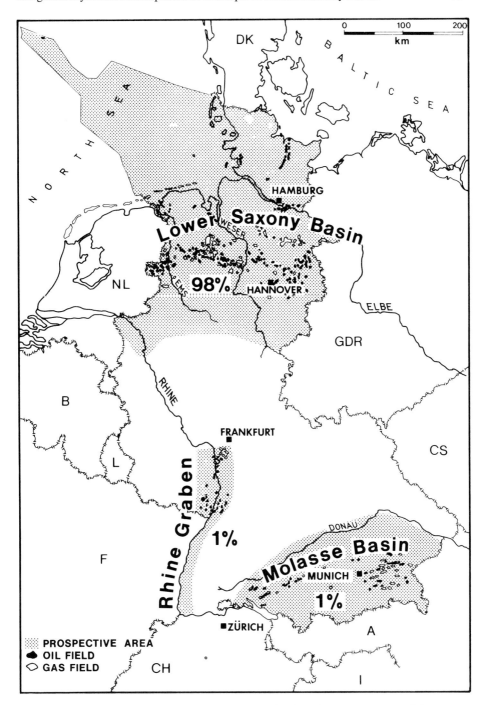

Fig. 1. Prospective areas in West Germany and distribution of HC reserve potential

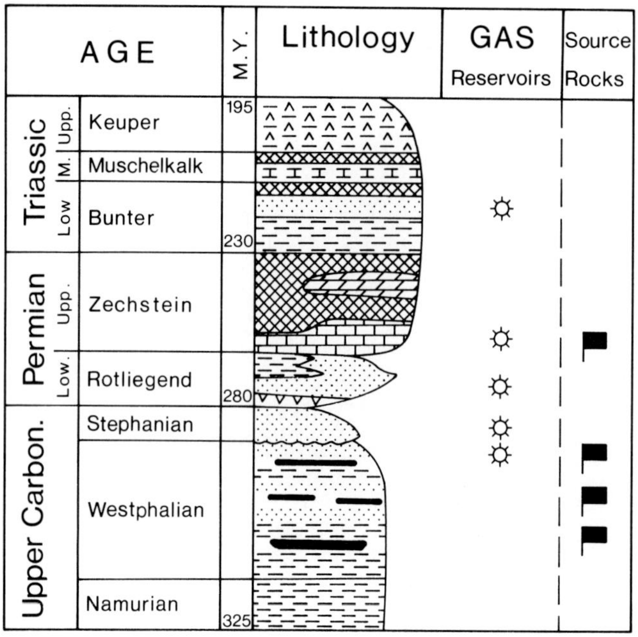

Fig. 2. Age and lithology of the main gas reservoirs in West Germany

clastics and carbonates (Fig. 2) at depths between 3000 and 6000 m, thus total reserves are about 270 billion m^3 gas (Plein 1978; Sannemann et al. 1978; Lübben 1987). But it is believed that additionally some 250–350 billion m^3 gas may still be found in the conventional exploration potentials of West Germany (Novak and Keshav 1977; Plein 1979). To mobilize these speculative reserves only modern integrated exploration methods will lead to success in a highly matured exploration area such as West Germany. However, if exploration is based only on the classical exploration methods, only minor success rates can be expected.

2 Classical Exploration Concept

The procedure of the so-called Classical Exploration Concept of the late 1970s (see Fig. 3) still has validity and advantages in less explored areas of the world. In highly matured exploration areas like West Germany the chance of success would sink below 10% within the next 10 years if this concept were still to be followed. Therefore, an integrated concept which is extremely adaptable to the present economic situation in production and sales is necessary.

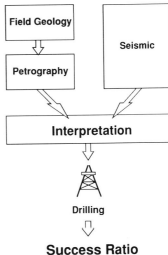

Fig. 3. Flow chart of a conventional exploration concept

3 Integrated Exploration Concept

The integrated exploration concept makes use of the abundance of data available for the last two generations of hydrocarbon exploration such as: extensive field geology and considerable petrographic and stratigraphic knowledge of the subsurface, where about 400 km have been cored by the German oil industry and a dense grid of about 240 000 km of 2D reflection seismic profiles has been recorded. But since exploration has reached depths of about 4500–6000 m, attention is now focused on:

1. Permeability alterations in clastics sediments due to diagenesis;
2. Seismic resolution below a highly tectonized and/or inverted overburden;
3. Reliable hydrocarbon generation and migration history by quantitative basin modelling;
4. Prospect selection by probabilistic methods.

Integrated exploration methods are considered to be part of the future standard tools for German hydrocarbon exploration. The following selected case histories will show how the exploration risk in relation to the present economic situation can be minimized by using less conventional exploration methods and how these methods have been applied successfully in German exploration.

3.1 Landsat Imagery Analysis

Landsat imagery data can also be used to identify major reactivated fault systems which are often difficult to recognize with routine exploration methods only. Large-scale N to NE and NW trending lineaments and major basement block boundaries in association with deep crust and complex movement history have formed the sedimentary basins of NW Germany. These lineaments show a complex cross-cutting and therefore are often very difficult to map with seismic data only. The following case study will show that the characteristics of these elements may control the distribution of hydrocarbon traps and how Landsat data was used to support the exploration. Field A (Fig. 4) was discovered early in 1970 and remained the only producing Permian and Carboniferous gas field for more than 15 years in this area. The reservoir behavior shows clearly that the field has been producing from a highly fractured and faulted reservoir formed by deep-rooted, basement-involved faults, a so-called flower structure. Other wells were

Fig. 4. Landsat imagery and its structural interpretation

drilled in this area but with poor porosities, although drilled on good structures. By using Landsat imagery interpretation together with seismic, gravity and sedimentological data, it was possible to define two major lineaments; the NW-trending lineament, representing the right-lateral wrench fault system, leading for example to the development of the Rotliegend graben system, and a NNE-trending lineament with left-lateral wrenches. This concept now allowed us to predict the structures which were intersected by these lineaments observed to have an area of weakness expressed as a so-called pop up structural element due to late Cretaceous compressional movements and therefore likely to be fractured. Two other structures were subsequently drilled, resulting in the discovery of field B with some 5 billion m^3 of natural gas and structure C which was also highly fractured but had no commercial gas due to a sealing problem. In conclusion, Landsat imagery interpretation provides significant information which should be routinely integrated into the interpretation of all other available data sets.

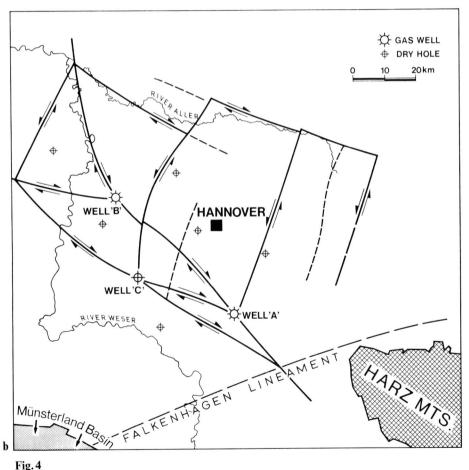

Fig. 4

3.2 Non-Seismic Methods

Before reflection seismic methods were applied worldwide, other methods such as gravity and magnetic methods were in use in German hydrocarbon exploration. During the 1930s and 1940s northern Germany was covered with a widespread grid of gravity data. This regional survey was aimed at mapping the borderline of the abundance of salt plugs with their typical oil traps in the Lower Saxony Basin (Betz et al. 1987). In general, these salt plugs consist of Permian salt and had their movement or piercement between Triassic and Tertiary time (Trusheim 1957; Sannemann 1963; Brink 1984). For economic reasons gravity maps and reflection seismic methods, were now used to define a drillable location preferentially on top of a salt plug. Drilling through thick (3000–4000 m) monotonous salt sections saves more than one-fourth of the drilling costs. But the more salt domes were drilled, the more obvious it became that the assumed salt plugs might actually consist of thick sedimentary non-halite intercalations, creating many technical and therefore financial problems. In some cases the salt plug did not exist at all because an updoming, fractured and faulted structure, a so-called flower structure, was interpreted, due to poor seismic resolution, as a salt plug. The following case history is shown in Fig. 5 where the interpretation of the seismic profile indicates a faulted monoclinal structure just below the salt plug. Before drilling well D it was decided to run a 3-dimensional gravity modelling over the salt body (Blaume 1989). After several modelling steps a best fit between model gravity and recorded Bouguer gravity was achieved. The thickness and width of the modelled salt body have been transferred to the seismic section (Fig. 6), resulting in a different shape of the salt plug and a different structure below the salt plug. It is also obvious that the time-depth conversion, which resulted from having more sediments below the salt plug overhang than before, will definitely influence the existence of a closure. Well A was drilled as a discovery through the predicted salt sequence followed by appraisal wells E and F drilled through the salt plug overhang. Finally, the combined interpretation of gravity and seismic data allows the construction of a physical 3D salt plug model which is shown in Fig. 7. As a result it can be summarized that a complex structural problem can be satisfactorily resolved by combining seismic and gravity methods at relatively low cost.

3.3 Advanced Log Interpretation

Incorrect predictions of facies and diagenesis are often the reason for well results such as: "gas-bearing but uneconomic". Since exploration on Permian and Carboniferous reservoirs has reached depths below 4000 m, more than one-third of all well failures are due to the fact that reservoir permeabilities in non-reservoir facies are too low. As already mentioned, the German oil industry has an extensive petrographic data base of cored sections available,

Integrated Hydrocarbon Exploration Concepts in the Sedimentary Basins

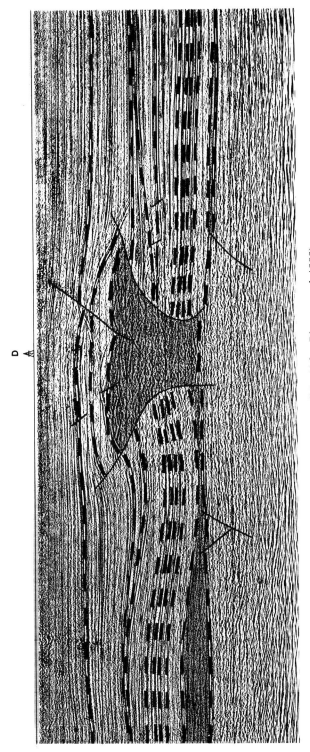

Fig. 5. Migrated seismic section with interpretation before drilling well D. (After Blaume et al. 1989)

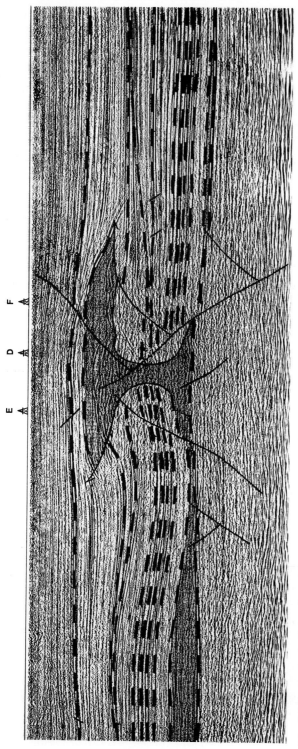

Fig. 6. Migrated seismic section with interpretation after 3-D modelling as preparation for well E and F (After Blaume et al. 1989)

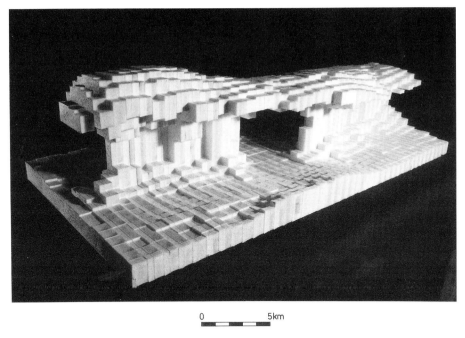

Fig. 7. 3-D salt plug model. (After Blaume et al. 1989)

but these cores represent only a small geological and time event in each well. Due to the extremely high coring costs, only some 20% of the Permian and Carboniferous reservoirs have been cored, but in general 100% of the reservoirs are logged. Therefore, efforts are focused on the use of open-hole logs in order to obtain detailed lithology (Mayer and Sibbit 1980) and facies information from uncored sections (Wolff and Pelissier-Combescure 1982; Delfiner et al. 1984). The disadvantage until recent times was that only a rough and generalizing petrophysical data base was used for the lithofacies analysis, resulting in a too coarse lithological description, which showed its limits in such complex and diagenetically altered reservoirs as in northern Germany.

The following example depicts how a detailed multi-well data base for the NW German Zechstein has been created (Stowe and Hock 1988) and meanwhile successfully used in exploration and production by integrating various disciplines (Fig. 8). Lithology and facies types can now be identified from well logs. Of course, the minimum thickness of a distinct facies type to be identified depends on the vertical resolution of the logging tools used. From the reconstructed geological section using the facies and litho-columns (Fig. 9), it can be determined how even in a monotonous mudstone sequence two diagenetic processes control the reservoir properties. After the deposition of a calcareous mudstone in a basinward direction of a carbonate platform, an early diagenetic dolomitization created favorable reservoir condi-

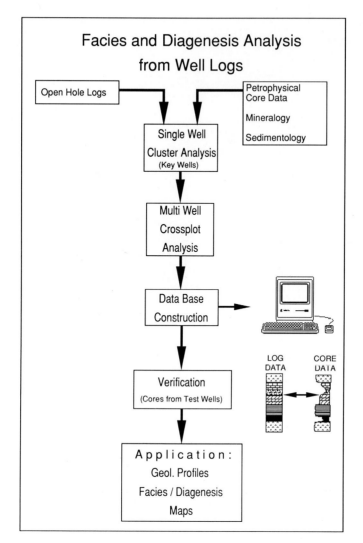

Fig. 8. Flow chart of a multi-well log analysis

tions over the whole area. Late diagenetic recrystallization in the mudstone sequence, progressing from well G to K, plugged all the initial pore space created previously with calcite cement. Well G was an economic failure; well H had uneconomic gas production. Wells I and K are now discoveries. This demonstrated that in cases with only a small amount of core information, a detailed interpretation of depositional and diagenetic processes would not be possible without advanced log interpretation.

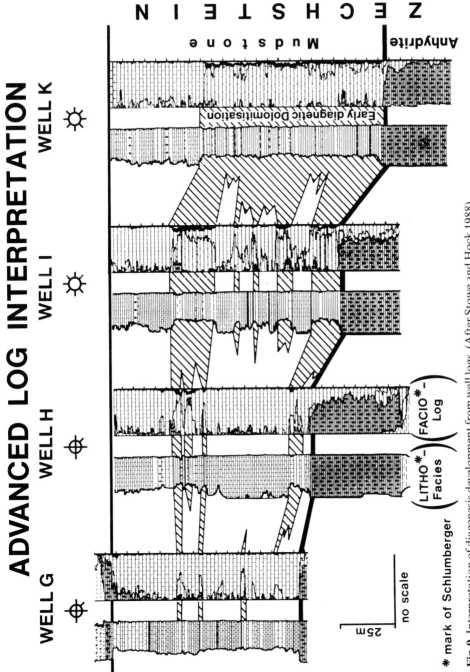

Fig. 9. Interpretation of diagenesis development from well logs. (After Stowe and Hock 1988)

3.4 Quantitative Basin Modelling

Since the invention of reflection and refraction seismic methods in the 1920s a big step forward was taken in the search for hydrocarbon accumulations in the subsurface. But all the geophysical methods could only answer the question "where". A quantification of the hydrocarbon generation mechanism was not possible. Continental sedimentary basins are the result of physical, chemical and tectonic processes between the surface and the upper crust. A regional subsidence led to the accumulation of sediments. Through geologic times organic material within the sediment has been thermally degraded to oil or gas. To quantify this process three main groups of questions have to be answered: burial history, thermal history and hydrocarbon generation, migration and accumulation history (Lerche 1989). Only two decades ago the determination of the maturity of organic matter and the correlation between temperature and time (Lopatin 1971; Waples 1980) gave the oil industry a first predictive tool (Fig. 10). Today, we know that the three main groups of questions have to be subdivided into: paleogeography, sediment (type, source, accumulation rate), depositional environment, mineralogic changes, tectonic movements, hydrodynamics, thermodynamics and organic geochemistry, to give only a few examples (Welte and Yükler 1981; Tissot and Welte 1984). As most of these factors vary interdependently in minor and major consequences for the whole system, it is obvious that only modern

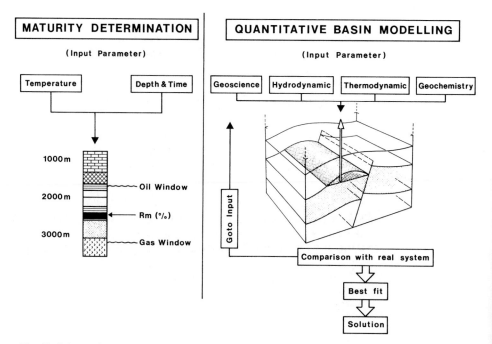

Fig. 10. Schematic comparison of maturity determination and quantitative basin modelling

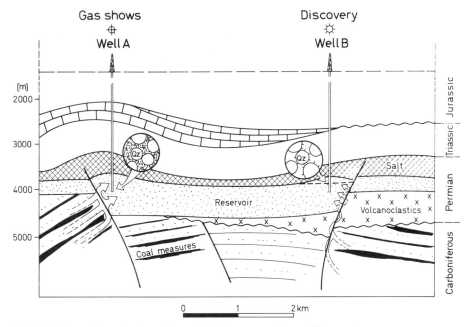

Fig. 11. Schematic geologic section with the interpretation of diagenetic processes

and powerful computers (e.g. Bethke et al. 1988) are able to handle and model these multi-dimensional problems.

The next case study will demonstrate how basin modelling can help to reduce the exploration risk by optimizing the proposed drilling site. Permian clastic reservoirs at depths of about 4500 m are often plugged by illite. Having modelled these processes, an explanation could be given for the diagenetic changes in this reservoir. Figure 11 shows that compaction and organic maturation in the deeper coal-bearing Carboniferous generated aggressive fluids which migrated along strata and faults in an up-dip direction. The left case shows that the relatively short and direct way to the adjacent Permian reservoir resulted in a strong illitization, plugging most of the pore space. On the right side, due to a long migration path through red beds and volcanics, acidic fluids lost part of their reactive potential and their permeability-reducing effect when reaching the reservoir. Then, with continuous subsidence, gas was generated from Jurassic times from the Carboniferous source rocks, which followed similar pathways as the former acidic fluids and filled the remaining pore space in the Permian reservoir. The trapped gas now protected the reservoir from further mineralogic changes.

3.5 Probabilistic Methods

All the methods and case studies described above have had only one objective: to minimize the exploration risk and to reduce uncertainty. In Germany

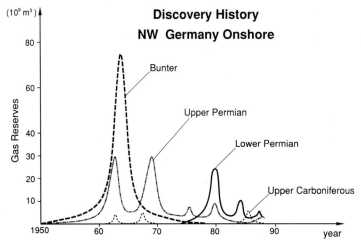

Fig. 12. Discovered reserves of the main gas plays of West Germany (source WEG)

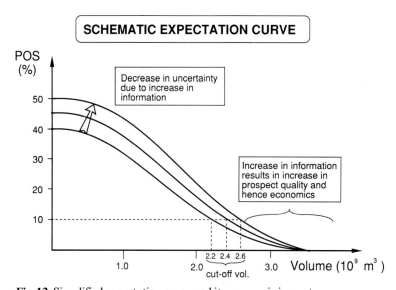

Fig. 13. Simplified expectation curve and its economic impact

generally one has to consider two types of risk: the geologic risk, which governs the existence of gas or oil, and the economic risk, which determines whether the discoveries can be produced. A third risk, which is becoming increasingly important, should be mentioned here: the environmental risk, especially in very populated countries like Germany. At a time of depressed oil prices the recognition and quantification of potential risk is essential. The use of probabilistic methods, which has been applied in the oil industry since

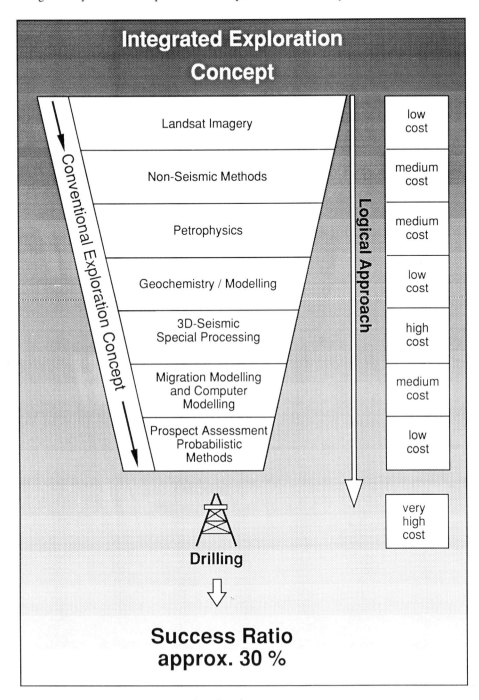

Fig. 14. Flow chart of the integrated exploration concept

the 1970s (Newendorp 1975; Megill 1985; Rose 1987) is a very helpful tool when decisions have to be made and expenditures optimized. The exploration history of NW Germany shows that the largest fields of the prominent gas plays were found in an early exploration cycle, whereas the smaller fields were found in the later stages of exploration (Fig. 12). Therefore, a mixed exploration strategy has to focus on a small number of high-risk, high-potential projects with the aim of finding a new play and furthermore on less risky prospects in the remaining "old" plays. The probabilistic approach illustrated schematically in Fig. 13 demonstrates that the prospect quality can generally be improved by increasing information, which is expressed in this graph by a higher probability of success (POS) and/or more resistance towards economic cutoffs.

4 Conclusion

Only the application of an integrated exploration concept in a highly matured hydrocarbon province like West Germany makes it possible to maintain a relatively high exploration rate of around 30%. Figure 14 summarizes the philosophy of using the least expensive techniques at the beginning and increasing the amount of information by maximizing cost effectiveness. The decision to use a further technical exploration effort in order to minimize risk and uncertainty should be carefully checked with an exploration assessment by using the probabilistic method. However, successful exploration programs should also include opportunities to find a large field or a new play outside the conventional pathway. Tailor-made combinations of the integrated methods, which fit geological, environmental and economic conditions, will result in substantially lower overall exploration costs. This cost-saving effect allows additional investment in future high-risk, high-potential prospects.

Acknowledgements. The author wishes to thank the management of BEB Erdgas und Erdöl GmbH for permission to publish this paper. I am also obliged to my colleagues who contributed with numerous critical comments and discussions.

References

Bethke CM, Harrison WJ, Upson C, Altaner SP (1988) Supercomputer analysis of sedimentary basins. Science 238:261–267
Betz D, Führer F, Greiner G, Plein E (1987) Evolution of the lower Saxony Basin. Tectonophysics 137:127–170
Blaume F, Brink H-J, Onasch H-J, Ries H (1989) Seismisch gravimetrisches 3D-Modelling am Beispiel eines norddeutschen Salzstockes. Erdöl Erdgas Kohle 105:4
Brink H-J (1984) Die Salzstockverteilung in Nordwestdeutschland. Geowiss Unserer Zeit 2:160–166

Delfiner PC, Peyret O, Serra O (1984) Automatic determination of lithology from well logs. In: 59th Annu Tech Conf Exhib Soc Petrol Eng, Houston, Sept 16–19

Lerche I (1989) An assessment of quantitative basin analysis. Energ Explor Exploit 7:63–67

Lopatin NV (1971) Temperature and geologic time factors in coalification. Izv Akad Nauk Uzb SSR Ser Geol 3:95–106

Lübben H (1987) Gegenwärtige Situation und zukünftige Perspektiven der deutschen Erdöl- und Erdgasgewinnung. Erdöl Erdgas Kohle 103:11

Mayer C, Sibbit A (1980) GLOBAL, a new approach to computer processed log interpretation. In: 55th Annu Fall Tech Conf Exhib Soc Petrol Eng, Dallas, Sept 21–24

Megill RE (1985) Evaluating and managing risk – a collection of readings. SciData, Tulsa, p 152

Newendorp PD (1975) Decision analysis for petroleum exploration. Penn Well Bocks, Tulsa, p 668

Novak H-J, Keshav NC (1977) Zur Erdöl- und Erdgasexploration in der Bundesrepublik Deutschland. Erdöl Erdgas Z 93:296–303

Plein E (1978) Rotliegend-Ablagerungen im Norddeutschen Becken. Z Dtsch Geol Ges 129:71–97

Plein E (1979) Das deutsche Erdöl und Erdgas. Jahresh Ges. Naturkd Württ 134:5–33

Plein E (1985) Die Entwicklung und Bedeutung der Erdöl-/Erdgasfunde zwischen Weser und Ems. Oldenburg Jahrb 85:267–311

Rose PR (1987) Dealing with risk and uncertainty in exploration: how can we improve? AAPG Bull 71:1–16

Sannemann D (1963) Über Salzstock-Familien in NW-Deutschland. Erdöl Z 11:3–10

Sannemann D, Zimdars J, Plein E (1978) Der basale Zechstein (A2-T1) zwischen Weser und Ems. Z Dtsch Geol Ges 129:33–69

Stowe I, Hock M (1988) Facies analysis and diagenesis from well logs in the Zechstein carbonates of northern Germany. In: SPWLA 29th Annu Log Symp Trans vol 2, Pap HH, San Antonio, June 5–8

Tissot B, Welte DH (1984) Petroleum formation and occurrence. Springer, Berlin Heidelberg New York, pp 495–608

Trusheim F (1957) Über Halokinese und ihre Bedeutung für die strukturelle Entwicklung Nordwestdeutschlands. Z Dtsch Geol Ges 109:111–151

Waples DW (1980) Time and temperature in petroleum formation – application of Lopatin's method to petroleum exploration. AAPG Bull 64:916–926

Welte DH, Yükler A (1981) Petroleum origin and accumulation in basin evolution – a quantitative model. AAPG Bull 65:1387–1396

Wolff M, Pelissier-Combescure J (1982) Faciolog-automatic electrofacies determination. In: SPWLA 23rd Annu Log Symp Trans, Houston, July 6–9

Ziegler P-A (1982) Geological atlas of western and central Europe. Elsevier, Amsterdam, 130 pp, 40 plates

Sedimentological and Petrophysical Aspects of Primary Petroleum Migration Pathways

Ulrich Mann[1]

CONTENTS

Abstract		152
1	Introduction	153
2	Petroleum Formation	154
3	Primary Migration: Present State of Knowledge	155
3.1	Distribution of Water, Kerogen, and Petroleum within the Source Rock	155
3.2	Possible Modes of Primary Migration	158
3.3	Current Theories and Concepts	158
3.4	An Unknown Petrophysical, Pressure Driven Flow Process	159
4	Primary Migration Type I: Via the Pore Network	161
4.1	The Migration Pathway Inside the Source Bed	161
4.2	A Quantitative Approach to Pore Network Definition: Pore Size Distribution	161
4.2.1	Experimental Methods and Calculation of Pore Radii	163
4.2.2	Effects of Lithofacies and Diagenesis	163
4.2.3	Effects of Compaction, Petroleum Formation, Petroleum Expulsion and Kerogen Type on the Pore Network	166
4.3	Pore Size Distribution and Related Petrophysical Properties	168
5	Primary Migration Type II: Via Fractures	169
5.1	Horizontal or Vertical Fractures?	169
5.2	Tectonic Fractures	171
6	Perspective: Revealing Primary Petroleum Migration Pathways	174
References		175

Abstract

The movement of petroleum from its source rock to a carrier bed called primary migration is mechanistically and quantitatively poorly understood compared to petroleum formation.

In the present article current theories and concepts of primary migration are reviewed with special emphasis being placed on sedimentological and

---

[1] Institute of Petroleum and Organic Geochemistry at the Research Centre (KFA) Jülich, FRG.

petrophysical considerations. In particular the migration pathway through the pore network of the rock matrix is discussed with reference to field examples where hydrocarbon expulsion has occurred.

It seems likely that the factors determining the dimensions of the three migration pathways organic, pore and fracture network – have to be quantified in order to evaluate primary petroleum migration mechanism. The respective key factors are:
1. Organofacies for the organic network;
2. Lithofacies, diagenesis and compaction for the pore network;
3. Permeability versus petroleum generation rate for the fracture network.

1 Introduction

Successful exploration for oil and gas means to the petroleum geologist the quantitative understanding of *petroleum formation, petroleum migration* and *petroleum trapping* within the limits of a certain volume of predominantly sedimentary rocks. While the processes of petroleum formation and petroleum trapping are relatively well known, petroleum migration is less well understood. Especially the first stage of petroleum migration, i.e. the movement of hydrocarbons within the source bed, and expulsion, which is called primary migration, still represents a fairly unknown process.

The objective of this chapter is to review selected sedimentological and petrophysical aspects of primary migration and to shed some light on the beginning of the migration pathway, immediately after the formation of petroleum. To this end the main emphasis will be put on the pore network of the source rock which can be quantitatively described by its pore size distribution.[1]

Most research done so far on the primary petroleum migration process was conducted on fine-grained clastic source rocks such as clay shales, calcareous shales and marlstones. Most findings apply also to carbonate/evaporite source rock sequences, even though this is not explicitly pointed out in the following.

Section 2 provides a summary of the basic ideas of petroleum formation. Primary migration of petroleum and petroleum formation are closely related processes, and therefore petroleum formation must be understood first, before specific details of the migration process are discussed.

---

[1] The author was fortunate to have had experience with pore size distribution curves and their interpretation (Mann 1979) at a time when questions about primary migration and related organic-geochemical phenomena arose. I had already studied such phenomena during my Ph.D. time at the Institute of Sedimentary Research of the Ruperto-Carola University at Heidelberg, under the guidance and the professional leadership of Prof. Dr. German K. Müller.

2 Petroleum Formation

Petroleum formation in source rocks is relatively well understood (Tissot and Welte 1978). With increasing burial and temperature oil and gas are formed from high-molecular organic material, the kerogen, within low-permeability clay shales or carbonates and evaporites (Fig. 1a). Only these fine-grained sedimentary rocks tend to be enriched in organic material. Initial input of organic matter, sedimentation rate and microbial activity govern this enrichment process. Little chemical transformation occurs in the immature zone of organic matter which is equivalent to the diagenetic stage of petroleum geochemists. As temperature increases to the catagenetic stage, hydrocarbon molecules and particulary aliphatics are produced from the kerogen, the amount and composition of which depend on the kerogen type. This is the principal stage of oil formation. With further increase of burial and temperature, condensate and wet gas is formed by cracking of carbon-carbon bonds. During the final stage, metagenesis, no significant amounts of hydrocarbons are generated from kerogen except methane. However, large amounts of methane result from the cracking of hydrocarbons still present in the source rock.

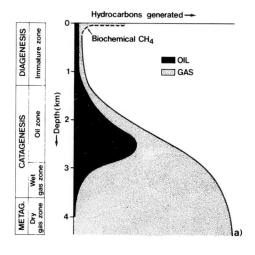

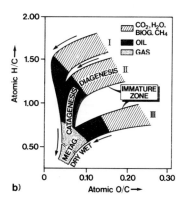

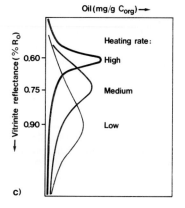

Fig. 1a–c. Petroleum formation: **a** Summary of hydrocarbon formation as a function of burial (After Tissot et al. 1974). **b** Scheme of an evolutionary pathway for principal kerogen types (After Tissot 1973). **c** Variation of the oil generation zone versus vitrinite reflectance (maturation stage) due to different heating rates. (After Yalçin and Welte 1988)

The petroleum potential of a source rock depends on the amount of petroleum that the kerogen is able to generate. This is directly related to the amount and type of the original organic matter as well as to the conservation/transformation conditions. It is clear that according to the chemical composition, kerogen composed of marine or limnic autochthonous organic matter will generate more petroleum (oil plus gas) than a kerogen comprised of continental plant debris. In fact, besides the amount of organic matter its hydrogen content determines the genetic potential of the source rock and also the oil/gas ratio of the generated petroleum charge. Tissot (1973) presented the general evolutionary pathway for the three standard kerogen types I, II, and III with their different hydrogen contents on a van Krevelen diagram (Fig. 1b).

Early research in organic geochemistry led to the maturity concept of immature, mature and overmature source rocks. The maturity level is determined by measurement of vitrinite reflectance, a method adapted from coal petrography. For each of the three standard kerogen types, fixed vitrinite reflectance values are used to define the "oil window" (Tissot et al. 1974), i.e. the beginning and end of oil generation. More recent studies on hydrocarbon generation and comparisons between case histories revealed that the kinetic reaction rates differ between vitrinite reflectance evolution and petroleum generation (Tissot and Espitalié 1975; Yükler and Kokesh 1984). Consequently, it is generally accepted today that petroleum formation can be described by a kinetic model (Jüntgen und Klein 1975; Tissot and Espitalié 1975; Ungerer and Pelet 1987). Also, the adjustment of the model to variable geologic heating rates in different sedimentary basins became possible by this progress. The shift of the oil peak on the maturity scale shown in Fig. 1c demonstrates the effect of different heating rates on the timing of oil generation.

3 Primary Migration: Present State of Knowledge

Movement of petroleum from the fine-grained source rock to the porous and permeable reservoir rock is generally called "migration". It is subdivided into *primary migration*, the movement within the narrow pores of the source rock up to the expulsion into a carrier bed, and into *secondary migration*, the movement through wider pores along one or more carrier beds up to the trap. The loss of petroleum from the trap is called *dismigration*. Because of density differences, gas is found above oil and oil above water in the reservoir rock which must be sealed by a low permeable cap rock (Fig. 2).

3.1 *Distribution of Water, Kerogen, and Petroleum within the Source Rock*

Organic-petrographical studies on source rocks revealed a highly variable distribution of organic matter: concentrations in various types of microlayers

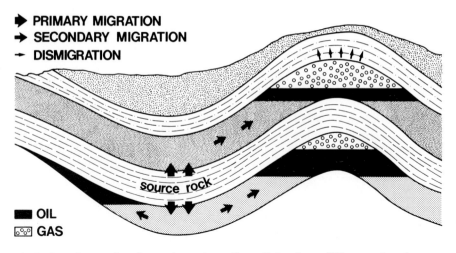

Fig. 2. Petroleum migration: schematic outline of the three different migration stages: primary migration, secondary migration, dismigration

or laminations occur (Plate 1a), i.e. bioturbated textures, thin or thick organic-rich layers or relatively evenly dispersed organic material.

It may be understood intuitively that the distribution of organic matter or kerogen predicts to a large degree the migration pathway inside a source bed. This is certainly the case if primary migration takes place along an organic network, within the pore network of the rock matrix, or along fractures (see below).

The small-scale distribution of kerogen and of oil, gas and water is very inhomogeneous. This is due to the highly variable pore sizes within the same rock as well as to interactions at the solid/liquid interfaces (Fig. 3).

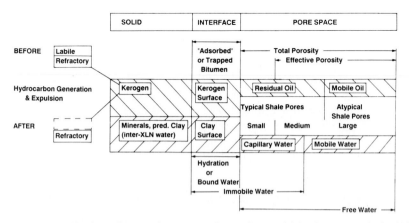

Fig. 3. Distribution of water, kerogen and petroleum within the source rock. (After Mann 1989)

Aspects of Primary Petroleum Migration Pathways

Plate 1a–b. *Effects of maturation in an oil shale-type source rock as expressed by macroscopic view (Lower Toarcian, Hils syncline, NW Germany).* **a** Immature stage at 0.53% vitrinite reflectance: fine laminations reflect the dominating mineralogic component, i. e. the varying contribution of coccolith shells to the clay mineral matrix. **b** Mature stage at 0.88% vitrinite reflectance: more or less vertical fractures, filled with calcite and bitumen (for microfabric, cf. plate 3)

During catagenesis, volumes of both organic and inorganic phases undergo severe changes as expressed by a macroscopically completely different view (Plate 1b): available pore space and the ratio of labile to refractory kerogen is continuously reduced. Due to conversion of the labile kerogen to petroleum, hydrocarbon saturation increases and water saturation decreases at first, whereas later on, during extensive petroleum expulsion and hydrocarbon cracking reactions, oil saturation decreases and water saturation increases again.

It can be assumed that at least four molecular layers of capillary water cling to the pore walls (Burgeff 1988). Furthermore, water from clay dehydration or other lattice water has to be taken into account separately and treated differently from the "normal" formation water.

For the petroleum phase, we have to differentiate between the lighter compounds and the more asphalt-like bitumen (Plates 1b, 3a). Many studies from the basin scale (e.g. Talukdar et al. 1987; Espitalié et al. 1987) to the core scale (e.g. Vandenbroucke 1972; Leythaeuser et al. 1984; Mackenzie et al. 1987; Mann 1989) have shown that primary migration favours light compounds over heavy compounds, and saturated over aromatic hydrocarbons.

3.2 Possible Modes of Primary Migration

From a theoretical point of view, primary migration can occur as discrete oil or gas phases and as individual oil droplets or gas bubbles. Furthermore, transportation as colloidal and micellar solutions has been proposed by several authors (Baker 1959; Meinschein 1959; Cordell 1972). Data by Price (1973) and McAuliffe (1966, 1980) on the molecular solution limit this transporting mechanism to the most soluble, lighter, hydrocarbon fractions. The most important form of primary migration during the main phase of oil (and gas) formation seems to be discrete hydrocarbon phase movements. Physicochemical, geochemical and geological considerations make individual droplets and bubbles or colloidal and micellar solutions highly unlikely as an effective means of transport during primary migration (Tissot and Welte 1984).

3.3 Current Theories and Concepts

At present, most researchers agree that primary migration takes place in its own phase, separate from water. Primary migration by solution in water or by molecular diffusion is assumed to be efficient (Durand 1988) only for the most water-soluble hydrocarbons such as methane, ethane and benzene.

Three concepts for primary migration have been proposed so far. They are named according to the respective migration pathway.

1. The *organic network concept* (Yariv 1976; McAuliffe 1980): the kerogen accumulates hydrocarbons for some time and then its oil-wet network serves as the pathway for hydrocarbons to leave the source bed.

2. The *pore network concept* (Momper 1978; Honda and Magara 1982; Mann 1989): hydrocarbon movement takes place predominantly through the largest available pore throats, also called "atypical" shale pore throats.

3. The *fracture network concept* (Snarsky 1962; Tissot and Pelet 1971): here it is supposed that the internal pore pressure increases so far that it exceeds the mechanical strength of the rock and opens up fractures. Then fluids can easily flow immediately out of the source rock.

The organic network concept provides a migration network with a very high oil saturation, and thus petroleum does not need to overcome a high capillary resistance as exhibited by narrow pore throats. Very recent analytical results seem to indicate, however, that the kerogen network functions predominantly as a first distributary system of petroleum to the pore network, and mainly during the early catagenetic stage (Ropertz 1989; Mann et al. 1989b). Expulsion finally takes place predominantly through the largest available pore throats at a more advanced stage of catagenesis after a minimum oil saturation is reached.

All three concepts need pressure as the driving mechanism. Some aspects of how this pressure is achieved are given below. However, the precise functioning of the source rock, i.e. the description of the primary migration process by water and petroleum flow equations has only been partly worked out to date.

3.4 An Unknown Petrophysical, Pressure Driven Flow Process

The existence of a two-phase flow, oil or gas plus water, needs the extension of Darcy's law on the concept of relative permeabilities, although this very recent approach from petrophysical reservoir description will certainly not satisfy such low permeability media like source rocks due to poorly understood mechanisms at the solid-liquid interface(s). However, as part of this concept, it can be confirmed up to now that a minimum hydrocarbon saturation of the pore volume is necessary for petroleum flow, and that a theoretical value of 25% for migration on modelling (Welte 1987) is probably close to the actual value. Based on detailed analyses, averaged saturation values for the Lower Toarcian Posidonia Shale (Lower Jurassic) are 24% for the calcareous shale facies with narrow pore throats and low pore volume, and 14% for the marlstone facies with wider pore throats and a relatively high pore volume (Ropertz 1989).

To obtain flow in porous media, a pressure gradient is needed and to obtain this pressure gradient, pressure has to be generated within the source

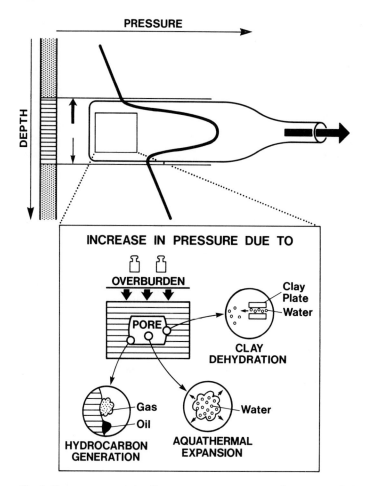

Fig. 4. Pressure versus depth across an overpressured source rock interval, with possible causes for formation of overpressuring. Pressure buildup provides the chance to pass narrow pore throats as expressed by the bottleneck. (Direction and intensity of primary migration as indicated by *arrows*)

rock. There are four factors which may contribute to varying degrees (Fig. 4, insert):

1. *Volume increase of organic matter* due to hydrocarbon generation. Ungerer et al. (1983) estimated this increase to be up to 15% of the initial volume, Goff (1983) found similar values between 10 and 20%.

2. *Overburden* as calculated by the weight of the sedimentary load. Some or all of this load can be transferred from the solid grains to the pore liquids.

3. *Thermal expansion of water* (Barker 1972). This is not easy to understand because water is present throughout the entire petroleum formation stage, and it is unclear whether a closed system exists at all.

4. *Clay mineral dehydration* contributes to a further increase of the water volume.

To summarize these pressure sources, which generally cause an overpressured interval, explorationists in many oil companies still apply the so-called Buried Bottle Model (Fig. 4). This model determines a higher pressure in the source bed than in the surrounding carrier beds (as usually measured during well site tests) which should provide the necessary driving force for petroleum to pass the narrow pores of the shale as expressed by the bottle neck.

4 Primary Migration Type I: Via the Pore Network

4.1 The Migration Pathway Inside the Source Bed

Shales, marls and argillaceous carbonates generally convey an impression of being homogeneous, dense and impermeable, except for an occasional fracture or other discontinuities. Scanning electron micrographs of core samples often reveal a remarkable open fabric, with a network of regular and irregular pores, vugs, secondary crystals and laminations. Detailed optical and electron-optical (scanning electron microscopy as well as transmission electron microscopy) analyses are the only ways to gain insight by direct observation into the actual migration pathway, and to reveal which features and minerals contribute to shale permeability and porosity.

Generally, most sedimentary non-clay minerals (quartz, calcite, pyrite, dolomite, siderite, halite, feldspars, barite) provide a certain amount of permeability because they contribute grains in the silt- and sand-size range. Secondary crystals of quartz or calcite also provide their own permeable halos (Plate 2b–d). Clusters (Plate 2a) or laminations of microfossils (Plate 1a) or lenses of silt are generally the areas of relatively large pores. Complete dissolution and removal of calcite from a calcareous shale or marl leads to a "clean" clay-shale fabric (Plate 2f) with a considerably enlarged pore network. Furthermore, permeability and porosity can be provided by corroded calcite crystals (Plate 2e) which may be related to organic-inorganic interactions with carboxyclic acids from the conversion of kerogen to petroleum (Mann et al. 1989a).

4.2 A Quantitative Approach to Pore Network Definition: Pore Size Distribution

If petroleum movement takes place via the sedimentological inherited and diagenetically modified pore network, it is necessary to quantitatively define the volume and shape of this network. The most commonly used approach,

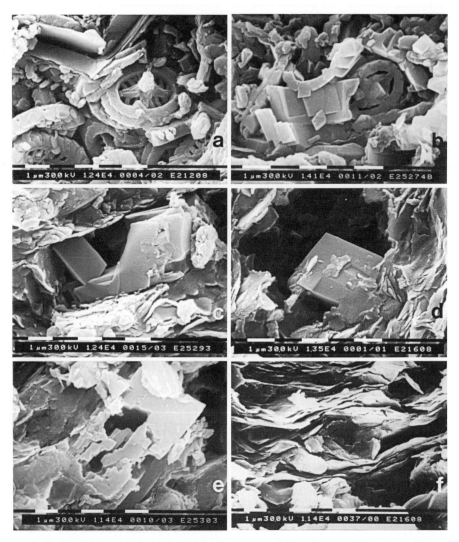

Plate 2a–f. *Primary petroleum migration type I: via the pore network (Lower Toarcian, Hils syncline, NW Germany). Permeability of the pore network is provided by:* **a** clusters of coccolith shells and large clay mineral flakes; **b** growth of individual coccolith shield crystals (calcite); **c, d** individual, large, newly formed, authigenic calcite crystals; **e** corroded calcite crystals, probably due to organic-inorganic interaction with carboxylic acids; **f** a "clean" clay-shale fabric due to total dissolution of calcite

first applied on coarser-grained reservoir rocks (Purcell 1948), is the determination of the pore size distribution.

In source rocks, however, the interpretation of pore size distribution is further complicated by aspects related to the content and type of organic matter. This section discusses the effects of the lithological composition of the sedimentary rock, possible diagenetic alterations as well as the effects of petroleum formation and expulsion.

4.2.1 Experimental Methods and Calculation of Pore Radii

Determination of pore radius distribution curves requires the saturation of an evacuated rock plug with a non-wetting liquid, generally mercury. Then a volume of mercury is forced into a pore under an external pressure p. Assuming clylindrical pores, applying the Washburn (1921) equation, and using $\theta = 140°$ (contact angle) and $\gamma = 484$ erg cm^{-2} (surface tension) results in:

$$r = \frac{1530}{p} \quad \text{for p in kg cm}^{-2} \text{ and r in µm;}$$

$$\left(r = \frac{106.7}{p} \quad \text{for p in PSI} \quad \text{and r in µm}\right).$$

This dictates that as pressure increases, mercury will intrude into progressively narrower pore throats.

The volume forced into the pores is usually monitored in a penetrometer which is the calibrated, mercury-containing stem, above a glass cell containing the sample (plus mercury). As intrusion occurs, the level in the capillary stem decreases. The mercury level is monitored continuously via the capacitance of a circular plate condenser, whereby the mercury in the stem represents the variable inner plate and a metal sheet the outer plate. Via the transformation of capacitance to voltage, the amount (=pore volume) of each pore throat range is obtained and generally shown as a pore volume curve which visualizes the summation of the intruded volume versus the respective pore throats.

4.2.2 Effects of Lithofacies and Diagenesis

Effects of *lithofacies* can be easily recognized from the pore size distribution curves as demonstrated by the following three rock samples (Mann and Müller 1985). The analyzed sediment samples, collected from the outer ridge of the Japan Trench, represent silty siliceous clays or claystones in which the siliceous component consists of diatoms, radiolaria and sponge spicules as well as vitric ash.

In detail, sample (1) is a hemipelagic sediment of Pliocene age from 145.47 m depth, sample (3) is a pelagic sediment of Cretaceous age from 375.46 m depths, and sample (2) is derived from the transition zone between pelagic and hemipelagic sediments and is of Middle Miocene age from 350.38 m depth. Their pore size distribution curves (Fig. 5) represent these lithofacies features very well. Predominantly detrital clay minerals form the

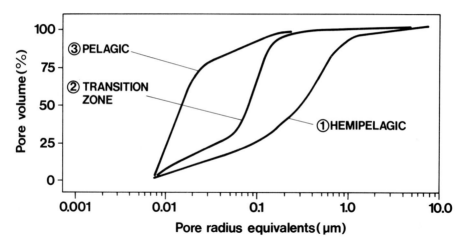

Fig. 5. Effect of lithofacies change on pore radius distribution. (After Mann 1979). For explanation, see text

large interparticle pores of the hemipelagic sediment. The pelagic one, on the other hand, consists of authigenic nontronite formed from volcanic glass (Mann and Müller 1980; Figs. 5 and 6). Due to the authigenic origin of the nontronite, all the individual mineral flakes have a relatively constant small size, which predetermines the possible pore system. The result is a well-sorted pore radii distribution with comparably small pores (Mann and Müller 1985). According to the nature of the third sediment sample (transition zone), we obtained a distribution of intermediate pore sizes.

Similar to the alteration of the chlorite/kaolinite clay mineral ratio (Mann und Müller 1980), the change of the lithofacies character as expressed by the pore size distribution can be used to indicate the drift of the position of Site 436 across the Pacific Plate to its present site on the rise of the Japan Trench due to the continuous change of the sediment source.

Diagenetic effects on the pore network of source rocks are a commonly observed phenomena, especially in calcareous shales and marlstones. Due to calcite dissolution, transportation and reprecipitation elsewhere, zones with very different permeability, porosity can be created which allow an increased or restricted hydrocarbon flow. The following is an example from the case history of the Lower Toarcian Posidonia Shale (Lower Jurassic) which represents the most prolific petroleum source rock of central and western Europe as in the Lower Saxony Basin, and the Paris Basin.

In the Hils syncline, which is located at the southern rim of the Lower Saxony Basin, Mann et al. (1989a, b) and Leythaeuser et al. (1989) studied four diagenetically modified lithofacies types within this strata: (A) a calcareous claystone facies; (B) a marlstone facies; (C) a calcite-cemented marlstone facies, now a limestone; and (D) a marlstone facies with secondary

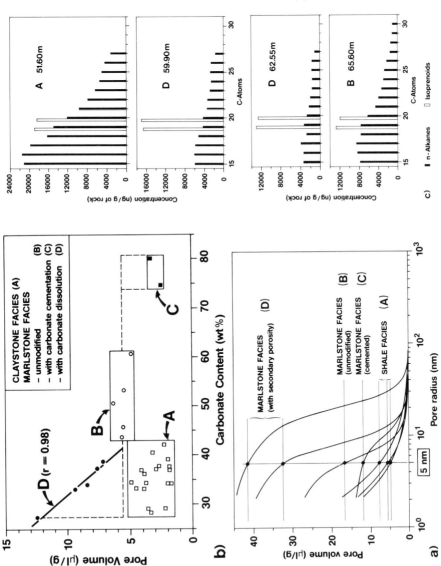

Fig. 6a–c. Effects of lithofacies and diagenesis on pore network and effect on hydrocarbon redistribution (After Mann et al. 1989a) **a** Pore volume of the micro- plus mesopores versus carbonate content. **b** Pore size distributions of the different lithofacies types. **c** n-Alkane concentrations of the different lithofacies types

porosity due to removal of calcite and therefore now a calcareous clay shale. Pore volumes of the micro-plus mesopores as analyzed by nitrogen adsorption and carbonate contents are directly related to these four facies (Fig. 6a). Detailed analyses by mercury intrusion revealed specific pore size distribution curves (Fig. 6b) which are responsible for a more or less intensive redistribution of hydrocarbons as outlined below.

From a sedimentological viewpoint, the pore size distributions reflect the dominating mineralogic component, i.e. the varying contributions of coccolith shells to the clay mineral fabric. Since the coccoliths are larger than the clay minerals, a greater number (= more pore volume) of pores as well as wider pore throats can be provided in the marlstone and in the diagenetically modified marlstone facies. By comparing the relative amount of pore throats greater than 5 nm, one can estimate that the permeability of facies (C) is about double that of facies (A), the permeability of facies (B) is about three times as much, and the permeability of facies (D) is about seven times as much. With the help of these permeability estimations, Mann et al. (1989a, b) were able to explain the different hydrocarbon concentrations encountered within rocks of the four facies. Highest concentrations of n-alkanes are exhibited by the low-permeable rock samples (= clay shale) and lowest concentrations occur in the most permeable rocks (Fig. 6c). Because all facies types had the same potential for hydrocarbon generation, hydrocarbons must have been expelled from the source rock, but to a variable extent controlled by petrophysical criteria.

In addition, the marlstone facies with secondary porosity exhibits phenomena of "water washing", i.e. the partial removal of more water-soluble compounds like phenanthrene and naphthalene (Price 1976, Radke et al. 1982). The different effects of water washing observed in oil pools in combination with biodegradation are well documented in the literature (see Tissot and Welte 1984, pp. 463 ff).

4.2.3 Effects of Compaction, Petroleum Formation, Petroleum Expulsion and Kerogen Type on the Pore Network

Both *compaction,* originating from increasing overburden pressure associated with deeper burial, and petroleum formation and expulsion during catagenesis, show similar effects on the pore network of a source rock. Very rarely can they be separated from each other.

To achieve a volume reduction of the pore network by compaction, length and area dimensions of the network, i.e. the pore radii, have to be reduced. This means a continuous, parallel shift of the sedimentologically inherited pore size distribution curve to smaller pores, if the volume effect is disregarded (examples in Mann 1979).

Petroleum or bitumens can be present either as free hydrocarbons in the pore space, adsorbed by (or even within) the (clay) minerals, or adsorbed on

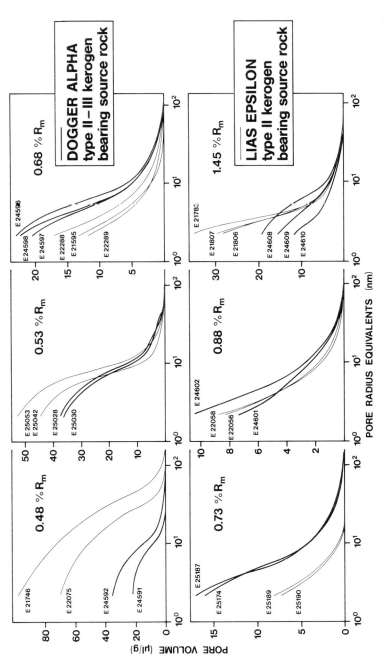

Fig. 7. Evolution of the pore network during catagenesis for two adjacent source rocks, Dogger alpha (Lower Aalenian) and Lias epsilon (Lower Toarcian) as exemplified by their pore size distribution. Catagenetic stages (caused by variable proximity to a deep intrusive heat source and not by different burial) are specified by vitrinite reflectance values

(or trapped within) organic matter. Intuitively, it is easy to understand that the free bitumens can migrate more easily from the source rock, whereas the adsorbed ones cannot. This might be partly related to so-called chromatographic effects, observed in many case histories. It can be accepted today that saturated hydrocarbons are expelled more efficiently than isoprenoids and aromatics, and most probably aromatics more efficiently than asphaltenes. Consequently, with increasing maturation due to petroleum formation and expulsion, more bitumen of higher viscosity restricts the volume of the pore network of the mineral matrix. Depending on the kerogen type present in the source rock, this blocking of pores by bitumen starts earlier (type II) or later (type I or III). One can recognize this effect by comparing the pore volume and pore size of adjacent Dogger alpha and Lias epsilon rock samples (types II–III and type II kerogen-bearing source rock respectively) of the same compaction stage as shown in Fig. 7 (after Mann et al. 1989a). With advanced maturity stages, the analyst observes a pore volume increase on non-extracted rock samples (Fig. 7, from 0.88 to 1.48% Rm). This enlargement is due to the removal of the bitumen by chemical cracking reactions.

Theoretical considerations by Larter (1988) and preliminary analytical results by Mann et al. (1989a) seem to indicate that petroleum expulsion alone can change fabric and pore network of the rock matrix because the labile part of the solid phase kerogen is converted to fluid oil and gas and expelled from the source rock; only the refractory part is left. Conventional approaches to simulate shale compaction ignore this effect and relate the density increase in sedimentary rocks solely to burial and compaction.

4.3 Pore Size Distribution and Related Petrophysical Properties

Rock bulk densities multiplied by the analytically obtained pore volume provide directly the desired porosity value, provided the pore size distribution has really been analyzed up to complete mercury saturation of the pore space. If some pore throats are narrower than the diameter of the mercury molecule, the distribution has to be extrapolated. Typical porosity values for shales during catagenesis are in the range of 2–20% (Mann 1987a).

Permeability data are needed very much for investigating hydrocarbon transport through source rocks and identifying dominant processes. Due to very low values, extremely low flow rates over a long time span have to be monitored. New analytical know-how ("pulsed permeability analysis") seems to overcome this problem.

Permeability values obtained so far were therefore based predominantly on calculations via pore size parameters. During catagenesis, calculated permeability values for the tight Lower Toarcian "Posidonia" oil shale vary between 10^{-2} and 10^1 nD (Mann 1987b), and between 2 and 8×10^1 nD for the more silty marine "Katharina" Shale from the Upper Carboniferous

(Mann 1989). For the petroleum geochemist, who checks hydrocarbon expulsion effects versus permeability, a relative comparison of estimated permeabilities as outlined in Sect 4.2.2 is often sufficient. In this case the contribution (= pore volume) of a selected range of large pores is compared between rock samples, or a ratio is calculated over the contribution of small pores (Mann 1989). This is correct because large pores are the main carriers of permeability (Khanin 1968).

Another approach to permeability estimations is the comparison of the maximum pore radius (Khanin 1968). This approach has also been used before to estimate the effective methane diffusion coefficient (Nesterow and Uschatinskij 1972).

5 Primary Migration Type II: Via Fractures

In general, microfractures in source rocks are a likely means of primary oil migration if present in the source rock. Microfractures may be caused by tectonic disturbances such as lateral tension or differential uplift and burial. Pore fluid helps by creating pressure, whereby fracturing reduces effective stresses.

Several authors suggested that oil generation may induce fracturing in source rocks (Tissot and Pelet 1971; Momper 1978; du Rouchet 1981). Especially at high petroleum generation rates the available pore network, as provided by lithofacies plus diagenetic modifications, may volumetrically no longer be sufficient for a fast drainage of the generated oil and gas. Additionally, high viscous bitumen blocks the pore throats (Mann et al. 1989b). As a result, the pore pressure exceeds the mechanical strength of the rock and fractures may open up. Therefore, fracturing, if associated with oil generation must depend on the highest possible expulsion rate through the pore network which in turn depends on the permeability of the source rock (Özkaya 1988).

5.1 Horizontal or Vertical Fractures?

Both horizontal fractures, which are parallel to the bedding planes of the source rock as well as being relatively irregular, and vertical fractures have been observed in source rocks (Meissner 1978; du Rouchet 1981; Talukdar et al. 1987; Jochum 1988; Littke et al. 1988).

The idea of primary migration through vertical fractures provides two conceptual advantages for the understanding of the migration pathway. First, migration takes place in a vertical direction and therefore petroleum can quickly reach the reservoir. Second, tectonic models with an increasing sedimentary load during burial as initiating force for rock fracturing will create vertical fractures more easily than horizontal ones (Breckels and van

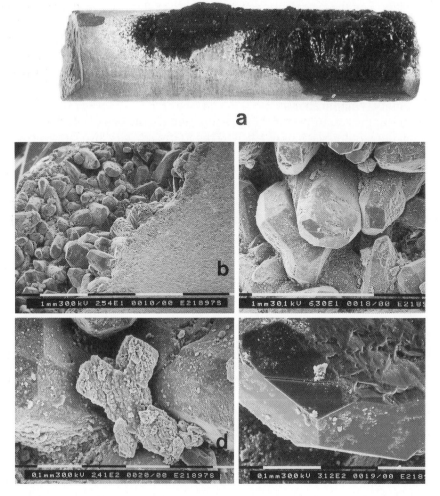

Plate 3a–e. *Primary petroleum migration type II: via the fracture network (Lower Toarcian, Hils syncline, NW Germany. View of microfabric of open fractures:* **a** Plug from vertical zone of slabbed rock specimen in Plate 1b with high-viscous to solid bitumen (*black*) and calcite (*white*) at the fracture walls (diameter of plug = 1 cm). **b, c** Close-up of fracture wall after removal of bitumen exhibits authigenic, barrel-shaped calcite. **d** Besides calcite, twinned baryte occurs sporadically. **e** Intense corrosion marks on the side faces (= slow growth) in contrast to very smooth top faces (= fast growth) of the calcite crystals prove that fracture cementation started from fracture walls and not from the center. Corrosion marks may be caused by organic acids

Eekelen 1982): In a sedimentary basin the maximum principal stress is vertical and equals overburden weight, whereas the total least lateral stress ranges between 0.6 and 0.9 of the vertical one. However, when the fabric of the source rock is taken into account, and an approach according to Griffith's (1921) theory of fracture development is used (Schmidt 1977; Costin 1981), horizontal fractures are the expected results of overpressuring in a shale or similarly laminated or layered rocks (Özkaya 1988). This implies longer migration pathways because petroleum movement takes place over a much longer distance up-dip within the strata of the source bed. There is also the question of how horizontal fractures can be kept open for a longer time period.

In the clay shale of the Lower Toarcian Lias epsilon Mann et al. (1989a) identified fractures filled with calcite and high-viscous to solid bitumen (Plate 3a). Fracture walls are primarily covered by authigenic, barrel-shaped calcite (Plate 3b,c) and sparsely by twinned baryte crystals (Plate 3d). Intensive corrosion marks on the side faces (= slow growth) in contrast to very smooth top faces (= fast growth) of the crystals (Plate 3e) indicate that fracture cementation started from fracture walls and took place during or after the impregnation by petroleum, i.e. during or after primary migration. The corrosion marks are interpreted as the result of an interaction with organic acids. Completely cemented fractures can be checked as a possible migration pathway by identifying fluorescent fluid inclusions of petroleum in between individual calcite crystals (Plate 4a), or within microfissures of broken crystals (Plate 4b).

5.2 Tectonic Fractures

Petroleum geochemists are generally biased and try to relate fractures observed in a source rock to hydrocarbon generation. Very often petroleum expulsion may have taken place along a preexisting, tectonically induced fracture network just because it provided the widest pore throats and best permeability. Tectonic disturbances such as lateral tension or differential uplift and burial are common geological processes which cause fracturing in most cases. Uplift of a sedimentary rock pile alone is sufficient to cause intensive shear fractures in relatively impermeable rocks (Gretener 1984) such as a source rock. Shear fractures, however, can be easily diagnosed because of the more or less 45° angle they form in the direction of stress release (Plate 4c,d).

Careful source rock examinations may not only reveal whether fractures are used for migration purposes, but also whether they are related to petroleum formation.

Plate 4a–d. *Primary petroleum migration type II: via the fracture network (Lower Toarcian, Hils syncline, NW Germany). Identification of completely cemented fractures as petroleum migration pathway by:* **a** fluorescent petroleum inclusions, trapped in between crystal boundaries (length of illustrated view is 0.25 mm); **b** fluorescent petroleum inclusions, trapped in fissures of broken crystals (length of illustrated view is 1 mm). *Identification of fractures as tectonically induced shear fractures by:* **c, d** their more or less 45° angle formed in the direction of stress release (length of illustrated view is 4 mm)

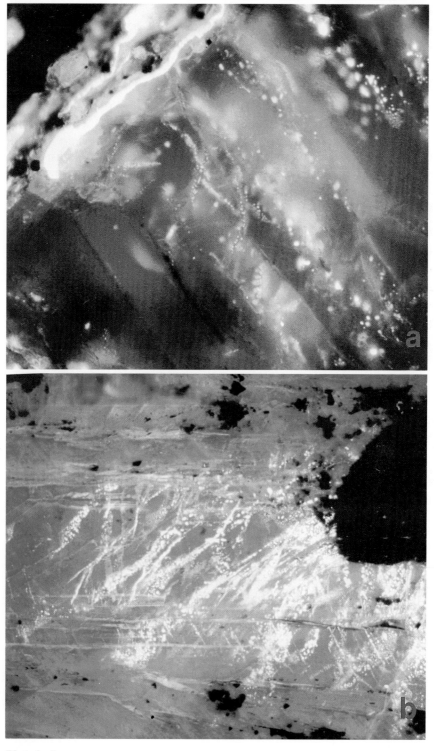

Plate 4 a, b

Aspects of Primary Petroleum Migration Pathways

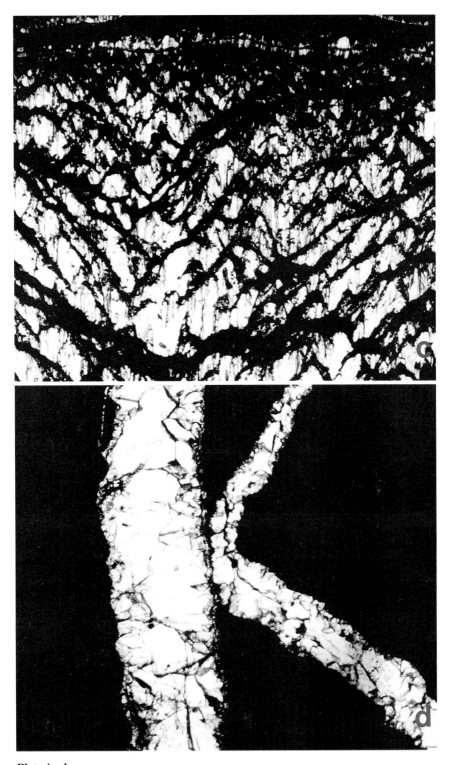

Plate 4 c, d

6 Perspective: Revealing Primary Petroleum Migration Pathways

Revealing petroleum migration pathways may be useful at various exploration stages, but then it is linked to a specific geologic scale. On a basin scale, it is possible to evaluate petroleum drainage, a key factor in the petroleum production of a sedimentary basin besides charge volume and trap availability (Demaison and Huizinga 1989). On a geologic structure-related scale, the prospective use of individual traps can be rated and compared (Blackwelder 1989). Revealing petroleum pathways during the primary migration stage means predominantly to work on a relatively small scale: a stratigraphic section from one well profile to identify potential vertical migration routes, or selected rock samples from two or more wells to reveal lateral migration directions.

Although a general scheme has been worked out for defining the direction of primary migration according to petrophysical criteria (England et al. 1987), such a simplification remains unsatisfactory as soon as somewhat more complicated prospects have to be considered. To decipher the correct migration pathway an interdisciplinary study which includes sedimentological, mineralogical, organic-petrographical, organic-geochemical and petrophysical analyses on source rock cores, is a prerequisite. However, the effort to obtain good core material from source rocks is in most cases a hopeless task: oil companies drill in anticlines with immature source rocks, whereas mature source rocks are generally present in synclines which makes sampling material very expensive. Nevertheless, primary migration pathways and petroleum expulsion effects can only be studied on core material, especially due to the need for petrophysical analyses which require a comparably large sample volume.

The heterogeneity of recent concepts describing the primary migration process (see Sect. 3.3) indicates the variety of possible factors involved which may participate to some degree as a control mechanism, and which more or less predominate. However, it seems likely that they might be combined some day to a set of equations providing an exact, quantitative description which can predict primary petroleum migration.

So far it has been recognized that organofacies is responsible for the composition and dimension of the organic network, which is the earliest and most easily available migration pathway, and which can take over the distribution of generated hydrocarbons to the pore network. The effectivity of the pore network as migration pathway depends on its permeability. This represents a sedimentologically inherited petrophysical factor, modified by compaction and diagenesis. If hydrocarbon generation is not too fast and the source rock still relatively permeable, an additional primary migration pathway, e.g. fractures, may not be required. More pathways will be needed to drain the source rock only if the source rock is relatively tight and the petroleum generation rate is high. Then the pressure buildup will be used for fracturing.

Thus, the final equations for the prediction of type and contribution of the actual primary migration mechanism(s) may contain the following key factors:

1. The *organofacies,* for the contribution of primary petroleum migration via the *organic network;*

2. *Lithofacies, diagenesis and compaction* for the contribution via the *pore network;* and

3. An evaluation of *permeability versus petroleum generation rate,* for the contribution via a *fracture network*.

Acknowledgements. I thank Dr. H.S. Poelchau for reviewing earlier drafts of the manuscript. Dipl.-Min. J. Jochum kindly provided Plates 4b,c and d; Dipl.-Geol. B. Ropertz Plate 4a; and Dr. R.G. Schaefer data of n-alkane distributions. Typing of the manuscript by Miss S. Bardon and Mrs. B. Schmitz is gratefully acknowledged.

References

Baker EG (1959) Origin and migration of oil. Science 129:871–874
Barker C (1972) Aquathermal pressuring – role of temperature in development of abnormal-pressure zones. Am Assoc Petrol Geol Bull 56:2068–2071
Blackwelder BW (1989) Hydrocarbon accumulation of rifted continental margin – examples of oil migration pathways, West African salt basins. Am Assoc Petrol Geol Bull 73:324
Breckels IM, van Eekelen HAM (1982) Relationship between horizontal stress and depth in sedimentary basins. J Petrol Technol 34 (September): 2191–2199
Burgeff M (1988) Immobiles Wasser in feinkörnigen Sedimentgesteinen und sein Einfluß auf Migrationsprozesse von Kohlenwasserstoffen. Diplomarb, Univ Freiburg, p 103
Cordell RJ (1972) Colloidal soap as proposed primary migration medium for hydrocarbons. Am Assoc Petrol Geol Bull 57:1618–1643
Costin LS (1981) Static and dynamic fracture behavior of oil shale, ASTM STP 745, Philadelphia, pp 169–184 (1981)
Demaison G, Huizinga BJ (1989) Genetic classification of petroleum basins. Am Assoc Petrol Geol Bull 73:349
Durand B (1988) Understanding of hydrocarbon migration in sedimentary basins (present state of knowledge). In: Mattavelli L, Novelli L (eds) Advances in organic geochemistry 1987. Pergamon, Oxford, pp 445–459
England WA, Mackenzie AS, Mann DM, Quigley TM (1987) The movement and entrapment of petroleum fluids in the subsurface. J Geol Soc London 144:327
Espitalié J, Marquis F, Sage L (1987) Organic geochemistry of the Paris Basin. In: Brooks J, Glennie K, (eds) Petroleum geology of N.W. Europe. Graham and Trotman, London, pp 71–86
Goff JC (1983) Hydrocarbon generation and migration from Jurassic source rocks in the E. Shetland Basin and Viking Graben of the northern North Sea. J Geol Soc London 140:445–474
Griffith AA (1921) The phenomena of rupture and flow in solids. Philos Trans R Soc London A 221:163–198

Honda H, Magara K (1982) Estimation of irreducible water saturation and effective pore size of mudstones. J Petrol Geol 4:407–418
Jochum J (1988) Untersuchung zur Bildung und Karbonat-Mineralisation von Klüften im Posidonienschiefer (Lias epsilon) der Hilsmulde. Diplomarb, RWTH Aachen, p 76
Jüntgen H, Klein J (1975) Entstehung von Erdgas aus kohligen Sedimenten. Erdöl und Kohle, Erdgas, Petrochemie, Ergänzungsband 1, Von Hernhausen, Leinfelden bei Stuttgart, pp 74–75; 52–69
Khanin AA (1968) Evaluation of sealing abilities of clayey cap rocks in gas deposits. Geol Nefti Gaza 8:17–21
Larter S (1988) Pragmatic perspectives in petroleum geochemistry. Mar Petrol Geol 5:194–204
Leythaeuser D, Mackenzie AS, Schaefer RG, Bjorøy M (1984) A novel approach for recognition and quantification of hydrocarbon migration effects in shale-sandstone sequences. Am Assoc Petrol Geol Bull 68:196–219
Leythaeuser D, Mann U, Rullkötter J, Welte DH (1989) Generation and migration of petroleum in Toarcian source rocks of northwest Germany – the Hils syncline as a model area. In: 1st Conf Eur Assoc Petroleum Geoscientists, West Berlin, 30 May to 2 June 1989
Littke R, Baker DR, Leythaeuser D (1988) Microscopic and sedimentologic evidence for the generation and migration of hydrocarbons in Toarcian source rocks of different maturities. In: Mattavelli L, Novelli L (eds) Advances in organic geochemistry 1987. Pergamon, Oxford, p 549
Mackenzie AS, Price I, Leythaeuser D, Müller P, Radke, Schaefer RG (1987) The expulsion of petroleum from Kimmeridge clay source-rocks in the area of the Brae oilfield, UK continental shelf. In: Brooks J, Glennie K (eds) Petroleum geology of N.W. Europe. Graham & Trotman. London, pp 865–877
Mann U (1979) Mineralogisch-sedimentologische Untersuchungen von Tiefseeablagerungen am ostasiatischen Kontinentalrand unter besonderer Berücksichtigung frühdiagenetischer Veränderungen an biogenem Opal. Diss, Univ Heidelberg, 146 pp
Mann U (1987a) Veränderungen von Porosität und Porengröße eines Erdölmuttergesteins in Annäherung an einen Intrusivkörper. Facies 17:181–188
Mann U (1987b) Geneseprojekt-Petropysikalische Analysendaten Lias Epsilon. KFA/ICH-5. Int Ber Nucl Res Cent Jülich 500487
Mann U (1989) Revealing hydrocarbon migration avenues. Geol Rundsch 78:1
Mann U, Müller G (1980) Composition of sediments of the Japan Trench transect, Legs 56 and 57, Deep Sea Drilling Project. In: Honza E et al. (eds) Init Rep DSDP 56, 57. US Gov Print Off, Washington, DC, pp 939–977
Mann U, Müller G (1985) Early diagenesis of biogenic siliceous constituents in silty clays and claystones, Japan Trench. N Jahrb Mineral Abh 153, 1:33–57
Mann U, Düppenbecker S, Langen A, Ropertz B, Welte DH (1989a) Evolution of the network of a clastic petroleum source rock during catagenesis (Lower Toarcian Posidonia Shale, Hils Syncline, northwest Germany). Am Assoc Petrol Geol Bull 73, 3:385
Mann U, Düppenbecker S, Langen A, Ropertz B, Welte DH (1989b) Petroleum pathways during primary migration: evidence and implications (Lower Toarcian, Hils Syncline, NW-Germany). In: 14th Int Meet Organic geochemistry, Paris, September 18–22
McAuliffe CD (1966) Solubility in water of paraffin, cycloparaffin, olefin, acetylene, cycloolefin and aromatic hydrocarbons. J Phys Chem 70:1267–1275
McAuliffe CD (1980) Oil and gas migration: chemical and physical constraints. In: Roberts; WH III, Cordell JR (eds) AAPG studies in geology 10. Problems of petroleum migration. AAPG, Tulsa, pp 89–107
Meinschein WG (1959) Origin of petroleum. Am Assoc Petrol Geol Bull 43:925–943
Meissner FF (1978) Petroleum geology of the Bakken Formation. In: Williston Basin Symp, Montana Geol Soc, pp 207–227

Momper JA (1978) Oil migration limitations suggested by geological and geochemical considerations. In: Roberts WH, Cordell R (eds) Physical and chemical constraints on petroleum migration. Am Assoc Petrol Geol Course Note 8, B1–B60

Nesterow II, Uschatinskij IN (1972) Abschirmeigenschaften der tonigen Gesteine über Erdöl- und Erdgaslagern in den mesozoischen Ablagerungen der Westsibirischen Tiefebene. Z anorg Geol 18, 12:548–555

Özkaya J (1988) A simple analysis of oil-induced fracturing in sedimentary rocks. Mar Petrol Geol 5:293

Price LC (1973) The solubility of hydrocarbons and petroleum in water as applied to the primary migration of petroleum. Ph D Thesis, Univ Cal, Riverside

Price LC (1976) Aqueous solubility of petroleum as applied to its origin and primary migration. Bull Am Assoc Petrol Geol 60:213–244

Purcell WR (1948) Capillary pressures – their measurement using mercury and the calculation of permeability thereform. Petrol Trans AIME 186:39–48

Radke M, Willsch H, Leythaeuser D, Teichmüller M (1982) Aromatic components of coal: relation of distribution to rank. Geochimica et Cosmochimica Acta 46:1831–1848

Ropertz B (1989) Petrophysikalische, organisch-chemische und organisch-petrographische Untersuchungen zur Verteilung des organischen Materials in Ton- und Mergelsteinen in zwei ausgewählten Katagenesestadien (Unteres Toarc, Hilsmulde, NW-Deutschland). Diplomarb, RWTH Aachen, p 123

Rouchet J du (1981) Stress fields, a key to oil migration. Am Assoc Petrol Geol Bull 65:74–85

Schmidt RA (1977) Fracture mechanism of oil shale – unconfined fracture toughness, stress corrosion cracking and tension test results. In: Proc 18th US Symp Rock mechanics, Col School Mines, Golden, pp 2A2-1–2A2-6

Snarsky AN (1962) Die primäre Migration des Erdöls. Freiberger Forschungsh C 123:63–73

Talukdar S, Gallango O, Vallejos C, Ruggiero A (1987) Observations on the primary migration of oil in the La Luna source rocks of the Maracaibo basin, Venezuela. In: Doligez B (ed) Migration of hydrocarbons in sedimentary basins. Technip, Paris, 681 pp

Tissot B (1973) Vers l'évaluation quantitative du pétrole formé dans les bassins sédimentaires. Rev Assoc Fr Tech Petrol 222:27–31

Tissot B (1987) Foreword. In: Doligez B (ed) Migration of hydrocarbons in sedimentary basins. Technip, Paris, 681 pp

Tissot B, Espitalié J (1975) L'évolution thermique de la matière organique des sédiments: application d'une simulation mathématique. Rev Inst Fr Petrol 30:743–777

Tissot B, Pelet R (1971) Nouvelles données sur les mécanismes de genèse et de migration du pétrole, simulation mathématique et application à la prospection. In: Proc 8th World Petroleum Congr, vol 2, pp 35–46 (1971)

Tissot B, Welte DH (1978) Petroleum formation and occurence. Springer, Berlin Heidelberg New York, 538 pp

Tissot BP, Welte DH (1984) Petroleum formation and occurence, 2nd edn. Springer, Berlin Heidelberg New York, 699 pp

Tissot B, Durand B, Espitalié J, Combaz A (1974) Influence of the nature and diagenesis of organic matter in formation of petroleum. Am Assoc Petrol Geol Bull 58:499–506

Ungerer P, Behar E, Diskamp D (1983) Tentative calculation of the overall volume expansion of organic matter during hydrocarbon genesis from geochemistry data. Implications for primary migration. Adv Organ Geochem 1981:129–135

Ungerer P, Pelet R (1987) Extrapolation of the kinetics of oil and gas formation from laboratory experiments to sedimentary basins. Nature (London) 327, 6117:52–54

Vandenbroucke M (1972) Etude de la migration primaire: variation de la composition des extraits de roche à un passage roche-mère/réservoir. In: Von Gaertner HR, Wehner H (eds) Advances in organic geochemistry 1971. Pergamon, Oxford, pp 547–565

Washburn EW (1921) Note on a method of determining the distribution of pore sizes in a porous material. Proc Natl Acad Sci USA 7:115

Welte DH (1987) Migration of hydrocarbons. Facts and theory. In: Doligez B (ed) Migration of hydrocarbons in sedimentary basins. Technip, Paris, pp 393–413

Yalçin MN, Welte DH (1988) The thermal evolution of sedimentary basins and significance for hydrocarbon generation. TAPG Bull 1, 1:12

Yariv, S (1976) Organophilic pores as proposed primary migration media for hydrocarbons in argillaceous rocks. Clay Sci 5:19–29

Yükler MA, Kokesh F (1984) A review of models in petroleum resource estimation and organic geochemistry. Adv petrol geochem 1:69–113

The Influence on Subrosion of Three Different Types of Salt Deposits

WOLFGANG SESSLER[1]

CONTENTS

Abstract	179
1 Introduction	180
2 The Cap-Rock Strata and the Tectonic Structures within the Salt of the Hänigsen-Wathlingen Salt Dome in Northern Germany	181
2.1 General Information	181
2.2 Stratigraphy and Insoluble Content of the Salt Body	182
2.3 The Cap-Rock Sequence of the New Ventilation Shaft	183
2.4 The Cap-Rock of the Riedel Shaft	183
2.5 The Tectonic Structures of Salt in the Shafts	183
2.6 Conclusions	185
3 Subrosion in the Salt Deposit of the Khorat Plateau in Thailand	185
3.1 General Information	185
3.2 Stratigraphy of the Salt Deposit and Overburden	185
3.3 Genesis of the Cap-Rock	187
3.4 Petrographic Description of the Cap-Rock Gypsum/Anhydrite	188
3.5 General Structure of the Salt Deposit	189
3.6 Regional Distribution of Salt and Cap-Rock Gypsum	191
4 Irregular Subrosion of Salt in the Deposit with Horizontal Bedding in the Fulda Basin in Germany	191
4.1 General Geology of the Fulda Basin Salt Deposit	191
4.2 Regular Subrosion of Salt in the Fulda Basin	193
4.3 The Rock Sequence of Borehole Tiefengruben 3	194
4.4 Interpretation of the Structure and Conclusions	194
References	196

Abstract

In the study of salt-bearing deposits, the geologist has to bear in mind that salt is readily soluble. Since water is ubiquitously present in porous sedimentary rocks, salt is frequently dissolved. The residues of the subrosion

---
[1] Wolfgang Sessler, Kali und Salz AG, Friedrich-Ebert-Str. 160, D-3500 Kassel, FRG.

processes are found in the cap-rocks or gypsum cap-rocks in salt domes. At the Hänigsen-Wathlingen salt dome within the northwest German Zechstein Basin it is shown that both the stratigraphy and the content of the insolubles of the cap-rocks depend on previous tectonic structures.

Apart from the diapiric salt structures, cap-rocks are also known from horizontally bedded salt deposits as shown in a deposit from the Khorat Plateau in Thailand. At the edges as well as in uplift areas the deposit is intensely subroded, and filled by insolubles such as anhydrite and clastic components of the preceding salt strata. Since the clastics occur as distinct and continuing layers, containing only minor amounts of salt, the stratigraphic section is still preserved even after subrosion has taken place.

Subrosion at the top of the salt body, called regular subrosion, is also common in the Fulda Basin of western Germany. In the horizontally layered salt strata of this area, a rare combination of regular and irregular subrosion at the bottom of the beds occurs. The top of the salt is covered by a thin sheet of layered cap-rock gypsum, overlain by a subrosion breccia consisting of a mixture of clastics and boulders of gypsum, and finally overlain by strata of brecciated overburden sediments. Beneath the salt horizons a collapse breccia composed of rock fragments from strata overlying the salt strata and displaying chaotic structures is found.

1 Introduction

Salt is a very soluble rock type. In nearly all salt deposits it contains a certain amount (normally a few percent) of insoluble or only partly soluble matter, e.g. clay or anhydrite. The insoluble matter is partly intergrown with halite or other salt minerals or it forms, in part, distinct layers within the salt sequence.

The subrosion of salt, i.e. dissolution below the surface by groundwater, seems to be quite common in many salt deposits and generates characteristic rock types mostly known from salt domes as "cap-rock" or "cap-rock gypsum". An example of this type of rock is described from the Hänigsen-Wathlingen diapir in northern Germany, including investigation results from outcrops in a newly constructed ventilation shaft.

In salt deposits with horizontal bedding we also observe strata which can be regarded as cap-rock because they are generated by the same mechanism of subrosion and show many petrographic similarities to the cap-rock of salt domes. The cap-rock strata above horizontally bedded salt deposits are either well-stratified beds with a good stratigraphic correlation to the insoluble beds within the salt or they are collapse breccia with a more or less disturbed stratigraphy.

Examples of these rock types are described from the Khorat Plateau in Thailand and from the Fulda Basin in Germany.

2 The Cap-Rock Strata and the Tectonic Structures within the Salt of the Hänigsen-Wathlingen Salt Dome in Northern Germany

2.1 General Information

Many publications on cap-rock of salt domes have used data gained either from boreholes or from shafts (Seidl 1921; Fulda 1935, 1938; de Boer 1979; Walker 1974; Martinez 1980; Batsche and Klarr 1980; Bauer and Sessler 1985; Bornemann and Fischbeck 1986). In some cases cap-rocks could also be studied in outcrops at the surface (Seidl 1921).

Most studies presume a connection between the internal structures and the composition of insoluble material of salt and the petrographic sequence of the overlying cap-rock, but usually the geological information on the uppermost part of salt domes and the related cap-rock is restricted due to the safety regulations of drilling and mining operations.

The following information on cap-rocks was obtained from two shafts only 190 m apart horizontally. The new ventilation shaft was built in 1980/1981 and mapped in detail during the lowering of the shaft. The other shaft, the "Riedel-shaft" was built in 1907/1908.

Table 1. Stratigraphy, thickness and insoluble content of the Zechstein beds, encountered in the Hänigsen-Wathlingen salt dome

Sequence	Name	Symbol	Average thickness (m)	Insoluble material Content	Percent
Zechstein 4	*Tonbanksalz*	Na4 tm	30	Fine-grained clastics	15–20
	Tonbrockensalz	Na4 δ	40	Fine-grained clastics	5–10
	Rosensalz	Na4 γ			
	Schneesalz	Na4 β	60	Anhydrite	1–2
	Aller-Basissalz	Na4 α			
	Pegmatitanhydrit	A4	1	Anhydrite	100
	Roter Salzton	T4	20–30	Fine-grained clastics	80
Zechstein 3	*Tonmittelsalz*	Na3 tm	20	Fine-grained clastics	15–20
	Kaliflöz Riedel	K3Ri	2–5		
	Schwadensalz	Na3 ϑ	40	Anhydrite	1
	Anhydritmittelsalz	Na3 η	25	Anhydrite	2–5
	Linien- und Bändersalz	Na3 β/ξ	45	Anhydrite	1–3
	Basissalz	Na3 α	2–3		
	Hauptanhydrit	A3	25–30	Anhydrite	100
	Grauer Salzton	T3	5–8	Fine-grained clastics	90
Zechstein 2	*Deckanhydrit*	A2r	1–2		
	Decksteinsalz	Na2r	1		
	Kaliflöz Staßfurt	K2H	10		
	Staßfurt-Steinsalz	Na2	600–700	Anhydrite	3

The rock sequence was also described in detail during the lowering of the shaft, but it had to be reinterpreted according to modern Zechstein stratigraphy (Table 1) and today's knowledge on the cap-rock of the Hänigsen-Wathlingen salt dome.

Information on the internal structure of the salt body was obtained from the lower parts of the two shafts, from boreholes in the mine, drilled in advance of the shaft-sinking operations of the new shaft, and from geologic maps of the mine openings below the upper 300-m mine level.

2.2 Stratigraphy and Insoluble Content of the Salt Body

The Hänigsen-Wathlingen diapir, located 20–25 km ENE of Hannover in northern Germany, consists mainly of Zechstein 2, 3 and 4 strata. The insoluble material of the different stratigraphic horizons (Table 1) varies in amount and composition. The halite of Zechstein 2, called *"Staßfurt-Steinsalz* (Na2)" contains approxionately 3–4% of disseminated sulfate, normally anhydrite, forming anhydrite/gypsum cap-rock as a result of subrosion. The thickness of the related cap-rock depends on the amount of dissolved *Straßfurt-Steinsalz*, but normally it is only a few meters (de Boer 1970).

In many salt domes in the Hannover region, the *"Hauptanhydrit* (A3)", the basal anhydrite of the Zechstein 3 cycle with a thickness of up to 30 m is the main source of cap-rock gypsum/anhydrite with gypsum thicknesses of more than 70 m (Bauer and Sessler 1985). In the Hänigsen-Wathlingen salt dome, the *Hauptanhydrit* is normally missing in the upper part of the diapir and cannot contribute large amounts of anhydrite to the cap-rock gypsum. The Zechstein 3 halite (*"Leine-Steinsalz"*, Na3) contains a large number of thin anhydrite layers in the lower part (*"Liniensalz"*) and some beds of anhydrite with a maximum thickness of each layer of 1–2 m in the *"Anhydritmittelsalz* (Na3 eta)". This part of the salt sequence is the major source of cap-rock gypsum in the Hänigsen-Wathlingen diapir and leaves behind a cap-rock gypsum with a maximum thickness of 25 m. The upper part of the Zechstein 3 halite, called *"Tonmittelsalz* (Na3 tm)" contains mostly grey claystones and siltstones as insoluble material. The residue of this salt will be a greyish or reddish blue siltstone or claystone. At the base of Zechstein 4 we find brecciated red siltstone of a considerable thickness of 20–30 m, called *"Roter Salzton* (T4)".

The halite above the *Roter Salzton* in the lower part of Zechstein 4 (*"Schneesalz"*, *"Rosensalz"*) is bedded by thin anhydrite layers, whereas the upper part of Zechstein 4 halite (*"Tonbrockensalz"*, *"Tonbanksalz"*) contains larger amounts of claystone and siltstone as inclusions in salt or intergrown with halite cubes and hoppers. The reddish and greenish grey claystones and siltstones of the cap-rock are the residual beds of *Tonmittelsalz*, *Roter Salzton* and *Tonbrockensalz/Tonbanksalz*.

2.3 The Cap-Rock Sequence of the New Ventilation Shaft

In the new ventilation shaft, a horizontally bedded sequence of gypsum/anhydrite with a thickness of 4–5 m covers the top of the salt. It consists of single layers of 10–50 cm thickness.

Above this gypsum/anhydrite with horizontal bedding we find dense and massive gypsum, partly coarse-grained and recrystallized, with fractures and cavities filled with dark grey plastic clay or sometimes coarse-grained, translucent gypsum. The fractures and cavities within the gypsum are oriented similarly to the bedding of the salt underneath.

The cap-rock gypsum is overlain by reddish and light greenish grey siltstone and claystone and contains pseudomorphs of halite cubes and hoppers filled with gypsum. It also contains boulders of gypsum in the lower part and only a few remnants of gypsum in the upper part. The fractures of the siltstone are normally filled with translucent gypsum. Similar to the top of the salt, the top of cap-rock is cut in a horizontal direction and is overlain by horizontally bedded Quaternary sand and gravel.

2.4 The Cap-Rock of the Riedel Shaft

In the Riedel shaft, the cap-rock sequence is different from the sequence in the ventilation shaft. The top of the salt is covered by a thin anhydrite layer, overlain by reddish claystone and blue and reddish blue siltstone at the top and bottom. Above this claystone, cap-rock gypsum was found. It is covered by red and light grey siltstone. The cap-rock of the Riedel shaft is also overlain by Quaternary sand and gravel with horizontal bedding.

2.5 The Tectonic Structures of Salt in the Shafts

In the lower part of the ventilation shaft Zechstein 4 halite ("*Schneesalz/Rosensalz*") dips vertically. The salt below shows an inclined bedding towards the east and southeast, whereas the salt above, including the basal clastic of Zechstein 4 ("*Roter Salzton*"), is overturned with an inclination of strata towards the southwest.

The inclined salt strata of the diapir are cut in horizontal direction by the top of salt, a subrosion plane which was completely flat in the ventilation shaft and showed no morphology.

The Zechstein salt sequence in the Riedel shaft also dips steeply, in general, but it is considerably thinned by tectoncis. The salt in the area of the two shafts consists mainly of two inverse synclines of folded Zechstein 4 halite and a thin anticline of Zechstein 2 and 3 halite in between. Further to the northwest and southeast, we find large inverse anticlines with complicated folding of Zechstein 2 and 3 halite (Fig. 6d in Schachl 1987).

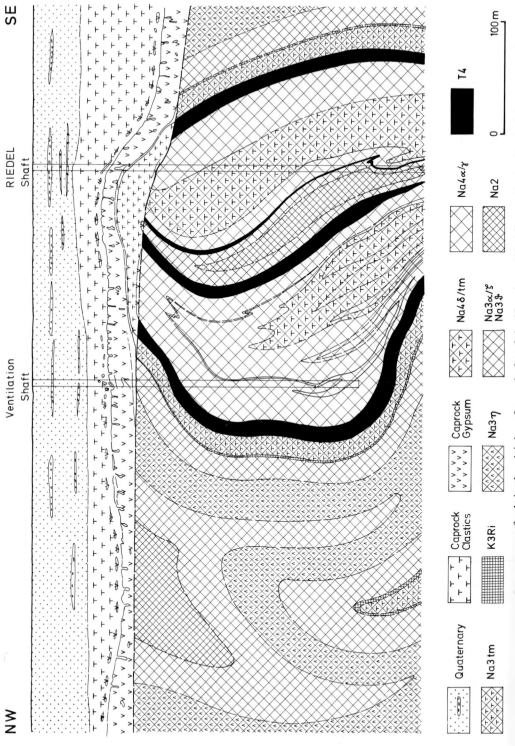

2.6 Conclusions

All the available geological information is summarized in a cross-section, including the interpretation of the tectonic structures in the salt and cap-rock sequence (Fig. 1). The cap-rock gypsum above the salt in the ventilation shaft and above the red siltstone in the Riedel shaft is a residue of Zechstein 3 halite. An anhydrite band of the "*Anhydritmittelsalz*" in the ventilation shaft, outcropping at the top of the salt into the anhydrite/gypsum with horizontal bedding confirms that this lowest part of the cap-rock gypsum is undoubtedly a residue of the "*Anhydritmittelsalz*", a halite sequence of Zechstein 3 with ten distinct anhydrite layers, each with a thickness between 10 cm and 1–2 m.

The thin anhydrite layer just above the top of salt in the Riedel shaft may be a residue of the "*Aller-Steinsalz*", whereas the red siltstone above is a residue of "*Tonmittelsalz*" and "*Roter Salzton*".

The majority of the cap-rock gypsum is a residue of Zechstein 3 halite, mainly "*Anhydritmittelsalz*". The red and light grey siltstone above the cap-rock gypsum in both shafts can be regarded as a residue of "*Roter Salzton*" and upper Zechstein 4 halite.

This uppermost section of the cap-rock is the oldest part of the cap-rock. It is strongly influenced by weathering and karstification of the gypsum cap-rock underneath, as can be seen by the brecciation, the content of gypsum boulders, and the fracture fillings of gypsum.

In the new ventilation shaft, the bedding of the older, upper parts of the cap-rock is parallel to the bedding of the salt, whereas the bedding of the youngest, lower part of cap-rock is oriented horizontally like the top of salt, the top of cap-rock, which is cut by erosion, and the bedding of the Quaternary sands.

3 Subrosion in the Salt Deposit of the Khorat Plateau in Thailand

3.1 General Information

The geology of the Khorat Plateau salt deposit in northeast Thailand (Fig. 2) had been described by Hite and Japakasetr (1979). In order to describe the subrosion phenomena in detail, the well-investigated area at the western border of the Khorat basin SE of Chaiyaphum was chosen (see area of investigation in Fig. 2). In this area more than 80 boreholes were drilled to investigate the salt deposit and its overburden.

3.2 Stratigraphy of the Salt Deposit and Overburden

The stratigraphy of the completely undissolved salt deposit is very simple: The deposit consists of three salt beds, each covered by clastic rocks, consist-

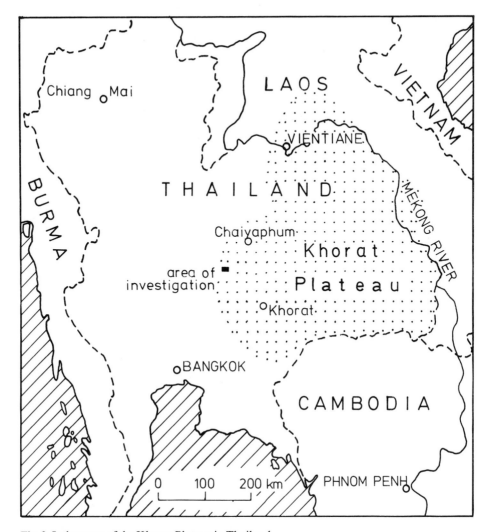

Fig. 2. Index map of the Khorat Plateau in Thailand

ing of siltstones and claystones. In the uppermost part of the lower salt bed a thick potash seam consisting mainly of carnallite ($KCl \times MgCl_2 \times 6H_2O$) and tachyhydrite ($2MgCl_2 \times CaCl_2 \times 12H_2O$) was discovered. This potash seam can be altered to high-grade sylvinite or to halite by the influence of water causing the subrosion of salt.

According to Jacobson et al. (1969), the salt of the Khorat Plateau belongs to the Maha Sarakam Formation of Upper Cretaceous age.

Anhydrite/gypsum beds were found on top of each halite bed or instead of halite in part of the investigation boreholes (Fig. 3).

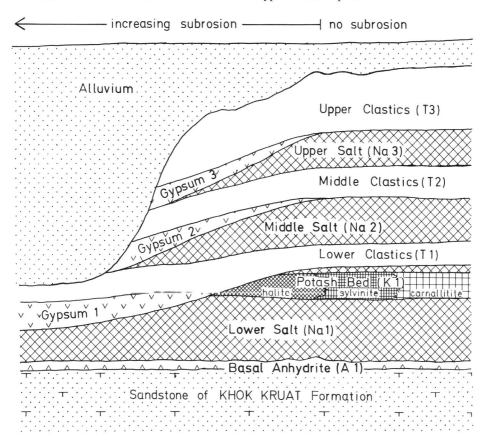

Fig. 3. Modified stratigraphy of the Maha Sarakam Formation of Upper Cretaceous age in the Khorat Plateau in Thailand

3.3 Genesis of the Cap-Rock

The anhydrite/gypsum beds were generated by the subrosion of salt, as can be seen by thin anhydrite stringers within the halite, outcropping at the top of salt into the anhydrite/gypsum.

Further evidence of subrosion of salt in the Khorat Plateau deposit is the similarity between the clastics within the salt containing 10–15% halite, and the clastic layers above the top of the salt without a visible salt content.

Because of good stratigraphic markers, the middle clastics at the top of the salt sequence two were chosen as examples (Fig. 4). One of the logs was chosen from a borehole with the complete salt of an undisturbed deposit a bit closer to the centre of the basin, whereas the other log shows the subroded remnants of middle clastics above the top of salt. Despite the influence of subrosion on one of the two logs, the stratigraphic markers of middle clastics are visible in both boreholes.

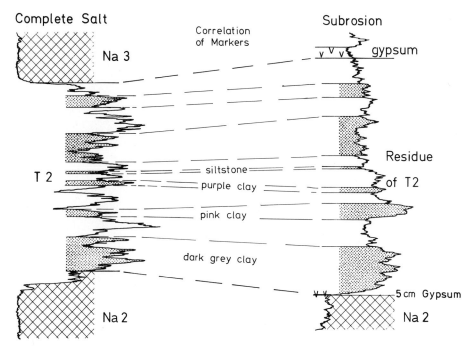

Fig. 4. Correlation of the stratigraphic markers of middle clastics in salt and in cap-rock

The Upper Cretaceous strata are overlain by unconsolidated Quaternary sand, gravel, clay and laterite.

3.4 Petrographic Description of the Cap-Rock Gypsum/Anhydrite

The anhydrite/gypsum normally consists of light grey to creamish white, thinly laminated anhydrite, partly or completely altered to gypsum with a thickness of the single layers up to 15 mm. Coatings of dark grey clay and carbonaceous material intercalate with the anhydrite/gypsum layers.

The anhydrite shows a sharp, wavy bedding with changes of the inclination in horizontal and vertical directions (Fig. 5). This type of anhydrite occurs on top of all three salt beds and is caused by the subrosion of salt, leaving behind the insoluble anhydrite layers of the dissolved halite on top of the salt.

Besides this finely laminated type of rock, the anhydrite/gypsum of the lower salt bed can be dense and massive with some vugs and cloudy and wavy bedding.

The anhydrite/gypsum of the upper salt bed in some boreholes is a dense, fine-grained rock, sometimes brecciated and fractured. The interspace between the anhydrite particles is filled with brown clay from the upper

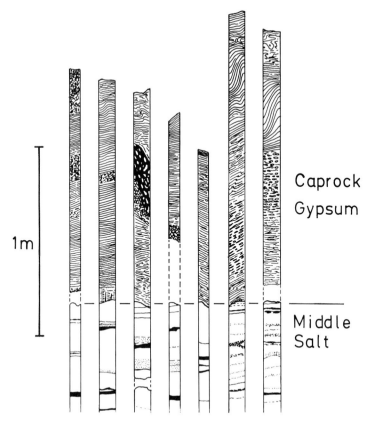

Fig. 5. Examples of the detailed structure of anhydrite/gypsum cap-rock of middle salt

clastics. Pseudomorphs of halite hopper crystals and halite cubes are also common. Intergrowths with larger crystals of translucent gypsum and anhydrite indicate recrystallization of the anhydrite/gypsum.

The anhydrite/gypsum of the upper salt bed shows the most intense influence of dissolution. Sometimes it is karstified with open cavities partly filled with idiomorphic calcite crystals. In some boreholes the anhydrite is more intensely karstified, consisting of isolated boulders within the cap-rock sequence, or is completely absent because of dissolution of the anhydrite.

3.5 General Structure of the Salt Deposit

The general structure of the salt deposit and its overburden can be seen in the cross-section (Fig. 6). The vertical scale of the cross-section is exaggerated fivefold compared to the horizontal scale.

The base of the salt deposit is comparatively flat in the investigated area with a gentle dip in the eastern direction towards the centre of the basin. The

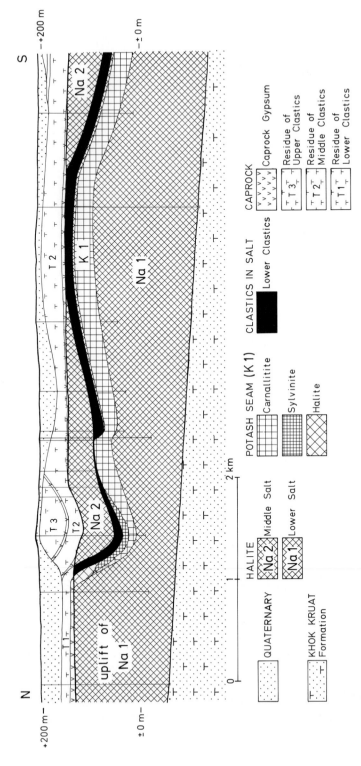

Fig. 6. Cross-section with the general structures of the deposit near the western boundary of the Khorat Plateau

thickness of the lower salt bed varies widely, indicating salt flow towards the uplifts with the bedding planes of salt dipping nearly vertically in the central part of the uplift areas. The potash layer on top of the lower salt bed and all the other Cretaceous strata above show a wavy structure parallel to the surface of lower salt bed caused by the varying thicknesses of this salt bed.

The cap-rock strata are eroded and covered with Quaternary sands and clays. In the uplift area in the north the residue of lower clastics is directly overlain by Quaternary sands of considerable thickness (more than 40 m in some places), whereas in the central part of the cross-section further south the Quaternary is only a few meters thick and residual beds of the middle and upper salt beds can be found.

3.6 Regional Distribution of Salt and Cap-Rock Gypsum

The regional distribution of Lower, Middle and Upper Salt (Na1, Na2, Na3) deposits and related anhydrite/gypsum residual beds as shown in Fig. 7 depends on the tectonics of the salt deposit and on the intensity of subrosion influenced by the thickness of Quaternary strata and the depth of salt below the surface.

In the uplift zone in the north and towards the outcrops of the deposit further west, only the Lower Salt bed and related anhydrite gypsum are present.

In the central part of the investigation area we mostly find anhydrite/gypsum of the Middle Salt bed, whereas in the eastern part we find a 2–3 km wide rim of anhydrite gypsum of the Upper Salt bed. The intensity of subrosion becomes more and more important towards the outcrops of the deposit in the west. Towards the centre of the basin the effects of subrosion diminish. This generalized picture is varied by the effect of salt uplifts within the deposit and by the erosion of the cap-rock.

4 Irregular Subrosion of Salt in the Deposit with Horizontal Bedding in the Fulda Basin in Germany

4.1 General Geology of the Fulda Basin Salt Deposit

The Fulda Basin represents an area of approximately 100 km^2, underlain by salt of Zechstein 1, called "*Werra-Steinsalz* (Na 1)". The salt, containing two potash seams, has a normal thickness of 220 m and shows horizontal bedding. The detailed, modern Zechstein stratigraphy of the Werra and Fulda Basin was published by Käding (1978) and shall not be described in detail here (Table 2).

The insoluble constituents of salt in the lower and middle section of the "*Werra-Steinsalz*" are sulfates, mainly anhydrite, whereas in the upper sec-

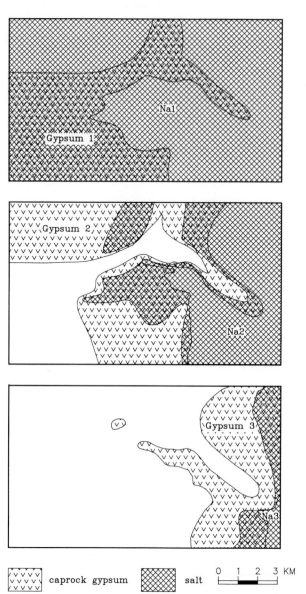

Fig. 7. Distribution of Lower, Middle and Upper Salt (Na1, Na2, Na3) and related anhydrite/gypsum cap-rock in the Khorat Plateau in Thailand

tion the salt is intergrown with claystones and siltstones or it contains distinct layers and bands of clastic material, intergrown with halite cubes or fibrous halite.

The salt is overlain by a sequence of fine-grained brown clastics, containing some anhydrite bands and a carbonate layer of 5 m thickness

Table 2. Stratigraphy and thickness of Zechstein in the Fulda Basin (Käding 1978)

Thickness (m)	Name	Symbol
3 – 4	*Friesland-Ton*	T6–T6r
	Friesland-Sandstein	S6
2.5– 4	*Ohre-Ton*	T5–T5r
	Ohre-Sandstein	S5
2 – 3	*Oberer Aller-Ton*	T4r
0.7– 1.8	*Aller-Anhydrit (Pegmatitanhydrit)*	A4
1 – 1.5	*Unterer Aller-Ton*	T4
1 – 1.5	*Aller-Sandstein*	S4
3 – 5	*Oberer Leine-Ton*	T3r
1.5– 4	*Leine-Anhydrit (Hauptanhydrit)*	A3
5 – 8	*Leine-Karbonat (Plattendolomit)*	Ca3
	Unterer Leine-Ton	T3
1 – 2	*Leine-Sandstein*	S3
15 – 35	*Staßfurt-Ton*	T2–T2r
3 – 7	*Oberer Werra-Anhydrit*	A1r
8 – 12	*Oberer Werra-Ton*	T1r
70	*Oberes Werra-Steinsalz*	Na1γ
2 – 3	*Kaliflöz Hessen*	K1H
35 – 55	*Mittleres Werra-Steinsalz*	Na1β
2 – 3	*Kaliflöz Thüringen*	K1Th
60 –100	*Unteres Werra-Steinsalz*	Na1α
3	*Unterer Werra-Anhydrit*	A1
5 – 8	*Anhydritknotenschiefer*	A1Ca
6 – 20	*Werra-Karbonat (Zechsteinkalk)*	Ca1
0.2– 0.5	*Unterer Werra-Ton (Kupferschiefer)*	T1
0 – 4	*Werra-Konglomerat (Zechsteinkonglomerat)*	C1

(*Plattendolomit*). The lower 50 m of the fine-grained clastic sequence is Zechstein strata (Zechstein 2–6), whereas the upper 30 m is the lowest section (*Bröckelschiefer*) of the Buntsandstein.

Above the Bröckelschiefer, we find sandstones and siltstones of Lower and Middle Buntsandstein in the Fulda basin.

4.2 Regular Subrosion of Salt in the Fulda Basin

Salt is usually subroded at the boundary of the deposit by groundwater downwards from the top, leaving behind a residue of the insoluble constituents of salt. In some cases the strata overlying the salt subside during the subrosion of salt and the stratigraphic sequence is well preserved. The strata of this origin are only slightly brecciated. If the rocks of the hanging wall sequence are competent enough, larger open cavities with an arched roof are formed by the subrosion of salt. When the cavities collapse, the rock of overburden and the residue of salt subrosion is mixed and more strongly brecciated. Subrosion breccia of this origin have a maximum thickness of 200 m in the Fulda Basin and the stratigraphy of the former, undisturbed rock sequence is hardly visible.

4.3 The Rock Sequence of Borehole Tiefengruben 3

The rock sequence of the borehole Tiefengruben 3 is interpreted in a profile shown in Fig. 8. The *Werra-Steinsalz* is subroded from the top to a thickness of approximately 80–90 m. The rock sequence above the top of the salt is fractured and brecciated, but most of the former stratigraphy can still be recognized by the gamma log and the drilled cores. The top of the salt is covered by layered gypsum of 0.7 m thickness which is partly brecciated and mixed with claystone fragments in the lowest part. The gypsum is overlain by a consolidated breccia of claystone with fragments and boulders of gypsum.

The salt in the borehole Tiefengruben 3 is not only subroded from the top, but also from the bottom. The *Werra-Steinsalz* with normal bedding is underlain by approximately 10 m of coarse crystalline, translucent halite without bedding with only a few inclusions of anhydrite stringers. The halite is undoubtedly of secondary origin and a product of recrystallization.

Below this coarse-grained halite we observe another breccia, containing only fragments of the rocks above, e. g. siltstone, anhydrite, carbonate from the *Plattendolomit* and sandstone of Buntsandstein.

Fragments of volcanic rocks or rocks from below the base of salt were not found in this breccia.

4.4 Interpretation of the Structure and Conclusions

One concept was that the structure with the subrosion of salt from the bottom was caused by a volcanic pipe, but despite careful investigation of drilled cores, no volcanic material or rock fragments from below the base of salt could be found.

As known from the stratigraphic position of the remnants of *Werra-Steinsalz* in the borehole, approximately 40–50 m of halite are missing at the bottom, whereas the distance between the lower boundary of salt and the base of salt is only 25 m.

From this fact it is assumed that there is a fault in the base of salt in the vicinity of the borehole.

The fault may have originated by the main tectonic events in the area during Late Jurassic or Early Cretaceous.

The fracture of the fault may have been the preferred path for undersaturated formation water from below, which first dissolved the salt and

The Influence of Subrosion on Three Different Types of Salt Deposits

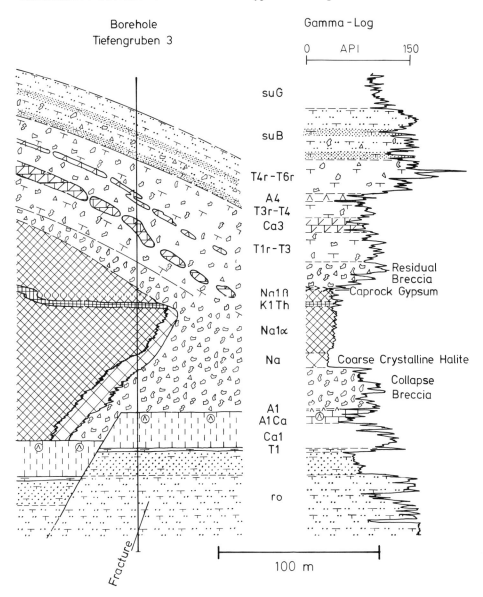

Fig. 8. Rock sequence and structure of borehole Tiefengruben 3 in the FUL Fulda Basin

was saturated with halite. In a second step the halite-saturated brine reprecipitated halite in some places, as can be seen from the occurrence of the coarse-grained, translucent halite at the lower boundary of salt.

All these events may have taken place within the salt deposit, away from the rim, long before the salt was subroded from the top.

In the coarse of time, the normal subrosion of salt slowly penetrated further into the interior parts of the deposit, and today the older, extraordinary structure with subrosion of salt from below is superimposed by the younger, normal subrosion at the top of the salt.

Acknowledgements. I thank Kali and Salz AG and the Thailand Department of Mineral Resources for the permission to publish the article. I am greatly indebted to Dr. Anant Suwanapal and his staff and to my colleagues of the Thailand Department of Mineral Resources for their support and hospitality during my field trips to Thailand.

Many thanks to Dr. Karl-Christian Käding for constructive criticism of the manuscript. Mr. Tiller did the drawings. Miss Everding and Miss Iba typed the manuscript.

References

Batsche H, Klarr K (1980) Beobachtungen und Gedanken zur Gipshutgenese. In: Coogan AH, Hauber L (eds) Proc 5th Symp Salt, vol 1 Northern Ohio Geol Soc, Cleveland, Ohio, pp 9–19

Bauer G, Sessler W (1985) Das Deckgebirge über zwei norddeutschen Salzstöcken in den Wetterschächten Kolenfeld und Riedel. Kali Steinsalz 9:125–131

Boer HU de (1970) Genese und Morphologie der Grenzfläche zwischen wasserführendem Deckgebirge und Zechsteinsalinar über dem Salzstock von Hänigsen-Wathlingen. Bergbau Wiss 17:442–446

Bornemann O, Fischbeck R (1986) Ablaugung und Hutgesteinsbildung am Salzstock Gorleben. Z Dtsch Geol Ges 137:71–83

Fulda E (1935) Handbuch der vergleichenden Stratigraphie Deutschlands: Zechstein. Bornträger, Berlin

Fulda E (1938) Steinsalz und Kalisalze. In: Beyschlag F, Krusch P, Vogt JHL (eds) Die Lagerstätten der nutzbaren Mineralien und Gesteine, vol 3, pt 2. F Enke, Stuttgart, 233 pp

Hite RJ, Japakasetr T (1979) Potash deposits of the Khorat Plateau, Thailand and Laos. Econ Geol 74:448–458

Jacobson HS, Pierson CT, Danusawad T, Japakasetr T, Inthuputi B, Siriratanamongkol C, Prapassornkul SM, Pholphan N (1969) Mineral investigations in northeastern Thailand. US Geol Surv Prof Paper 618

Käding KC (1978) Stratigraphische Gliederung des Zechsteins im Werra-Fulda-Becken. Geol Jahrb Hessen 106:123–130

Martinez JD (1980) Salt dome cap-rock – a Record of geologic processes. In: Coogan AH, Hauber L (eds) Proc 5th Symp Salt, vol 1. Northern Ohio Geol Soc, Cleveland, Ohio, pp 143–151

Schachl E (1987) Kali- und Steinsalzbergwerk Niedersachsen-Riedel der Kali und Salz AG, Schachtanlage Riedel. Zechsteinstratigraphie und Innenbau des Salzstockes von Wathlingen-Hänigsen. In: Kulick J, Paul J (eds) Int Symp Zechstein 1987, Exkursionsführer 1:69–100

Seidl E (1921) Schürfen, Belegen und Schachtabteufen auf deutschen Zechstein-Salzhorsten. Arch Lagerstättenforsch 26, 209 pp

Walker CW (1974) Nature and origin of cap-rock overlying Gulf Coast salt domes. Proc 4th Symp Salt, vol 1. Northern Ohio Geol Soc, Cleveland, Ohio, pp 169–195

Particle Size Distribution of Saliferous Clays in the German Zechstein

Robert Kühn[1]

CONTENTS

Abstract 197
1 Introduction 197
2 Methods 198
2.1 Elutriation Analysis 198
2.2 Plotting of Grain Size Distribution 200
3 Grain Size Distributions according to Section 2.1
    for Samples Studied 201
4 Discussion of Grain Size Distribution 206
References 208

Abstract

Grain size analyses of 26 grey and 7 red salt clays from the German Zechstein are presented in bilogarithmic nets according to the Rosin-Rammler equation. Salt clays are characterized by average particle diameter d', calculated from Rosin-Rammler (grain size for residue 36.8%), and by n (uniformity coefficient) with a subsequent discussion of grain size distribution with regard to genetic relationships.

The striking correspondence between grain size distribution of red clays of the Rotliegend and loess supports Sonnenfeld's hypothesis, in which a geological relationship between evaporitic periods during geological history and their immediately preceding glaciations, is suggested.

1 Introduction

The vast oceanic salt deposits reflect rather isolated events in the earth's history. Lotze (1969) showed that the evaporite zones of the geological

---

[1] Institut für Sedimentforschung der Universität Heidelberg, Postfach 103020, D-6900 Heidelberg, FRG.

formations shifted systematically from north to south and that this shifting can be explained by the supposition of a drift of the earth's axis in geologic time. Sonnenfeld (1984, 1985) discussed a possible relationship between salt deposits and preceding ice ages. Storage of ocean water in the massive continental ice caps leads to a general lowering of sea level. Shallow basins developed towards evaporitic conditions and the shelf areas underwent enhanced erosion processes.

The Permo-Carboniferous glaciation might have created particularly effective conditions with regard to the development of the subsequent Zechstein evaporites. Weathering solutions rich in sulfate ions infiltrated the chloridic potash horizons of the adjacent areas and caused their sulfatization. This pattern, however, cannot be observed for the Ordovician glaciation where no salt was deposited during the Middle Ordovician Interglacial, although it seems to reoccur during the Neogene.

If Sonnenfeld's hypothesis is correct, the clastic deposits associated with evaporites should be studied more closely. The clastic deposits shift towards purely saliferous deposits, i.e. the salt clays from the base to the top of the sections. They may at least be interpreted as reworked glacial sediments from the preceding ice age. This hypothesis is not applicable, however, to the evaporites of the Devonian.

Particle size distribution of such clastics are supposed then to be comparable to loess. This study presents a preliminary approach to the problem, and some data on particle size distribution of salt clays from the German Zechstein are presented in addition to the already known petrographic parameters.

2 Methods

2.1 Elutriation Analysis

To determine the grain size composition of the clay material, salt clays are elutriated and then subjected to customary elutriation analyses, whereby the salts go into solution. Kühn (1938) was the first to quantitatively elutriate an entire group of salt clays in a cylinder in this manner, at the same time determining grain size distribution. Details of the process, such as the careful rubbing of samples under the addition of small amounts of ammonia (after Neumaier 1935), or regarding the elutriation cylinder, are given in Kühn (1938), while height and duration of fall, etc. are given in Lohse (1958). Crushing of the sample in a mortar should be avoided, since this falsifies the primary particle size distribution. Drawbacks to this method, namely its time-consuming nature, in particular the washing of the high salt content, as well as the large amounts of distilled water necessary, can be partially overcome, according to Lamcke (1939), by desalinating the samples by Brintzinger's rapid electrodialysis before elutriation, and, through addition of silicate of soda, stabilizing them against coagulation. Following this, the quick-

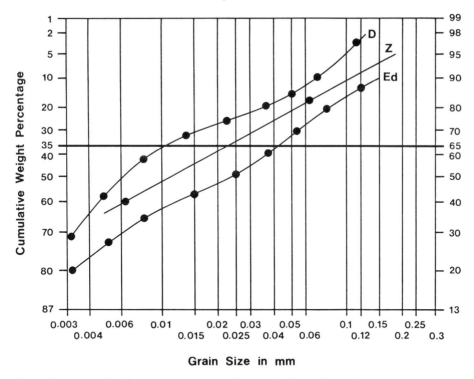

Fig. 1. Grain size distribution in a net according to the Rosin-Rammler equation: grey salt clay specimen No. 8 of Kühn (1938). Varying methods of analysis according to Lamcke (1939; see text)

Table 1. Table of values for Fig. 1 [a]

	Curve D dialyzed		Curve Ed electrodialyzed		Curve Z washed out by elutrition	
	(%)	(d)	(%)	(d)	(%)	(d)
	29	3.1	20	3.1	40	< 6.32
	41	4.9	27	5	82	<63.2
	58	7.9	35	8		
	69	14	43	15		
	73	23	50	25		
	80	37	61	38		
	85	50	70	54		
	91	70	79	78		
	97	98	87	120		
d'	0.01		0.041		0.023	
n	0.51		0.54		0.515	

[a] Grey salt clay No. 8 of Kühn (1938). Varying methods of analysis according to Lamcke (1939); wt % at grain diameter 10^{-3} mm.

er aerometric methods of analysis can be used for the determination of grain size distribution alone; without recovery of the grain size fraction (see also Lamcke 1939). In this manner, Lamcke was able to prove that grain size distribution among salt clays varies within one and the same specimen; the small differences between results of aerometric analyses (see Fig. 1) caused by varied processing, i.e. normal rapid analysis (curve D), electrodialysis (curve Ed), or through washing alone (curve Z), all lie within the natural grain size scattering of the material itself. Figure 1, together with the table of value (Table 1), shows the scattering for grain size distribution as well as for d' and n resulting from the individual type of pretreatment and processing of the materials studied. The method used here, simple elutriation, was the probable cause for the mean values, which are found between those for material pretreated by dialysis and electrodialysis.

2.2 Plotting of Grain Size Distribution

To plot size distributions, Puffe's net for the graphic evaluation of granulometric analyses based on the Rosin-Rammler equation was chosen, since evaluation is possible with even little material, and since such good results have been achieved in processing.

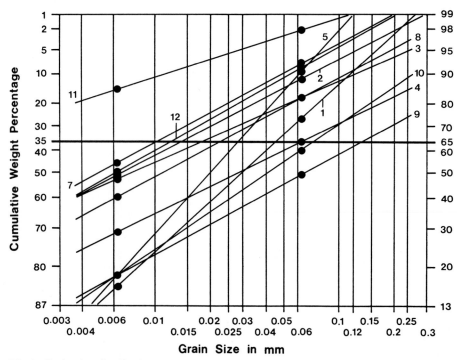

Fig. 2. Grain size distributions in the manner of Fig. 1: grey salt clays, Loeser and Haselgebirge, according to Kühn (1938)

Residue R in relation to particle $\emptyset$ can be represented by the formula: $R = 100 \times e^{-(d/d)^n}$. Granulometric curves appear as straight lines when residue R or D = 100-R are bilogarithmically plotted above the logarithms of the particle $\emptyset$. Special values d' and n are then read from the particle curves. The average grain diameter d' gives the particle size for which $R = 1/e = 36.8\%$. The constant n determines the gradient of the line. When n increases, the material becomes more uniform; n is thus called the uniformity coefficient. According to Puffe (1948), d' is a measurement of fineness; the smaller d', the finer the material; n is a measurement of uniformity; the greater n, the more uniform the material.

3 Grain Size Distributions According to Section 2.1 for Samples Studied

A summary of the grain size distribution of salt clays taken from the literature and illustrated after Kühn (1938), Niemann (1960) and Siebert (1952)

Table 2. Table of values for Fig. 2[a]

Specimen No.	Horizon and occurence	Wt % at grain diameter 10^{-3} mm			d' (mm)	n
		<6.32	6.32–63.2	>63.2		
1	Clay anhydrite T 3a Hansa, Hardsaltseam 4 curve to seam 7	16	57	27	0.045	0.87
2	As before, lower part	49	39	12	0.014	0.50
3	Lying part. Hansa, Blindshaft 3, 601 m level	48	34	18	0.018	0.42
4	Middle part. Hansa, Hardsaltseam 4 curve to seam 7	29	34	37	0.063	0.47
5	T3c with magnesite and pyrite. Hansa	18	73	9	0.028	1.08
7	Lower part. Glückauf I, Sondershausen	54	38	8	0.0103	0.52
8	Lower part of the middle section. Glückauf I, Sondershausen	40	42	18	0.023	0.515
9	Upper part of the middle section. Glückauf I, Sondershausen	18	33	49	0.14	0.52
10	T3c resp. upper part. Glückauf I, Sondershausen	18	42	40	0.075	0.65
11	Loeser above K 1H. Hattorf	85	13	2	~0.001	0.31
12	Haselgebirge Salt mine Hall/Tirol	50	41	9	0.012	0.54

[a] Grain size distributions of grey salt clays, Loeser and Haselgebirge according to Kühn (1938).

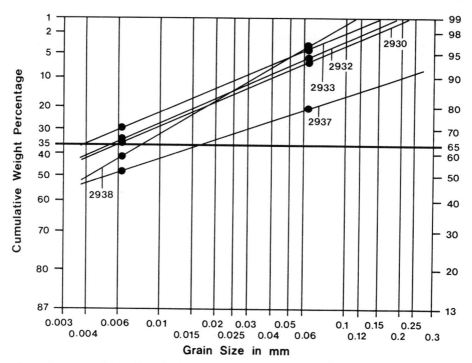

Fig. 3. Grain size distributions in the manner of Fig. 1: grey salt clays from layer T3a (lying, anhydritic part) according to Niemann (1960)

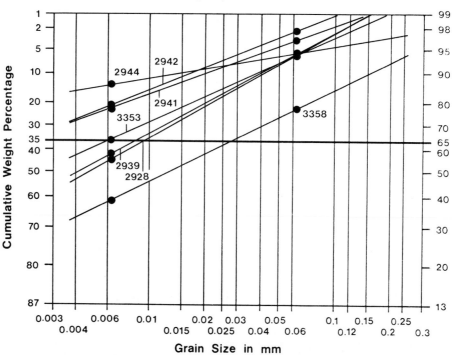

Fig. 4. Grain size distributions in the manner of Fig. 1: grey salt clays from layer T3b (middle, clayey, sandy part) according to Niemann (1960)

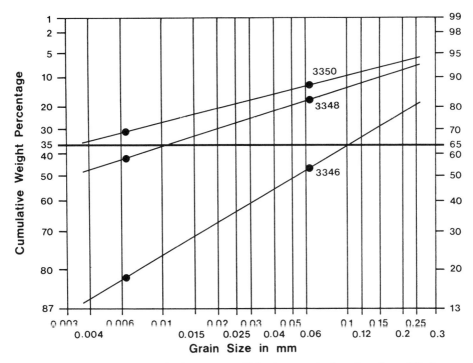

Fig. 5. Grain size distributions in the manner of Fig. 1: grey salt clays from layer T3c (hanging dolomitic part) according to Niemann (1960)

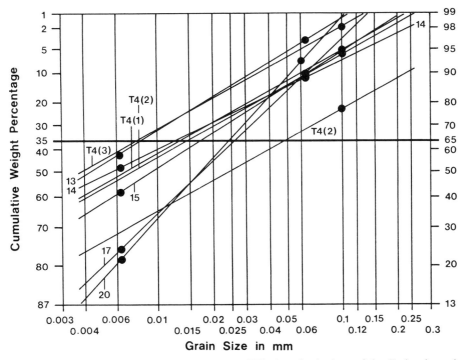

Fig. 6. Grain size distributions in the manner of Fig. 1: red salt clays of the Zechstein, red clay of the Rotliegendes and loess of Mauer according to Kühn (1938) and Siebert (1952)

Table 3. Table of values for Figs. 3–5[a]

Specimen No.	Horizon	Wt % at grain diameter 10^{-3} mm			d' (mm)	n
		<6.32	6.32–63.2	>63.2		
2938	T3a, i.e., lying, anhydritic part	55.9	40.5	3.6	0.008	0.57
2937	T3a, i.e., lying, anhydritic part	51.5	27.9	20.5	0.016	0.32
2933	T3a, i.e., lying, anhydritic part	70.6	26.0	3.4	0.0033	0.41
2932	T3a, i.e., lying, anhydritic part	65.1	29.1	5.8	0.0054	0.42
2930	T3a, i.e., lying, anhydritic part	64.3	29.3	6.5	0.0056	0.41
2928	T3b, i.e., middle, clayey-sandy part	58.9	35.4	5.6	0.084	0.515
2939	T3b, i.e., middle, clayey-sandy part	55.1	39.0	5.9	0.0095	0.54
2941	T3b, i.e., middle, clayey-sandy part	77.8	19.0	3.2	ca. 0.0015	0.35
3358	T3b, i.e., middle, clayey-sandy part	39.0	38.6	22.2	0.028	0.47
2942	T3b, i.e., middle, clayey-sandy part	79.3	18.5	2.1	ca. 0.0016	0.38
2944	T3b, i.e., middle, clayey-sandy part	86.3	8.3	5.4	–	0.155
3353	T3b, i.e., middle, clayey-sandy part	64.1	29.9	6.0	0.0059	0.43
3350	T3c, i.e., hanging, dolomitic part	69.8	17.5	12.7	0.0032	0.23
3348	T3c, i.e., hanging, dolomitic part	57.3	25.1	17.4	0.011	0.30
3346	T3c, i.e., hanging, dolomitic part	18.1	35.4	46.6	0.105	0.56

[a] Grain size distributions of grey salt clays from the Königshall-Hindenburg mine, according to Fig. 2 of Niemann (1960).

can be found in Figs. 2–6. For clarity, numerous illustrations were used. Evaluations for d' and n are given in the corresponding tables (Tables 2–4). To obtain exactly comparable ratios, the same grain size intervals were used. Grain size distributions of the four horizons of the red salt clay from Angersdorf after Siebert (1952), offer only an approximate picture, since only one sample per horizon was examined, and therefore n was taken over from the mean for the red salt clay of Hansa (Table 5).

Table 4. Table of values d' and n for Fig. 6[a]

Specimen No.	Horizon and occurrence	Wt % at grain diameter 10^{-3} mm			d' (mm)	n
		<6.32	6.32–63.2	>63.2		
13	T 4 lower part. Hansa, 601 m level 570 m western wall drifting 76	58	38	4	0.008	0.58
14	T 4 middle part. As before	51 (2:33 2–6.32:18)	37 (6.32–20:13 20–63.2:24)	12	0.012	0.46
15	T 4 upper part. As before	41	48	11	0.017	0.62
17	Red clay of the Rotliegendes. Lieth near Elmshorn	24	65	11	0.026	0.90
20	Younger Loess Wuerm Diluvium. Mauer, near Heidelberg	21	72	7	0.025	1.03

	Red salt clays of Angersdorf, near Halle	One grain section only			
		<0.1	>0.1		
T 4 (1)	Lying part; stratified 1.5 m	94.7	5.3	0.15	0.55 assumed
T 4 (2)	Above T 4 (1) stratified 0.85 m	77.8	22.2	0.05	
T 4 (3)	Middle part, above T 4 (2), not stratified 2.3 m	98.1	1.9	0.0094	
T 4 (4)	Upper part, above T 4 (3), not stratified 6.85 m	94.3	5.7	0.016	

[a] Grain size distributions of red salt clays from the Zechstein, red clay of the Rotliegendes and loess of Mauer, according to Kühn (1938) and Siebert (1952).

Table 5. Grain size distributions for the red salt clay horizons of Angersdorf, for grains >0.1 mm diameter

Horizon	Coarse residues >0.1 mm ⌀ from 500 g substance (g)	Average	Weight % >0.1 mm
T 4 (1) lying part, stratified 1.5 m	11– 42	26.5	5.3
T 4 (2) above T 4 (1), stratified 0.85 m	103–119	111	22.2
T 4 (3) middle part above T 4 (2), not stratified 2.3 m	2– 17	9.5	1.9
T 4 (4) upper part above T 4 (3), not stratified 6.85 m	2– 55	28.5	5.7

4 Discussion of Grain Size Distribution

Without going into detail regarding the widely studied mineralogical composition of the salt clays (Kühn 1938, 1968; Lotze 1957; Niemann 1960; Schulze et al. 1962; Lohse 1963; Sonnenfeld 1984), it should be mentioned that the main clastic constituents are quartz and mica (muscovite/illite), with additional anhydrite, which is abundant in the underlying grey salt clay. Abundant magnesite is found in the overlying part of the section, towards the marginal areas of the basin, as in the samples of the Königshall-Hindenburg mine, dolomite occurs instead of magnesite.

Small amounts of chlorite and koenenite occur as neoformations. The red salt clay seems to contain slightly more clastic components (quartz and mica), and also more koenenite than the grey salt clay. The grain size distribution of the clastic components is most probably influenced by the minerals of neoformation.

For reasons mentioned in Sect. 2.1, grain size analyses, and thus the characteristic results for d' and n, scatter significantly even among samples from the same horizon. To this are added the differences from the geological positions of the samples: samples from the Hansa and Glückauf I potash mines are probably from the central to deepest basin areas; samples from the Königshall-Hindenburg mine, as well as the Löser samples are likely to stem from areas closer to the basin margin. To consider these scatterings, the average of the results for d' and n are compared in Table 6.

The comparison between Rotliegend clays and loess shows striking similarities between d' and n. It may thus cautiously suggested that the Rotliegend clays have a similar origin; they were probably reworked from deposits originally formed during the Permo-Carboniferous glaciation. This could support Sonnenfeld's hypothesis mentioned above, although such studies should be extended to a much broader range of samples.

The grey salt clays of the basin margin exhibit a finer d', understandable through the greater influence of at least partially aeolian transport. For the same reason, n values for grey salt clays of the central basin areas (Hansa, Glückauf I) are greater than those for basin margin samples (Königshall and clay of the Löser). Material from the central basin is more uniform.

Detailed study of the grain size analyses of the grey salt clay (Kühn 1938; Niemann 1960) has shown an increase of the finer material from the footwall to the hanging wall, whereas in the highest part (T3c) grain size increases again. Thus, fluvial transport is assumed. Based upon the thickness distribution of salt clays in the Zechstein basin, Niemann (1960) concluded that the clayey-clastic components were transported from the south. The slight aeolian activity is supported by the findings of pollen- and quartz-bearing desert varnish. A total correspondence is found among the red salt clays. Löser clay is extraordinarily fine, although values for n are relatively low, indicating material of a non-uniform nature. With regard to d' and n the Haselgebirge closely corresponds to the middle section of the grey salt clay.

Table 6. Averages of d′ and n on grey and red salt clays.
Grey salt clays:

Horizon T 3a Specimen No.	Hansa and Glückauf I 1, 2, 3		Königshall-Hindenburg 2938, 2937, 2933, 2932, 2930	
	d′	n	d′	n
	0.0218	0.577	0.008	0.426
Horizon T 3b Specimen No.	4, 8, 9		2928, 2939, 2941, 3358, 2942, 2944, 3353	
	d′	n	d′	n
	0.08	0.50	0.01	0.41
Horizon T 3c Specimen No.	5, 10		3350, 3348, 3346	
	d′	n	d′	n
	0.051	0.87	0.04	0.36
Loeser above K 1 H Specimen No.	11			
	d′	n		
	0.001	0.31		
Haselgebirge Specimen No.	12			
	d′	n		
	0.012	0.54		

Red salt clays, Rotliegendes, sandstone and loess:

Specimen No.	13, 14, 15, T4 (1–4)	
	d′	n
	0.0182	0.55
Red clay of the Rotliegendes Specimen No.	17	
	d′	n
	0.026	0.90
Loess Specimen No.	20	
	d′	n
	0.025	1.03

According to von Engelhardt (1973) the average grain size for loess lies between 0.01 to 0.06 mm (roughly equivalent to our average particle fraction 6.32 to 63.2 µm); the same is true for the loess from Mauer. This grain size maximum in salt clays is only indicated at samples 1, 5, 9, and 10, to which no significance could be attached here. The shifting of the clastic grain size distribution due to neoformations is certainly more significant: in the middle (clayey-sandy) horizons of the grey salt clay this shifting produces through the formation of chlorite a finer grain size distribution, while in the hanging wall coarser grain size is reached due to the coarseness of the carbonates formed (magnesite and dolomite).

For a discussion of the genesis of grey salt clays based on their mineralogical content and geochemical relationship, see Braitsch (1962).

A similar discussion regarding the formation of the red salt clay is given by Kühn (1966) including synsedimentary reworking, which occurred frequently during the Leine and Aller series.

References

Braitsch O (1962) Entstehung und Stoffbestand der Salzlagerstätten. Springer, Berlin Göttingen Heidelberg
Engelhardt, v. W (1973) Die Bildung von Sedimenten und Sedimentgesteinen. Von Engelhardt W, Füchtbauer H, Müller G (eds) Sediment-Petrologie, pt3. Schweizerbart, Stuttgart, 378 pp
Kühn R (1938) Über den Mineralgehalt der Salztone, Thesis, Christian-Albrechts-Univ, Kiel
Kühn R (1966) Befahrung des Kalibergwerkes Siegfried-Giesen, Groß-Giesen bei Hildesheim. Zu den mineralogisch-petrographischen Verhältnissen. Fortschr Mineral 43:153–187
Kühn R (1968) Geochemistry of the German potash deposits. In: Mattox RB (ed) Saline deposits. Geol Soc Am Spec Pap 88:427–504
Lamcke K (1939) Die Vorbehandlung mariner Sedimente für die Schlämmanalyse. Geol Meere Binnengewässer 3:222–237
Lohse HH (1958) Erfahrungen bei der röntgenographischen Identifizierung semisalinarer und nichtsalinarer Minerale der Salzlagerstätten. Thesis, Christian-Albrechts-Univ, Kiel
Lohse HH (1963) Versuch einer Klassifikation der Salzgesteine des Zechsteins mit Hilfe iher wasserunlöslichen Bestandteile. Kali Steinsalz 3:402–410
Lotze F (1957) Steinsalz und Kalisalze, pt1 (Allgemein-geologischer Teil). Bornträger, Berlin
Lotze F (1969) Die Salz-Lagerstätten in Zeit und Raum. Arbeitsgemeinschaft für Forschung des Landes Nordrhein-Westfalen, 195. Westdeutscher Verlag, Köln Opladen
Neumaier F (1935) Über Vorbehandlungsverfahren der Sedimente zur Schlämmanalyse. Zentralbl Mineral A 1935, pp 78–95
Niemann H (1960) Untersuchungen am Grauen Salzton der Grube Königshall-Hindenburg, Reyershausen bei Göttingen. Beitr Mineral Petrogr 7:137–165
Puffe E (1948) Graphische Darstellung und Auswertung von Siebanalysen auf Grund der Rosin-Rammler-Gleichung. Erzmetall 1:97–103
Reidemeister C (1911) Über Salztone und Plattendolomite im Bereich der norddeutschen Kalisalzlagerstätten. Thesis, Christian-Albrechts-Univ, Kiel

Schulze G, Greulich C, Seyfert H (1962) Untersuchungen am Grauen Salzton im Spaltendiapir des oberen Alltertales. Angew Geol 8:1–9

Siebert G (1952) Vergleichend feinstratigraphische Untersuchungen der Jüngeren und Jüngsten Salzfolge (III u. IV Großzyklus) im Magdeburg-Halberstädter Becken mit einer speziellen Bearbeitung des Roten Salztones von Angersdorf bei Halle. Thesis, Freie Univ, Berlin

Sonnenfeld P (1984) Brines and Evaporites. Academic Press, New York London

Sonnenfeld P (1985) Comment evaporites, marine or non-marine? Am J Sci 285:661–667

Mineralogical and Petrographic Studies on the *Anhydritmittelsalz* (Leine Cycle z3) in the Gorleben Salt Dome

R. Fischbeck[1]

CONTENTS

Abstract	210
1 Introduction	210
2 The *Anhydritmittelsalz* in the Hannover and Gorleben Areas	212
3 Petrographic Studies of the Ninth *Anhydritmittel* in the Gorleben Salt Dome	213
4 Discussion	217
References	219

Abstract

An unusual mineral composition of sylvitic kieserite-anhydrite was found with an anhydrite layer of the *Anhydritmittelsalz* (Leine cycle z3 Zechstein, NW Germany). The mineral composition is considered to have formed diagenetically. Macroscopic and microscopic evidence suggest that both mineralogical composition and the textures were formed by postsedimentary alteration of the primary evaporites in different postsedimentary stages.

1 Introduction

The field of sediment petrography is in its variety like a bouquet of wild flowers. An especially beautiful and somewhat exotic flower from this bouquet is the group of evaporites. I was first introduced to these particulary interesting sedimentary rocks by Prof. German Müller, my academic teacher, to whom I am most grateful.

It is characteristic of the evaporites to show both the typical structure of sediments: stratification and bedding as well as the texture of crystalline rocks. Lotze (1957) called them "crystalline sediments".

Evaporites or salt rocks consist mainly of readily soluble minerals precipitated from evaporating aqueous solutions when the solubility product

---

[1] Bundesanstalt für Geowissenschaften und Rohstoffe, Stilleweg 2, D-3000 Hannover 51, FRG.

of each mineral was exceeded. In the ideal case, an entire sequence of salts would be formed by a continually and completely evaporating marine basin; such a sequence usually follows a pelitic basal layer, and begins with carbonates (calcite, aragonite→magnesite), followed by sulfates (gypsum→anhydrite), and chloride (halite), ending with potassium magnesium minerals.

Within increasing solubility the salt minerals in this evaporite cycle become more sensitive with respect to the conditions of their formation to climatic changes and tectonic movements. Thus, conditions of formation of the most soluble minerals are, in practice, seldom attained.

Therefore, complete evaporite cycles are comparatively rare; they are commonly interrupted at an early stage of evaporation, i.e. the sulfate stage. Often, thick deposits of evaporites originate from several either partly complete or incomplete evaporite cycles.

Both within and beyond the boundaries of northern Germany, more than 1000 m of marine evaporites were deposited during the Late Permian (Zechstein). These consist of seven cycles, and are, from bottom to top, as follows: Werra (z1), Staßfurt (z2), Leine (z3), Aller (z4), Ohre (z5), Friesland (z6), and Mölln (z7). The seventh cycle, the Mölln, was discoverd by Best (1988, 1989), who demonstrated its existence in the evaporite facies of the southern North Sea through borehole logging.

Not all of the cycles were deposited throughout the entire basin, however. Cycles were named after the areas where they were found in their most complete form.

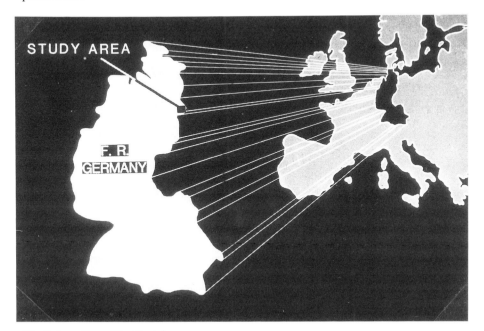

Fig. 1. Location of the Gorleben area

Cycles z1 to z3 contain of potassium magnesium salts (potash seams). The Staßfurt cycle (z2) corresponds most closely to an ideal evaporative cycle. Cycles z5, z6, and z7 exist only in very incomplete form.

This chapter deals with a conspicuous anhydrite layer of unusual mineral content (sylvitic kieserite-anhydrite) from the *Anhydritmittelsalz*, a salt sequence of the z3 cycle characterized by serveral anhydrite strata (the so-called *Anhydritmittel*).

The Gorleben salt dome is located in eastern Lower Saxony, near the Elbe, about 160 km northeast of Hannover (Fig. 1), extending 15 km NE-SW, with a width of up to 4 km. The salt dome has been investigated by borehole samples since 1979. These studies are intended to ascertain the adequacy of the dome as a repository mine for radioactive waste (Jaritz 1983).

Drilling exhibited strata of Zechstein cycles z2 to z4. On the basis of numerous cores, Bornemann (1982) set up a detailed stratigraphic succession. In general, these data agreed with the known stratigraphic Zechstein profile for the Hannover area. However, as demonstrated below, detailed comparison of the successions has yielded numerous differences. The *Anhydritmittelsalz* (Leine cycle z3) serves as an example.

2 The *Anhydritmittelsalz* in the Hannover and Gorleben Areas

The sequence of the *Anhydritmittelsalz* occurs in the entire northwest German Zechstein basin with occasional extension beyond, and consists of pure rock salt as well as rock salt contaminated with clay and anhydrite. The salt contains intercalations of anhydrite (*Anhydritmittel*) of varying thickness at varying intervals.

Schwerdtner (1959) and Hofrichter (1960) studied the *Anhydritmittelsalz* (Leine z3) in the area around Hannover, which forms a part of the northwest German Zechstein basin. According to Hofrichter (1960), the *Anhydritmittelsalz* is 30–35 m thick containing eight (0.01 to 1.5 m) anhydrite layers (*Anhydritmittel*). Its lower boundary coincides with the base of the first *Anhydritmittel*. The top of the eighth forms the upper boundary to the following layer, the *Schwadensalz*, a salt sequence within the Leine (z3), characterized by an alteration of pure and impure salt; the anhydrite and clay impurities give the salt a smokescreen-like appearance.

The two thickest *Anhydritmittel*, over 1 m thick each, are the fourth and the sixth. The first and second may locally contain kieserite. Most of the anhydrite beds start with a clay layer. Locally, in the rock salt immediately below the eighth, a 1-m-thick sylvinite seam occurs, specified as *Mittelflöz* by Herde (1953).

Bornemann (1982); Bornemann and Liedtke (1983); Bornemann and Fischbeck (1987) and Bisping (1984) dealt with the stratigraphy of the *Anhydritmittelsalz* in the Gorleben dome, which according to these authors has a thickness of 60 m there, and consists of alternating layers of red to brown-orange, fine to coarse crystalline rock salt, and nine 0.02 to 2.5 m thick anhydrite layers (*Anhydritmittel*). In the Gorleben salt dome the lower boundary of the *Anhydritmittelsalz* is identical to the base of the first *Anhydritmittel*. Contrary to the Hannover area, the upper boundary is not identical to the top of the last *Anhydritmittel*, which up to the contact with the *Schwadensalz* is overlain by a 2.5-m-thick orange to brown-orange, sulfate-rich rock salt layer. Only the seventh and ninth *Anhydritmittel* are more than 1 m thick, while all others are less than 0.5 m, with the second to fourth forming a characteristic triad.

In both the Hannover and Gorleben areas the *Anhydritmittelsalz* consists of a sequence of alternating rock salt layers and anhydrite beds (*Anhydritmittel*). Despite the basic similarity of the *Anhydritmittelsalz* in the two areas, correlations of individual *Anhydritmittel* are not possible. There are indications for a correlation of the eighth *Anhydritmittel* of the Hannover area with the ninth of the Gorleben area though both represent the final *Anhydritmittel* of the *Anhydritmittelsalz* in the respective areas, and show association with a conspicuously large amount of potassium magnesium minerals. In the Hannover area, locally a sylvinite seam underlies the *Anhydritmittel*, while in the Gorleben area a sylvitic kieserite-anhydrite is found. Adjacent rock salt layers also contain an elevated amount of potassium magnesium minerals.

3 Petrographic Studies of the Ninth *Anhydritmittel* in the Gorleben Salt Dome

Due to its thickness of 1.8 m and its high content of kieserite and sylvite, the ninth *Anhydritmittel* in the Gorleben salt dome represents a particularly conspicuous and interesting key horizon. It was encountered in the boreholes Go 1004, Go 1005, and GoHy 3130. In Go 1004 it was discovered three times. The layer was first penetrated at a depth between 867.5 and 871.0 m where it has undergone considerable tectonic deformation with parts actually missing. It is crossed by fractures up to 10 cm wide filled with fibrous carnallite and halite. Deeper in this borehole at a depth between 1463.8 to 1467.4 m the *Anhydritmittel* is complete. The angle of dip is steep (up to 75°).

The layer was then found a third time at a depth between 1531.4 and 1533.2 m. Again it is complete with a smooth dip of 20° and no tectonic deformation can be observed.

The ninth *Anhydritmittel* was penetrated initially by Go 1005 where, except for a 10-cm-thick basal layer, an almost complete alteration to fine-grained polyhalite was found. Individual layers are still recognizable, however, their outlines are blurred. Polyhalitization represents an additional local alteration, which has, as yet, been observed in the Gorleben area in this boring only.

In borehole GoHy 3130 the ninth *Anhydritmittel* ca. 29 m below the top of the salt dome was discovered. Crushed by drilling only small core sections are well enough preserved to enable recognition of the characteristic bedding.

The *Anhydritmittel* studied consists of a sylvitic kieserite-anhydrite, and can be subdivided into five layers, which are not separated by distinct bedding planes but show a transition zone of 1–2 cm thickness. The lowermost part is formed by compact anhydrite, grading upwards into alternating zones consisting largely of either anhydrite or kieserite. The upper part is formed by alternating beds consisting largely of either sylvite or anhydrite and kieserite. These typical lithological varieties were found in all boreholes of the Gorleben salt dome.

The individual layers are described from bottom to top as follows:

Layer 1: thickness 10 cm.
Anhydritic rock, gray, fine-grained, crossed by hair-like, dark lines (clay carbonate material), with anhydrite as the main mineral with lath-shaped crystals showing a mutually interpenetrating, irregular, radial texture. The interstices between the anhydrite crystals are filled with the following minerals (in order of decreasing frequency): kieserite, halite, sylvite, and carbonate.

Layer 2: thickness 25 cm.
Kieserite-anhydrite rock, grayish white with red patches (irregularly scattered brick-red sylvite flakes), fine-grained. The main component is anhydrite; kieserite and sylvite are minor constituents. Carbonate, present in the bottom of the layer, decreases in concentration to the top. Small amounts of fine-grained, acicular polyhalite occur as a new mineral.

Layer 3: thickness 45 cm.
Kieserite-anhydrite rock, light gray with red patches, fine-grained. The red sylvite flakes with a grain size between 1 mm and 1 cm are irregularly distributed. The following minerals are found in order of decreasing frequency: anhydrite, kieserite, sylvite, polyhalite, small amounts of halite and carnallite. Thus, the mineral content differs only slightly from that of the layer 2. There is, however, a distinct difference in the textural arrangement of the individual minerals. Both anhydrite and kieserite are enriched in certain zones, often forming rosette-like aggregates. In the kieserite zones considerably corroded anhydrite idioblasts occur.

Layer 4: thickness 65 cm.
Kieserite-anhydrite rock, gray white with red patches, fine-grained. Kieserite and anhydrite are enriched in certain layers, alternating on a scale of a few centimeters. Both minerals often form rosette-like aggregates (Fig. 2).

Strongly corroded anhydrite idioblasts up to 1 cm in size are found in the kieserite zone, and are occasionally enriched in layers; their longitudinal axes are aligned parallel to the bedding. Anhydrite zones are cut by numerous thin fractures of about 1 mm perpendicular to the bedding, and do not continue

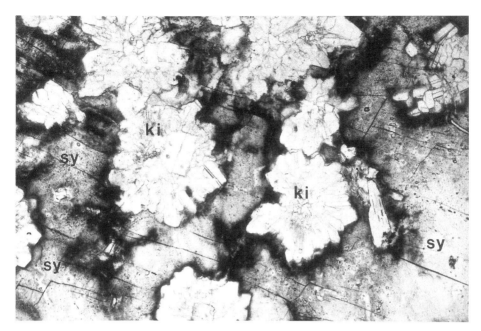

Fig. 2. Rosettes made up of kieserite surrounded by sylvite; the sylvite contains fine inclusions of hematite, which have a dark dusty appearance; *ki* kieserite; *sy* sylvite. Plane polarized light; width of photo 3 mm

into the neighboring kieserite zones. They are filled by colorless halite or sylvite.

At the contact between anhydrite and kieserite, the layer shows horizontal ruptures (1 cm) filled by sylvite and halite. On the bottom of these cavity fillings, the sylvite is particularly rich in hematite inclusions.

Locally, anhydrite is replaced by polyhalite. Very small inclusions of carnallite occur in the halite and sylvite.

Layer 5: thickness 35 cm.
Kieserite-anhydrite rock, gray white, certain layers brick-red, fine-grained.

The gray white kieserite anhydrite layers alternate on a centimeter scale with brick-red layers containing sylvite. In the gray white layers, kieserite grains form an interlocking mosaic, in which numerous anhydrite idioblasts are irregularly embedded. Both anhydrite and kieserite show replacement by fine acicular polyhalite (Fig. 3). Occasionally, anhydrite crystals display a mutally interpenetrating texture, and are mostly strongly corroded. Isolated anhydrite idioblasts also occur in the sylvite layers.

Halite
The kieserite-anhydrite rocks of layers 1–5 contain only very small amounts of halite. Halite is mostly found filling either fine vertical fissures, or, together with sylvite, horizontal fractures.

Fig. 3. Porphyroblastic anhydrite partially replaced by polyhalite. The thin section also contains rounded grains of kieserite; *an* anhydrite; *po* polyhalite; *ki* kieserite. Crossed polarizers; width of photo 3 mm

Sylvite
The sylvite is always brick-red in color caused by very fine hematite inclusions. The sylvite content increases from bottom to top, from layer 1 to 5. Sylvite fills the interstices between the anhydrite and the kieserite crystals, and also fills horizontal fractures up to 1 cm in width. Locally, sylvite has probably undergone recrystallization, or must be assigned to a second sylvite generation. It contains numerous, partly acicular, partly rounded birefringent inclusions. Locally, sylvite occurs free of inclusion, except for hematite, as rims around anhydrite idioblasts (Fig. 4).

Anhydrite
Often the lath-shaped anhydrite crystals show a compact texture. Sometimes they are mutually interpenetrating, or display a rosette-like arrangement. A second anhydrite generation, consisting of idioblasts 1 mm to 1 cm in size, shows peripheral alteration to polyhalite. In the contact area between anhydrite and kieserite, rather coarse anhydrite crystals commonly occur; their longitudinal axes are aligned perpendicularly to the anhydrite kieserite boundary.

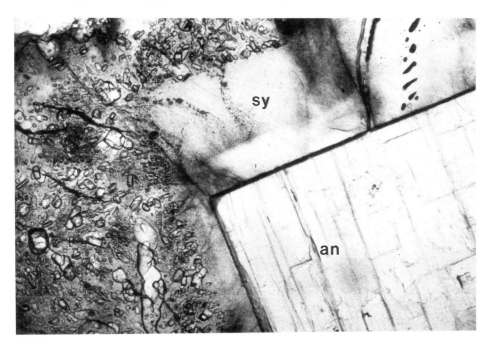

Fig. 4. Idioblastic anhydrite surrounded by sylvite; the sylvite forms an inclusion-free rim around the anhydrite; *an* anhydrite; *sy* sylvite. Partly crossed polarizers; width of photo 1 mm

Kieserite
Kieserite forms typical, rounded grains, some of which are twinned. In layers 1 and 2, it occurs in the interstices between the anhydrite crystals. In layers 3 to 5, kieserite forms discrete patches in which the individual crystals sometimes show a rosette-like arrangement.

Polyhalite
Polyhalite forms clusters of fine needles replacing anhydrite and occasionally kieserite. Small amounts of polyhalite first appear in layer 2 with the amount increasing in each successive layer.

Carnallite
Carnallite occurs in traces only in layers 2 to 4.

4 Discussion

The minerals encountered in the ninth *Anhydritmittel* are of diagenetic origin, and were formed during multiphase alterations by the solutions migrating through the salt dome. On the basis of both macroscopic and microscopic analyses, the mineralogy and complicated textures of the rocks led to the idea of the possible stages of alteration. The compositions of both solutions involved and of the initial rock remain unknown. Possibly, the ninth *Anhydritmittel* was originally an anhydritic rock with a considerable content of readily soluble salt minerals, for example, alternating layers of anhydrite and halite with sylvite. The successive stages of alteration are thought to have been as follows:

1st stage: The readily soluble salts halite and sylvite were almost completely dissolved, as was also possibly part of the anhydrite. Subsequently, anhydrite idioblasts were formed, and the sylvite partially recrystallized, forming pure clear rims around the anhydrite idioblasts.
2nd stage: Again major parts of the rock were dissolved. The strongly corroded anhydrite idioblasts represent relics of this solution stage, which preceded the formation of the kieserite surrounding the idioblasts.
3rd stage: Anhydrite, and locally, kieserite as well, were transformed to fine acicular polyhalite.

The sequence of transformation stages as described above resulted in the individual layers of the ninth *Anhydritmittel* displaying the same mineralogical and petrographical composition throughout the area covered by the boreholes. It can therefore be assumed that the transformation of the ninth *Anhydritmittel* happened at the same time throughout the Gorleben area.

The period of the transformation can be roughly estimated: alteration was only possible after consolidation of the *Anhydritmittel*, proven by the filled fractures representing the solution migration channels. Transformation, however, must have been completed before the commencement of diapirism and the folding of these layers. Otherwise, the alteration would not have been stratabound, nor would it be traceable over considerable distances along with the bedding.

Additionally, the ninth *Anhydritmittel* in borehole Go 1005 was almost completely altered to polyhalite. Nevertheless, the five characteristic layers can still be recognized by their blurred outlines. This transformation took place at a later time, and is only known from this locality. In this particular case, due to the uplift of the salt, the ninth *Anhydritmittel* was pushed into a position extremely close to the edge of the salt dome. In its present position, the distance between the polyhalitized *Anhydritmittel* and the edge of the salt dome is only about 120 m vertically and only about 30–50 m horizontally. It can therefore be assumed that this polyhalitization took place over a rather long period of geologic time following the formation of the salt dome, and was probably caused by interstitial water stemming from rocks adjacent to the salt dome.

References

Best G (1988) Die Grenze Zechstein/Buntsandstein in Nordwestdeutschland und in der südlichen deutschen Nordsee nach Bohrlochmessungen (Gamma-Ray und Sonic-Log). Geol Jahrb Hessen 116:19–22

Best G (1989) Die Grenze Zechstein/Buntsandstein in Nordwest-Deutschland nach Bohrlochmessungen. Z Dtsch Geol Ges 140:73–85

Bisping D (1984) Stratigraphie und Fazies der jüngeren Ronnenberg-Gruppe und der Riedel-Gruppe (Zechstein 3) im Salzstock Gorleben. Thesis, Univ Münster, 133 p

Bornemann O (1982) Stratigraphie und Tektonik des Zechsteins im Salzstock Gorleben auf Grund von Bohrergebnissen. Z Dtsch Geol Ges 133:119–134

Bornemann O, Fischbeck R (1987) Zechstein 2–4 des Salzstocks Gorleben. Exkursionsführer I, Int Symp Zechstein 87:145–160

Bornemann O, Liedtke L (1983) Ingenieurgeologische Arbeiten zur Planung des Endlagers für radioaktive Abfälle im Salzstock Gorleben. Ber Nat Tag Ing Geol, pp 195–211

Herde W (1953) Die Riedel-Gruppe im zentralen Teil des nordwestdeutschen Zechsteingebietes (Stratigraphie, Genese und Paläogeographie). Thesis, Univ Göttingen, 127 pp

Hofrichter E (1960) Zur Stratigraphie, Fazies und Genese der Ronnenberg-Gruppe und Anhydritmittel-Zone (Zechstein 3) in Nordwestdeutschland. Thesis, Univ Kiel, 133 pp

Jaritz W (1983) Das Konzept der Erkundung des Salzstocks Gorleben von übertage und die Festlegung von Schachtstandorten. N Jahrb Geol Paläontol Abh 166:19–33

Lotze F (1957) Steinsalze und Kalisalze, Borntraeger, Berlin, xii + pt 1, 2nd edn. 466 pp

Schwerdtner W (1959) Zur Stratigraphie und Tektonik der Staßfurt- und Leine-Serie im Salzhorst Höfer. Dipl Arb, FU Berlin, 81 pp

Lacustrine Paper Shales in the Permocarboniferous Saar-Nahe Basin (West Germany) - Depositional Environment and Chemical Characterization

ANDREAS SCHÄFER[1], ULLRICH RAST[2], and ROGER STAMM[1]

CONTENTS

Abstract	220
1 Introduction	221
2 Geological Setting of the Saar-Nahe Basin	221
3 Lake Environment	223
4 Paper Shales	225
5 Model of the Saar-Nahe Basin Paper Shales	235
References	237

Abstract

Lacustrine paper shales in Carboniferous and Permian rocks of the Saar-Nahe Basin, West Germany, are described in terms of their sedimentology, petrography and geochemistry.

The lakes formed mostly on flood plains associated with fluvial environments. They were exposed to a tropical climate. Thus, the bedding of the lacustrine shales shows an even lamination from layers of light silt to dark clay. The graded silts are correlated with the suspensions swept into the lakes by heavy rainfalls which are frequent in tropical summers. The clays are rich in particulated organic matter and most often associated with calcite individuals, so that biogenic carbonate precipitation by blooms of floating algae has occurred. The suspension input to the lakes was low during tropical winters. The lakes were shallow and eutrophic, lacked oxygen in the pelagic zone, and had no benthos.

Calcite quite often was converted to ankerite and siderite in the reducing conditions of the sediment. Thus, the organic matter is well preserved.

The chemistry of the sediments, characterized by major, trace and rare earth elements, is quite distinctive for each of the three types of paper shales, the siliciclastic ones, those containing carbonate and the pure carbonates.

---

[1] Geologisches Institut Universität Bonn, Nußallee 8, 5300 Bonn, FRG.
[2] Bayerisches Geologisches Landesamt, Heßstr. 128, 8000 München 40, FRG.

1 Introduction

Paper shales are thinly and evenly laminated shales originating from deposition in calm, undisturbed waters. Most often they contain a high amount of particulated organic matter and organic compounds, thus enhancing the concentration of trace and minor metals. They are known from modern and ancient marine and lacustrine environments and have potential as a petroleum source rock (Zimmerle 1985).

Lacustrine paper shales from the Permocarboniferous of the contential Saar-Nahe Basin have been previously described by Stapf (1970), Rast (1975), Boy (1977), Schäfer and Stapf (1978), Schäfer and Sneh (1983), Schäfer (1986), Stapf (1989), and Schäfer and Stamm (1989).

In this chapter a depositional model will be presented to explain the bedding characteristics, the presence of carbonates and the inorganic chemical constituents. The Saar-Nahe lacustrine paper shales have much lower organic content than true oil shales such as those known from major lacustrine deposits outside Europe, e.g. those from the Eocene Green River Basin, USA (Taylor 1987) or those from the Cenozoic Mae Sot Basin, Thailand (Gibling et al. 1985). Also, the Saar-Nahe Basin paper shales are by no means as rich in organic matter as the shales from the Eocene Lake Messel (Kubanek et al. 1988; Schaal and Ziegler 1988). According to their organic content, they are more in the range of the Eocene Nördlinger Ries Crater Lake (Jankowski 1981) or the Oligo-Miocene Lake Rott (von Königswald 1989).

The above examples vary widely in size and stratigraphic range, but they all represent a unique lake sedimentation which has always been very significant in the fossil record. Despite the fact that lake sediments are of only minor significance in the total volume of sedimentary rocks, their extraordinary use for stratigraphic correlations, for studies in environmental reconstructions and their potential as petroleum source rocks justify particular attention to their geology.

2 Geological Setting of the Saar-Nahe Basin

The Saar-Nahe Basin (Fig. 1) is a continental, intermontane sedimentary basin containing Variscan molasse deposits at least 7.5 km thick (Schäfer 1986, 1989). The basin has an area of about 300 x 100 km, but is mostly covered by Mesozoic and Tertiary rocks so that only one-third of its surface area is exposed immediately south of the Schiefergebirge (Fig. 2).

The basin fill consists of a fluvio-lacustrine grey facies and coal beds in the Namurian, Westphalian, Stephanian and Lower Rotliegend (whereby the coals have major thicknesses in the Westphalian and are intensely mined), and an alluvial red facies together with volcanic rocks in the Upper Rotliegend. Minor tectonic deformations produced a discordance between the

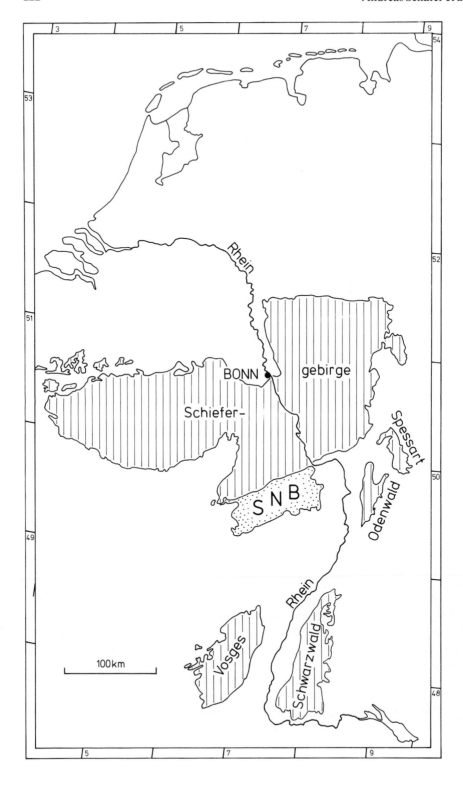

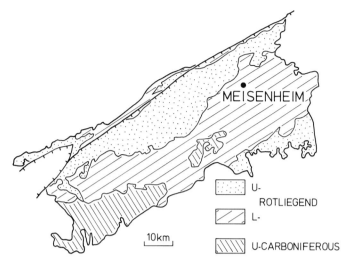

Fig. 2. Generalized stratigraphy of the Saar-Nahe Basin, also showing the locality of the well Meisenheim 1 which provided a number of insights into the sedimentology of the Permocarboniferous basin fill

Westphalian and Stephanian, and a gentle folding of all beds occurred by the end of Upper Rotliegend.

In the Upper Carboniferous and the Lower Rotliegend the basin had a tropical climate, whereas in the Upper Rotliegend arid conditions prevailed (Schäfer and Stamm 1989).

The subsurface geology of the Saar-Nahe Basin is well known because of a large number of wells from both the coal and oil industry (Schäfer 1986). One of these is the oil well Meisenheim 1, the core of which has been a major source of information for stratigraphy, sedimentology and petrography; the paper shales discussed below were sampled from this well (Fig. 3).

3 Lake Environment

Especially during Stephano-Autunian time, a large number of lakes at different times existed in the Saar-Nahe Basin (Stapf 1989; Schäfer and Stamm 1989). Extended lakes covered at least the presently exposed surface of the Saar-Nahe Basin (Stapf 1970); oxbow lakes and drowned flood basins were associated with meandering rivers (Schäfer 1986). The lake beds were of variable size and most existed only for short periods. Repeated fluvial processes filled them. Sands and silts formed lake deltas of variable sizes and shapes,

Fig. 1. Surface exposures of the Saar-Nahe Basin (*SNB*). The basin is one of the major internal molasse basins of the Central European Variscides and its true size is only known from the subsurface

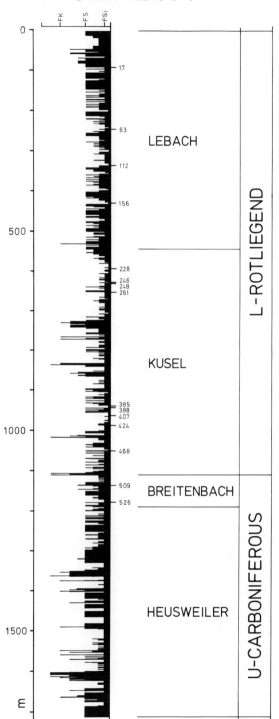

Fig. 3. Computer-generated lithology log of the cored well Meisenheim 1 given in full detail (grain size indicated by: *FK* fine-grained conglomerate; *FS* fine-grained sand; *FSi* fine-grained silt). The 15 paper shale samples studied here were taken from lacustrine beds of different stratigraphic levels

which quite often were only minor crevasse-splay deltas and not large-scaled systems (Rast and Schäfer 1978); Gilbert-type deltas were not observed. The deltaic sequences eventually reached a thickness of about 10 m, which is assumed to be the approximate depth of the former lakes as well; often the sediments deposited in the subsequent fluvial environments were much thicker. This fluvio-deltaic sedimentation style has been discussed using numerous well profiles from the oil and coal industry (Schäfer 1986).

According to their extent and their depth, the lakes should have been morphometrically eutrophic (Schäfer and Stamm 1989). The paper shales were formed in protected, probably deeper areas of the lakes and hence are well bedded, whereas along shorelines or on marginal mud flats more massive mud rocks with carbonate nodules, mud cracks, plant rootlets, onkoids and stromatolites were formed (Schäfer and Stapf 1978).

4 Paper Shales

4.1 Bedding

The bedding of the lacustrine sedimentary rocks is highly significant (Fig. 4) because it shows lamination couplets each consisting of a light silty layer of about, on average, 1 mm and a dark clayey layer of about 0.1 mm thickness. The variation of the average thicknesses of light/dark couplets in four samples is 0.35/0.04, 3.48/0.09, 1.19/0.20, and 1.04/0.18 mm, each representing the mean of several measurements. Compared to the silty layers, the clayey layers have a relatively constant thickness. Thus, the thicknesses of the light layers vary, the thicknesses of the dark layers are relatively constant and vary only in a much smaller range.

The light layers are graded, fining upward, and show sharp erosive bases to the dark layers below. Together they represent one sedimentary couplet (Rast 1975). The siliciclastic silts often contain randomly distributed coarser clasts.

The dark layers contain clay minerals and particulated organic matter which provides the dark color. The organic matter stems from residues of predominantly algae and plant remains, as fluorescence microscopy indicates. In addition, the dark layers contain fish and other vertebrates, as fragments or rarely as complete skeletons (Boy and Fichter 1988). The dark layers also contain calcite individuals, which sometimes are as large as the thickness of the dark layers.

The sequence of the lamination is not regular: Several light layers of subsequent couplets are thicker, several are thinner; the transition between either extremes is gradual. The variable thickness of the light layers was due to variable suspension input to the lakes (Sturm and Matter 1978).

As the bedding of the paper shales is well preserved and not disturbed by any benthos, a lack of oxygen at least in the deeper parts of the lakes is most

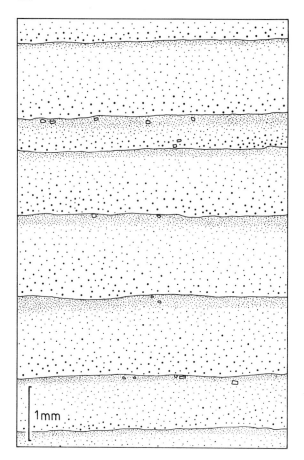

Fig. 4. Tracing of a photograph of a thin section of a paper shale (sample 219 of the transition of Kusel to Lebach Group) showing the principle of lacustrine deep-water suspension sedimentation, where graded silts erode carbonaceous clays to provide a characteristic light/dark lamination

probable. The organic richness of the dark layers was dependent on the biological productivity of the photic zone of the water mass above, whose organic remains were not or not fully oxidized after sedimentation (Zangerl and Richardson 1963).

4.2 Petrography

The paper shales contain fragments of quartz, feldspar and carbonate (Fig. 5). Clay minerals are dominantly illite; a small portion of the clay fraction is chlorite and kaolinite; smectite is rare (Schäfer 1986; Schäfer and Stamm 1989).

Analcite was detected in several horizons of the well profile; it is a good indicator of mineral formation from volcanic ash (Zimmerle 1985; Kubanek et al. 1988).

The paper shales also contain three types of carbonate, each with variable abundance and composition (Table 1). The carbonates occur as calcite

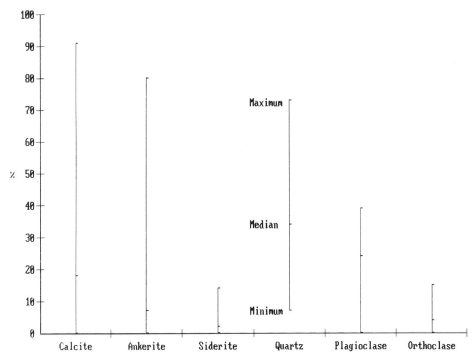

Fig. 5. Minerals in the selected 15 paper shale samples. Mineral identification and their relative quantities are determined by X-ray diffractometry. The minima and maxima of the data together with their median are shown

Table 1. X-ray diffraction analysis of 65 samples from the depth range 160–670 m of the well Meisenheim 1 (within the Upper Kusel and Lower Lebach Group) showing the relative abundance of the three carbonates; the 10% hydrochloric acid-soluble portion could roughly be estimated as the content of "total carbonate" (Schäfer and Stamm 1989)

	Minimum	Maximum	Mean	
Soluble	6.01	71.89	21.08	wt %
Calcite	0	57	16	rel %
Siderite	0	66	7	rel %
Ankerite	0	100	12	rel %

preferably in the dark layers (showing sharp reflection patterns in the X-ray diffractograms) or as siderite and ankerite in the fine-grained groundmass of the whole rock. Siderite and ankerite (both showing broad reflection patterns in X-ray diffractometry) occur as secondary carbonates formed under reducing conditions within the sediment (Matsumoto and Iijima 1981; Gibling et al. 1985).

4.3 Chemistry

Sixty-three paper shale samples from the well Meisenheim 1 (see Table 1) were analyzed for major elements related to carbonate formation (Table 2).

The data reflect possible carbonates which are abundant in the paper shales (Schäfer and Stamm 1989): The variation of Ca is to due to the variable content of calcite, whereby Fe_2 and Mn are related to the occurrence of siderite and only Mg and Ca to ankerite; Fe_3 is of no major importance in these reducing sediments. The relatively low average value of Sr and the major enrichment of it in ankerite seem to confirm calcite as the primary carbonate phase. Based on the content of total organic carbon (C_{org}), the paper shales may be assigned to the "grey shales" category defined by Odom (in Potter et al. 1980, p. 199). Their low amount is argued to be due to loss during the geothermal history of the Saar-Nahe Basin (as the data of Teichmüller et al. 1983 might suggest).

In addition, 15 paper shale samples from the well Meisenheim 1 (Fig. 3) were analyzed for their chemistry by neutron activation (analytical procedure in Wänke et al. 1973 and Rast 1981; Table 3). The data pertaining to the major elements Mg and P obtained by atomic absorption and photometry, the data of C_{org} by combustion and the mineral data by X-ray diffractometry are from Schäfer and Stamm (1989).

The range and the average contents of the minerals in those 15 samples are given in Fig. 5. Major elements, calculated as oxides together with the acid-soluble residue (R), and the amount of total organic carbon (C_{org}) are presented as a concentration profile in Fig. 6. Trace elements (Fig. 7) and the chondrite normalized rare earth elements (Fig. 8) are given as concentration profiles as well. All three concentration profiles show a grouping, according to their content in CaO, into three varieties. In addition, a correlation matrix was calculated for all minerals and elements (Fig. 9).

The samples investigated have been grouped into "siliciclastic paper shales" with CaO = 1–2%, "paper shales containing carbonate" with CaO

Table 2. Some carbonate-related major and minor elements of 63 samples from the depth range 160–670 m of the well Meisenheim 1 (right column: elements of "marine shale" from Turekian and Wedepohl 1961; C_{org} from Curtis 1980)

	Minimum		Maximum	Mean	"Marine shale"
Fe_3	0.182	–	3.889	1.765%	4.720% (Total Fe)
Fe_2	1.671	–	17.435	3.464%	
Mg	0.110	–	0.990	0.258%	1.500%
Ca	0.060	–	12.650	1.597%	2.210%
Mn	0.033	–	0.635	0.089%	0.085%
Sr	3	–	1767	399 ppm	300 ppm
C_{org}	0.15	–	3.90	1.37%	3.2%

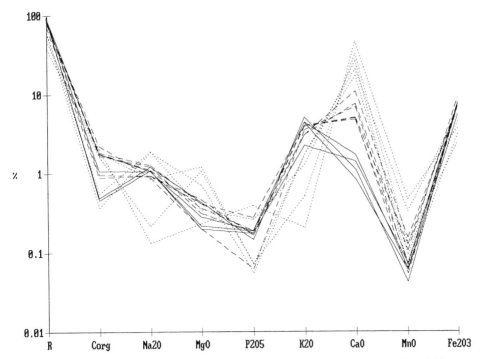

Fig. 6. A concentration profile for the major elements (as oxides) of the selected 15 paper shale samples together with their total organic carbon (C_{org}) and 10% hydrochloric acid residue (R in wt%). The data are grouped by the amount of 1–2% CaO (*solid lines*), 4–11% CaO (*dashed lines*) and >16% CaO (*dotted lines*)

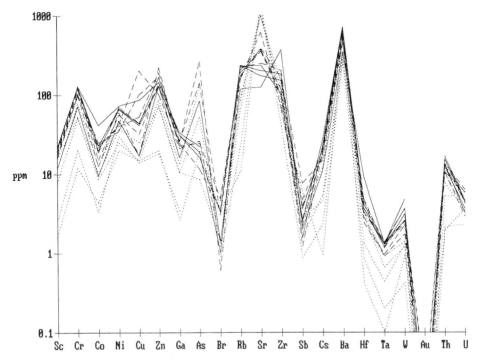

Fig. 7. A concentration profile for the trace elements of the selected 15 paper shale samples; grouping as given in Fig. 6

Table 3. Complete list of minerals, main, trace and rare elements analyzed from 15 paper $P_2O_5 = \%$; Cl to U = ppm)

Sample	17	63	112	156	228	246	248
Depth	95.00	246.20	336.40	429.10	594.30	626.80	628.30
Calcite	18.	7.	13.	0.	0.	36.	35.
Ankerite	0.	0.	9.	0.	80.	15.	56.
Siderite	5.	14.	2.	13.	0.	2.	0.
Quartz	37.	41.	35.	58.	10.	21.	7.
Plagioclase	33.	26.	31.	29.	6.	21.	2.
Orthoclase	7.	12.	7.	0.	2.	5.	0.
Analcite	0.	0.	3.	0.	2.	0.	0.
Pyrite	0	0	1	1	1	1	0
Residue	83.690	90.010	82.210	90.560	80.210	72.120	47.910
CaO	5.036	1.707	4.729	0.839	20.467	10.870	32.037
MgO	0.315	0.282	0.414	0.216	1.210	0.464	1.061
Na_2O	1.221	1.057	1.275	1.073	0.848	0.933	0.216
K_2O	3.977	4.374	4.007	4.338	1.711	3.025	0.517
MnO	0.124	0.062	0.073	0.059	0.329	0.153	0.128
Fe_2O_3	7.336	6.950	6.850	5.348	5.691	6.735	2.474
C_{org}	1.690	1.070	1.830	1.770	1.820	0.890	1.630
P_2O_5	0.175	0.188	0.275	0.186	0.071	0.165	0.072
Cl	300.000			200.000		610.000	
Sc	19.300	22.000	18.500	21.200	15.700	16.100	2.550
Cr	119.000	127.000	116.000	128.000	53.000	100.000	20.500
Co	15.200	19.800	22.800	22.900	10.100	21.600	3.320
Ni	56.000	65.000	40.000	67.000	37.000	42.000	30.000
Cu	32.000	41.000	52.000	43.000	25.000	200.000	14.000
Zn	143.000	126.000	190.000	166.000	70.000	101.000	18.300
Ga	24.400	30.900	26.000	30.500	10.400	20.900	3.420
As	19.700	16.300	141.000	11.400	19.000	26.300	
Se	0.800			1.000	0.500	0.100	
Br	3.000	3.300	3.000	3.500	1.000	1.370	0.892
Rb	220.000	226.000	206.000	234.000	91.000	155.000	23.800
Sr	247.000	234.000	373.000	200.000	1060.000	611.000	1300.000
Zr	200.000	100.000	100.000	180.000	80.000	67.000	40.000
Sb	2.190	1.560	4.670	2.490	2.610	2.740	0.873
Cs	18.800	22.200	19.900	26.600	12.000	18.500	2.310
Ba	590.000	720.000	630.000	720.000	530.000	564.000	263.000
La	41.200	44.200	40.400	44.700	19.900	32.600	
Ce	96.000	89.500	86.400	104.000	44.500	70.600	
Nd	45.000	36.000	35.000	50.000	20.000	29.800	6.040
Sm	8.600	7.910	6.500	8.140	4.670	6.470	1.400
Eu	1.580	1.470	1.320	1.530	0.891	1.330	0.262
Tb	1.120	0.925	0.883	0.991	0.680	0.794	0.183
Dy	5.750	4.840	4.760	5.560	3.400	4.510	0.790
Yb	3.400	2.920	2.440	3.370	2.020	2.320	0.654
Lu	0.490	0.440	0.365	0.488	0.320	0.248	0.083
Hf	4.520	3.630	3.310	4.190	1.320	2.440	0.850
Ta	1.360	1.290	1.330	1.300	0.440	0.963	0.200
W	3.600	3.650	2.340	4.800	1.300	2.270	0.423
Au	0.005	0.006	0.020	0.006		0.002	0.001
Hg	0.100		0.200	0.100			0.120
Th	15.200	15.300	12.700	16.600	7.580	10.400	1.950
U	4.200	4.430	5.760	4.600	3.340	5.860	3.570

Lacustrine Paper Shales in the Permocarboniferous Saar-Nahe Basin

shales of the well Meisenheim 1 (minerals in rel %; pyrite: *1* present, *0* absent; soluble to

261	385	388	407	424	468	509	526
653.90	940.00	943.80	965.20	988.80	1051.60	1138.80	1179.70
20.	12.	34.	22.	5.	23.	3.	91.
6.	22.	11.	28.	7.	0.	0.	0.
0.	0.	0.	0.	2.	5.	5.	0.
46.	34.	28.	24.	32.	37.	73.	9.
24.	24.	27.	24.	39.	30.	19.	0.
4.	8.	0.	2.	15.	5.	0.	0.
0.	0.	0.	0.	0.	0.	0.	0.
1	1	1	0	0	0	0	1
81.900	85.350	60.670	59.250	86.800	78.730	88.840	37.590
7.443	4.924	26.021	16.928	1.119	6.855	1.399	45.887
0.365	0.414	0.348	0.713	0.431	0.199	0.199	0.232
1.100	0.925	1.887	1.860	1.206	0.888	1.091	0.132
3.038	3.856	1.277	1.723	5.025	3.472	2.229	0.205
0.105	0.053	0.065	0.093	0.042	0.069	0.072	0.466
8.895	6.850	3.146	4.519	7.579	7.808	7.279	3.761
1.780	0.960	0.560	0.370	0.490	2.230	0.460	2.020
0.188	0.175	0.260	0.055	0.145	0.062	0.175	0.390
	100.000	1000.000				300.000	
17.400	18.400	7.820	11.700	20.600	16.200	11.900	1.670
97.800	105.000	51.400	78.000	129.000	100.000	71.800	12.000
25.000	18.800	8.490	12.800	41.200	12.500	9.380	4.320
35.000	70.000	25.000	48.000	72.000	60.000	47.000	20.000
100.000	44.000	15.000	17.000	85.000	17.000	17.000	15.000
170.000	130.000	74.000	87.000	145.000	223.000	130.000	20.000
22.000	25.200	10.300	13.400	32.900	22.400	16.200	2.610
264.000	52.800	24.500	22.200	83.800	19.700	8.500	132.000
1.500	0.500	0.500	0.500	1.800	0.200	0.800	0.400
4.500	0.600	1.210		1.400	4.840	1.400	3.400
168.000	190.000	66.900	84.300	230.000	174.000	118.000	10.900
383.000	357.000	992.000	961.000	172.000	353.000	125.000	1420.000
160.000	116.000	110.000	140.000	150.000	95.000	370.000	60.000
7.620	3.930	2.720	1.570	3.610	1.220	2.010	6.170
14.900	17.200	4.530	5.320	20.800	18.000	10.900	0.948
450.000	510.000	280.000	318.000	673.000	480.000	350.000	650.000
40.000	35.800	20.900	27.400	43.300	39.900	40.400	8.110
91.000	71.700	46.100	63.500	91.500	85.600	88.800	19.500
35.000	30.700	18.000	30.000	42.000	38.800	30.000	11.500
8.320	6.380	4.450	4.700	5.780	8.920	6.440	2.150
1.610	1.290	0.923	0.980	1.100	1.600	1.180	0.518
1.160	0.794	0.648	0.720	0.679	1.060	1.020	0.325
6.270	4.590	3.770	4.000	3.670	5.440	5.700	1.940
2.790	2.370	1.770	1.620	2.210	2.670	3.200	0.899
0.494	0.319	0.280	0.312	0.380	0.414	0.580	0.129
4.830	2.850	2.870	3.200	3.700	4.190	9.240	0.423
1.270	0.901	0.665	0.921	1.200	1.160	1.350	0.100
2.520	1.750	1.200	1.400	3.100	2.600	2.550	0.940
0.024	0.006	0.001	0.003	0.008	0.007	0.004	0.004
0.250	0.337						0.270
13.300	10.400	5.210	10.300	13.100	12.800	12.200	2.130
6.220	2.890	5.310	3.270	3.210	4.390	3.620	2.300

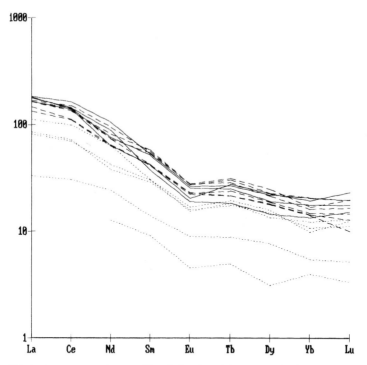

Fig. 8. A concentration profile of C1-chondrite normalized rare earth element (REE) data of the selected 15 paper shales samples; grouping as given in Fig. 6

= 4–11% and "pure carbonates" with CaO = > 16%. The three groups were marked not only by differences in their CaO content, but also by differences in the amounts of lithophile elements which are known to be concentrated in clay minerals, mica and, to a lesser extent, in feldspars: K, Na, Rb, Ga, Sc, Cs, Ba, REE, Zr, Hf, Ta, W, Th and U.

Nevertheless, some fractionation of Na and U occurs: U is enriched in carbonates and Na enters the weathering cycle. The enrichment of U by organic matter, though quite frequently observed in a variety of sedimentary environments (Rast 1981), seems to be of minor importance for the paper shales (Fig. 10): The Th/U ratio varies between 3.3 and 4.1 in siliciclastic paper shales, whereas the increasing carbonate content lowers the ratio considerably. In the paper shales containing carbonate and the pure carbonates a trend can be observed which lowers the Th/U ratio from 3.6 to 0.6, suggesting a preferred uptake of U in carbonates.

Several other elements, especially Zr, Hf and Ta which, despite their association with the silicates in general, are probably concentrated in heavy minerals. The REE (Fig. 8) show a "mature sedimentary" pattern which is typical of repeatedly reworked sediments and originates from simple mechanical mixing of various source rocks. There is no indication of enrich-

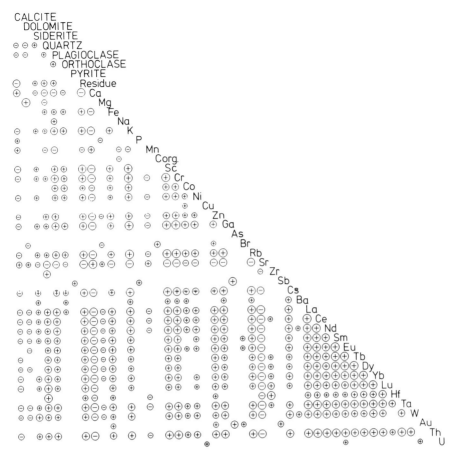

Fig. 9. Correlation matrix of minerals, major, minor and trace elements to show their possible interrelationship; limit of significance is 95%, *large* to *small circles* indicate a decreasing correlation in the range of $-1. < R < -0.53$ and $+0.53 < R < +1$

ment of heavy REE which is observable in some C_{org}.-rich sediments (Rast 1981).

Ga, Rb, Cs and Ba are closely correlated with Al-rich clay minerals. The transition metals Co and Cr are associated with silicates, although they might be enriched with chalcophile elements as well. This is due to their tendency to be absorbed by Fe-hydroxides which are nearly ubiquitous in sediments derived from the primary environment or from the weathering of reduced sediments.

Elements associated with carbonates are Sr, and to a much lesser extent Mn and U. Elements associated with sulfide phases are As, Se, Sb, Au, Hg, Zn and to a lesser extent Ni and Co. Sulfides appear to be more common in the siliciclastic paper shales than in the pure carbonates because of increased contents of the relevant elements in these rocks.

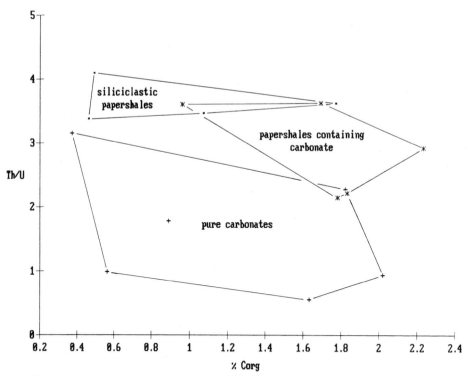

Fig. 10. Graph of the Th/U ratio versus Corg

In evaluating the data in terms of characterizing certain lithotypes, on the one hand, and the development of the depositional environment, on the other hand, attention has to be drawn to several elements or groups of elements. As mentioned earlier, the carbonate content differentiates the samples into three lithological groups: One with 1 to 2% CaO, another with 4 to 11% CaO and a third one with more than 16% CaO. The latter high value suggests that primary carbonates (calcites) formed in the environment by organic production rather than by diagenetic carbonate formation.

In the first group (low Ca-content) the chemical signature of clay minerals is quite obvious in the lithophile elements Na, K, Rb, Sr, Zr, Cs, Ba, REE, Hf, Ta, Sc, Cr, Ga and W; the data show a high consistency, with the exception of U which might have been fractionated due to its solubility and its affinity for carbonates and organic substances. However, there is one sample (No. 509) which does not fit into this scheme: The lithophiles Sc, Cr, Ga and W are depleted, while absolute amounts of Zr and Hf are increased. This is interpreted as a dilution of the clay material by quartz sand containing mainly zircon as the heavy mineral. Other elements like Fe, Mn, Co and chalcophiles like Zn, As and Sb show variations which may be attributed to differences in minor rock-forming minerals (e.g. sulfides) and organic substances which themselves reflect fluctuations in plant productivity and environmental conditions. Zn, As and Sb do not follow the variations of major rock minerals.

In the second group (medium Ca-content) the main rock mineralogy is much less consistent than in the carbonate-poor paper shales. Carbonate and organic production influences the distribution of the elements by dilution as well as by fractionation and differentiated uptake of specific elements in discrete phases while others remain constant.

In the third group (high Ca-content) a substantial depletion of all elements incompatible with the carbonate lattice is shown. This is a common feature of carbonates: all variations of the lithophile elements can be attributed to fluctuations of the silicate impurities (clay minerals) in them; the same is valid for sulfide impurities. Mn is increased due to uptake in the carbonate lattice; it can, however, also substitute for Mg in dolomite. But Mn is not enriched in all carbonate-rich samples: Mn might be introduced to the environment by some detrial component which is not homogeneously and/or continuously supplied. In principle, the non-carbonate impurities in the carbonates reflect a varying degree and abundance of clastic input to the lakes during the organic carbonate precipitation in them.

5 Model of the Saar-Nahe Basin Paper Shales

A model for the formation of paper shales in the Saar-Nahe Basin has to consider the paleolatitude and the climate of the Permocarboniferous of Central Europe (Schäfer and Stamm 1989).

According to paleomagnetic investigations in the Stephano-Autunian of the Permocarboniferous (Petersen in Schäfer and Stamm 1989), the Saar-Nahe Basin was placed about 10° north of the equator, in the tropics. From abundant caliche nodules in mud flats around the lakes it can be deduced that the lakes were in a savanna environment, rather than in tropical rain forests.

In the tropics, precipitation occurs mostly during summer. Therefore, the fluvial transports are most violent in this wet season. Sediments were transported to the lakes by rivers and flash floods, providing deltaic buildup with prograding shorelines from the coarse fraction and suspension sedimentation in the pelagic zone of the lakes from the wash load. Thus, the graded silts of the paper shales should be dominantly correlated with the summer wet season.

During winter, precipitation in the tropics and subtropics declines. The hot air temperatures remain for the dry season and provide a stable temperature bedding in the lakes. According to the model of Sturm and Matter (1978), the thermocline allows suspension transport as an undercurrent and suspension sedimentation of the fines far from the littoral zone. Additionally, a rich bioproduction together with occasional algal blooms are favoured. The same model is applicable in the Stephano-Autunian. Thus, the dark layers of the paper shales received their organic material from decaying algae and plants. As algae assimilate in the epilimnion, calcite crystallites were precipitated from the bicarbonate content of the lake water.

The lamination of the paper shales is due to the weather-dependent variation of dryness and precipitation. The dry season favours a close stacking of organic-rich layers as the sediment influx is reduced. The wet season causes a rich suspension of sediments derived from daily thunderstorms, resulting in repeated thicker silt layers with distinct graded bedding ("weather bedding", Rast 1975). An "annual bedding" is supposed to be hidden in a sequence of several couplets developing from thicker, light silt layers (summer) to thinner, light silt layers (winter), whereby the dark clay layers remain relatively constant in thickness.

Rich biomass production surely provided reducing conditions in the deeper parts of the lakes, which preserved the lamination of the deposits from benthos. The decay of the biomass, however, did not trigger gas-heave structures, which would normally destroy bedding. Only minor slump features and reworked laminae can be detected occasionally.

Calcite is the primary carbonate and was precipitated by algae (Schäfer and Stapf 1978; Stapf 1989). Pure carbonate rocks occur in several localities and may be formed on the occasion of a regional and/or time-dependent enrichment of bioproductivity and/or availability of biocarbonate in the lake water. Within the sediments, siderite and ankerite were formed (from the calcite?) and are assumed to be secondary. This was also found by Gibling et al. (1985), with the exception that siderite was absent from their samples. For the formation of siderite, the redox limit has to be at the sediment-water interface (Berner 1981). Ankerite replaces dolomite in Fe_2-rich reducing environments and obviously needs a higher Mg/Ca potential of the pore water for its formation.

The input of particulated organic matter preserved the reducing environment during diagenesis. The paper shales of the Saar-Nahe Basin, however, were quite rich in inorganic input, so that they only achieved the range of "oil shale C to D" according to the classification used by Gibling et al. (1985). In contrast, only the oil shales of Messel (Kubanek et al. 1988) might be "oil shale A" in richness. Probably, the diagenesis of the Paleozoic lake sediments reduced the former richness of organic material. Furthermore, the geothermal gradient of the Saar-Nahe Basin could have easily been about 50°C/km (Teichmüller et al. 1983) which possibly enhanced the diagenesis.

What type of environment does the chemistry of the paper shales define? The paper shales of the Saar-Nahe Basin were not formed as true black shales because of the importance of carbonate production and the lack of uniform enrichment of chalcophile elements (which is typical of such sediments). Instead, they were deposited in a water body which was at least intermittently stagnant, but showed a constant and dominant siliciclastic sedimentation. Apart from the carbonate-rich ones, most samples demonstrate a marked conformity of their data, indicating quite stable environmental conditions, with only gradual variation in carbonate production through time.

Acknowledgements. We thank Russell J. Korsch, B.M.R. Canberra/Australia, for thoroughly reading the manuscript!

References

Berner RA (1981) Authigenic mineral formation resulting from organic matter decomposition in modern sediments. Fortschr Miner 59:117–135
Boy JA (1977) Typen und Genese jungpaläozoischer Tetrapoden-Lagerstätten. Paläontographica Abt A 156:11–167
Boy JA, Fichter J (1988) Zur Stratigraphie des höheren Rotliegend im Saar-Nahe-Becken (Unter-Perm; SW-Deutschland) und seiner Korrelation mit anderen Gebieten. N Jahrb Geol Paläont Abh 176:331–394
Curtis CD (1980) Diagenetic alteration in black shales. J Geol Soc London 137:189–194
Gibling MR, Tantisukrit C, Uttamo W, Thanasuthipitak T, Haraluk M (1985) Oil shale sedimentology and geochemistry in Cenozoic Mae Sot Basin, Thailand. Am Assoc Petrol Geol Bull 69:767–780
Jankowski B (1981) Die Geschichte der Sedimentation im Nördlinger Ries und Randecker Maar. Bochumer geologische und geotechnische Arbeiten 6: Bochum, 315 pp
Königswald W von (ed) (1989) Fossillagerstätte Rott bei Hennef am Siebengebirge. Das Leben an einem subtropischen See vor 25 Millionen Jahren. Rheinlandia, Siegburg, 82 pp
Kubanek F, Nöltner T, Weber J, Zimmerle W (1988) On the lithogenesis of the Messel oil shale. Cour Forschsinst Senckenb 107:13–28
Matsumoto R, Iijima A (1981) Origin and diagenetic evolution of Ca-Mg-Fe carbonates in some coalfields of Japan. Sedimentology 28:239–259
Potter PE, Maynard JB, Pryor WA (1980) Sedimentology of shale. Study guide and reference source. Springer, Berlin Heidelberg New York, 310 pp
Rast U (1975) Sedimentologisch-fazielle Untersuchung eines Profils zwischen Duchroth und Odernheim/Glan. Dipl-Thesis, Univ Mainz, 164 pp
Rast U (1981) Vergleichende Untersuchungen zur Geochemie von Flachmeersedimenten und Ablagerungen limnischer Bildungsbereiche. Diss, Univ Mainz, 149 pp
Rast U, Schäfer A (1978) Deltaschüttungen in Seen des höheren Unterrotliegenden im Saar-Nahe-Becken. Mainzer geowiss. Mitt. 6:121–159
Schaal S, Ziegler W (eds) (1988) Messel – ein Schaufenster in die Geschichte der Erde und des Lebens. Kramer, Frankfurt, Senckenberg-Buch 64:315 pp
Schäfer A (1986) Die Sedimente des Oberkarbon und Unterrotliegenden im Saar-Nahe-Becken. Mainzer Geowiss Mitt 15:239–365
Schäfer A (1989) Variscan molasse in the Saar-Nahe Basin (W-Germany), Upper Carboniferous and Lower Permian. Geol Rundschau 78:499–524
Schäfer A, Sneh A (1983) Lower Rotliegend fluviolacustrine sequences in the Saar-Nahe Basin. Geol Rundschau 72:1135–1145
Schäfer A, Stamm R (1989) Lakustrine Sedimente im Permokarbon des Saar-Nahe-Beckens. Z Dtsch Geol Ges 140: (in press)
Schäfer A, Stapf KRG (1978) Permian Saar-Nahe Basin and Recent Lake Constance (Germany): two environments of lacustrine algal carbonates. Spec Publ Int Assoc Sediment 2:83–107
Stapf KRG (1970) Lithologische Untersuchungen der Altenglaner Schichten im saarpfälzischen Unterrotliegenden mit besonderer Berücksichtigung der Karbonatgesteine. Diss, Univ Mainz, 231 pp

Stapf KRG (1989) Biogene fluvio-lakustrine Sedimentation im Rotliegend des permokarbonen Saar-Nahe-Beckens (SW-Deutschland). Facies 20:169–198

Sturm M, Matter A (1978) Turbidites and varves in Lake Brienz (Switzerland): deposition of clastic detritus by density currents. Spec Publ Int Assoc Sediment 2:147–168

Taylor OJ (ed) (1987) Oil shale, water resources, and valuable minerals of the Piceance Basin, Colorado. The challenge and choices of development. US Geol Surv Prof Pap 1310, Washington, DC, 143 pp

Teichmüller M, Teichmüller R, Lorenz V (1983) Inkohlung und Inkohlungsgradienten im Permokarbon der Saar-Nahe-Senke. Z Dtsch Geol Ges 134:153–210

Turekian KK, Wedepohl KH (1961) Distribution of the elements in some major units of the Earth's crust. Geol Soc Am Bull 72:175–192

Wänke H, Baddenhausen H, Dreibus G, Jagoutz E, Kruse H, Palme H, Spettel B, Teschke F (1973) Multielement analysis of Apollo 15, 16, and 17 samples and the bulk composition of the moon. Proc 4th Lunar Sci Conf Suppl 4, Geochim Cosmochim Acta 2:1461–1481

Zangerl R, Richardson ES Jr (1963) The paleoecological history of two Pennsylvanian black shales. Fieldiana Geol Mem 4:1–252

Zimmerle W (1985) New aspects on the formation of hydrocarbon source rocks. Geol Rundschau 74:385–416

"Search for Poyang Lake and China"..."get 7 Hits"
Information Management in Environmental Sedimentology

WILFRIED SCHMITZ[1] and KLAUS WINKLER[2]

CONTENTS

Abstract		239
1	Scenario	240
2	The Information Problem	241
2.1	A World of Information at Your Fingertips	242
2.1.1	How to Use Your Fingertips	242
3	Back to the Facts	243
3.1	Online Data Bases in Use and Their Hosts	244
3.1.1	Advanced Retrieval Software	245
3.2	Desk-Top Data Bases Versus Online	246
3.2.1	CD-ROM - an Optical Mass Storage	246
3.2.2	Featured CD-ROM Titles	247
3.2.3	It is Your Choice: CD-ROM and/or ONLINE	247
3.2.3.1	Networking CD-ROM	248
4	What Comes Out?	249
4.1	What Happens with the Outcome?	249
5	Information Needs Communication	251
Appendix: Resources		252
Services		252
Books		253
Periodicals		253

Abstract

This chapter focuses on the so-called information problem in the interdisciplinary research subject "Environmental Sedimentology", arising from the increasing willingness of all scientists involved to pass beyond their traditional borders into adjoining scientific domains. Scientific curiosity and the widespread availability of personal computers progressively trigger the

---

[1] Institut für Sedimentforschung, Universität Heidelberg, Im Neuenheimer Feld 236, D-6900 Heidelberg 1, FRG.
[2] Informationsvermittlungsstelle, Universitätsbibliothek, Universität Heidelberg, Im Neuenheimer Feld 368, D-6900 Heidelberg 1, FRG.

evolution of data bases and the software to retrieve and manipulate information. We will elucidate applicable facilities for identifying and retrieving information from scientific data bases accessible today either from "off-site" information stores or from desk-top data bases. After a general view of relevant data bases both accessible as so-called online data bases and on optical mass storages (CD-ROMs), we feature some aspects of information retrieval software. For a suggestive application of retrieved information, we recommend data be imported on a desk-top data bases system capable of managing whole research project information. Finally, we indicate the applicability of electronic mailboxes for global information exchange along communication networks.

1 Scenario

10:00 a.m. A rainy day in February, 1989.

In less than a week, I shall be presenting a short report at a meeting of the steering commitee on our entry into the research project "Ecological Effects of Heavy Metal Pollution in the Dexing Copper Mine Region", in cooperation with scientists of the People's Republic of China. The main ecological problem encountered in the Dexing Mine Region is water pollution. The Dawu River flows through the region into the Lo An and Jishui Rivers and into Lake Poyang, which is the largest lake in China and an important fishing area and habitat for rare migratory birds. These river and lake waters are polluted mainly due to discharge from the mine of strongly acidic wastewater with a high content of metal pollutants.

Our Chinese colleagues, who visited us within the scope of the Cooperative Ecological Research Project[1] a few weeks ago, could give me only limited information about the research objects, above all the Poyang Lake. Little seems to be known about this lake, the Poyang Hu, as the actual ecosystem.

Yet, we are in urgent need of basic data, such as hydrological features of the lake, including data on water inflow and runoff, sediment transport, water quality, etc., to predict and evaluate the ecological impact of heavy metals on the river and lake ecosystems. Perhaps someone has already published his findings?

10:15 a.m. and still raining.

I hasten to my personal computer and change to the communication program. Starting, it dials through the communication network. One minute

---

[1] CERP, a joint project of the United Nations Educational, Scientific and Cultural Organization (UNESCO), Man and the Biosphere (MAB) Programme, People's Republic of China, Chinese Academy of Sciences and the Federal Republic of Germany, Federal Ministry for Research and Technology.

more and I will be in sunny California. A short acoustic signal, together with the rather meager message: "Connect 1200" appears – I am through. The software enters into dialogue with the "big" system at the other end of the line. It identifies me as an authorized user (only paying users get "Welcome" on their screens). I just have entered one of the biggest warehouses in the world of marketable data: DIALOG in Palo Alto, California.

And time begins seriously to pass.

I activate the data-base categories "Environment" with parts of "Pollution" and "Water". In this way, 14 data bases are connected for information retrieval. Quickly, I type my request into the computer: Select all records which contain the terms "Poyang Lake" and "China". Tension rises. Twenty seconds later, my vis-a-vis tells me 7 hits. Now everything is running nearly automatically. I read back the titles of the publications and, after this, the bibliographic data and abstracts. Bingo!

10:30 a.m. and it has stopped raining.

2 The Information Problem

In this age of information, as it has been described, it may seem strange to declare that information is a problem or that information users have a problem. However, no one will deny that there is difficulty in obtaining the right information in the right quantity at the right time.

So, if there is a problem with information, do we not have the means to deal with it? The answer is, as in many such cases, both yes and no.

The facilities for identifying and retrieving information are numerous and dissimilar today. Over 4000 data bases worldwide, managed by nearly 580 online services, simply called hosts, are available for public searching, covering every subject imaginable. Associated with this is the widespread availability of personal computers with the software to retrieve and manipulate information and the development of specialized telecommunication networks for access to "off-site" information stores.

Nevertheless, one has the dim feeling that publication has been extended far beyond our present ability to make real use of the records. Above all, this is obvious in chemistry. In no other field have we had such a huge increase of information in the last few decades. We need only to have a look at the *Chemical Abstracts Service* (CAS): they added nearly 0.5 million abstracts to the *Chemical Abstracts* data base during 1988. So, the CA FILE increases by one document per minute. Here, we have located the so-called information problem.

2.1 A World of Information at Your Fingertips

Advertising of the big hosts like *"A World of Information at Your Fingertips"* as composed by ESA-IRS, or similarly *"Getting the information you need without using DIALOG is like trying to find out where your high school sweetheart lives by going door to door"*, or *"Data bases – Your Knowledge Reservoir"* created by STN International makes us believe that there is no longer an information problem.

2.1.1 How to Use Your Fingertips

The starting point of electronic data bases was the development of typographical techniques 20 years ago, when they began to supply large-scale bibliographies such as *Chemical Abstracts*, making use of electronic data processing. This made it easier to manage the large number of sorting operations, e. g., the various registers. On this occasion, magnetic tapes were made available, at first merely used as "inhouse data bases". Later, with increasing development and spreading of specialized telecommunication networks, the idea was born to use large data bases as online versions.

The worldwide offer of data bases is not restriced to computer versions of bibliographies alone. There are various data bases of all possible designs with information on distinct fields of sciences and technology, economy and business, and patent literature as well as data on enterprises and companies. Numeric data bases do not primarily refer to literature; they contain directly the requested information, e. g., structural formulas and substance properties of chemical compounds. The increasing number of full-text data bases contain not only the bibliographic information, but the full text of a publication (except graphics).

With the hosts, one is now able to browse quickly through these huge data resources. Search-in data bases are far more efficient than conventional information supply and they offer more facilities as well. In traditional data resources, such as manuals, catalogues or card indices, one is usually only able to use a single criterion of order, whereas a data base offers the possibility of query for information in a very complex way. Different field names in the records (e. g., Title of Document, Author, Source, Corporate Source, Publication Year, Abstract Text, Controlled Terms, etc.) can be assigned individually . Furthermore, request terms can be combined with Boolean operators (AND, OR, NOT), and, in addition, with more differentiated data base-specific methods, such as the proximity operators: request terms, e. g., can be within a defined range of each other and in any specified order. After transmitting a request, the correct records can be selected and made available within seconds.

Access from the user's data terminal to data bases all over the world is managed by means of local and international telecommunication systems; in some cases (overseas) satellites are used. The basic requirement for data-base retrieval along these networks is not too difficult. Certainly, the ideal data terminal equipment is the PC. Together with efficient communication software, it makes working with online services much easier. In some cases, this combination even makes effective performance possible (e.g., offline preparations for information retrieval referred to structural formulas). The "way outside" routes via data circuit equipment (e.g., a modem or an acoustic coupler used for data transmission in public telecommunications networks).[2] Finally, dial-up contracts must be signed with local network coordinators and selected online services.

3 Back to the Facts

In 1984 we launched into information retrieval in online data bases. At first, only two data bases were feasible for questions concerning our domain, environmental analytics and techniques: the CA FILE[3] and the REGISTRY FILE[4], both provided by CAS, and offered in a linked computer system by STN International.[5]

At that time, asking a question and getting an answer was often like an exploration; seeking a collection of writings that would connect to form the literature surrounding a research question. By the time we developed a collection of terms sufficiently restrictive to exclude the majority of irrelevant documents, we often noticed that more than half of the desired documents had been excluded as well. If you expected to find the needle in the haystack with a broader formulation, you fairly often got to the haystack instead!

In the meantime, it has become routine, although tension still remains. This means, in practice, that one already tries to get information at the blueprint stage of a research or development project. One searches public data bases for facts from existing knowledge and institutions and their projects, which pertain to the same field. From the beginning, master theses and doctoral dissertations were supported in their initial stages with information retrieval in online data bases (e.g., in DISSERTATION ABSTRACTS ONLINE; including European dissertations from the host DIALOG). There is also a continuous update regarding the state of the art of long term research objectives. The update of analytical methods in environmental geochemistry (e.g., ANALYTICAL ABSTRACTS from DIALOG and ESA-IRS) is a main aspect as well.

---

[2] For example, packet-switched data services such as DATEX-P (Deutsche Bundespost), ITAPAC (Italy), TELENET and TYMNET etc. (USA), or VENUS-P (Japan).
[3] The CA FILE includes abstracts and index entries for papers, patents and other documents. In 1988, CAS produced the 12 millionth CA abstract.
[4] The REGISTRY FILE holds 9 million subjects (1988), searched mainly with structure diagrams. Answers include structure diagrams displayable on all terminals.
[5] STN International is the Scientific and Technical Information Network, offering direct access from Europe, North America, and Japan to major scientific data bases.

3.1 Online Data Bases in Use and Their Hosts

As mentioned above, we have more than 4000 data bases offered on the international hypermarket. The domains of "Natural Sciences and Technology", including "Biomedicine", reserve approx. 26% of all data bases, which are of interest to scientists. In practical operation with regard to environmental analytics and technology, one gets along with three to five "big" and approx. 20 "medium-sized" data bases.

The interdisciplinary research subject "Environmental Sedimentology" requires the willingness of all scientists involved to pass beyond their traditional borders into adjoining scientific domains. Whenever one works on problems, e. g. environmental analytics, one automatically finds superpositions with biology, biochemistry, inorganic and organic chemistry as well as medicine.

Thus, it is not only the "big" data bases already mentioned, like CA FILE and REGISTRY FILE of CAS, which are vital to our research, but also the data-base categories "GEOLOGY", "ENVIRONMENT", "POLLUTION", "MARINE SCIENCE", and "WATER". It is not merely coincidental that these categories match the DIALINDEX data-base categories. The host "DIALOG", for instance, offers the possibility to scan whole clusters of related data bases with the *OneSearch* feature for the information needed (related to *QUESTCLUSTER* from the host ESA-IRS).

For the category "GEOLOGY", we have the data bases GEOARCHIVE, GEOBASE (DIALOG) and GEOREF (DIALOG and via STN International). These three data bases cover more or less all types of information sources in geoscience. With GEOBASE, the extra aspect of ecology is also available.

Both the categories "ENVIRONMENT" and "POLLUTION" provide data bases offering interdisciplinary approaches to scientific, technical, and socioeconomic aspects of environmental problems. Outstanding data bases are ENVIRONLINE (DIALOG and DIMDI) and POLLUTION (ESA-IRS) respectively POLLUTION ABSTRACTS (DIALOG); the data bases OCCUPATIONAL SAFETY and HEALTH (NIOSH) and ANALYTICAL ABTRACTS are both available from DIALOG and ESA-IRS. The ANALYTICAL ABSTRACTS data base specifically extracts analytical aspects in chemistry in all its applications.

One of the most applicable data bases in the categories "MARINE SCIENCE" and "WATER" is BIOSIS (ESA-IRS) respectively BIOSIS PREVIEWS (DIALOG, DIMDI and STN International), which covers all life science subjects. This is followed by WATER RESOURCES ABSTRACTS (DIALOG), which offers a comprehensive range of water-related topics in the life, physical, and social sciences, as well as the engineering and legal aspects of the conservation, control, and the use and management of water. The two data bases AQUATIC SCIENCES and FISHERIES ABSTRACTS (ASFA) are mounted at DIALOG and OCEANIC (ESA-IRS) respectively OCEANIC ABSTRACTS (DIALOG) deals with the large-scale scope on the science, technology, and management of marine and freshwater environments and resources. They both contain ecology and environmental protection as substantial aspects.

Are you interested in your colleagues' activities on environmental research projects? The ENREP data base is an online directory of Environmental Research Projects in the Member States of the European Communities, collected on a national basis by focal points under the management of the Commission of the European Communities. The research projects cover all aspects of the environmental field and are of interest to those involved in this area. This data base is available free of charge from the European Commission Host Organization (ECHO).

UFORDAT stores comparable information on both running and completed research and development projects, as well as on research institutions in West Germany, Austria and Switzerland. If one is mainly interested in German language literature relevant to environmental science, one should couple with ULIDAT, a data base offered by STN International. The evaluation of these data bases supplied us with rather important additional information for our China project, regarding research institutions already working in China, and their priority research objects.

Much actual information on developments and results in scientific investigations occur as so called "Grey Literature". SIGLE (System for Information on Grey Literature in Europe), a bibliographic data base provided by the host STN International, contains publications such as research reports, discussion papers, dissertations, conference proceedings, etc. normally not available in book trade.

3.1.1 Advanced Retrieval Software

The so-called general-purpose communications packages dial up any online service automatically, and log on to the host automatically with your password masked. They allow the composition of search strategies offline and storing of search strategies on one's own computer. Some of them provide the chance to type ahead of the host computer while online, thus reducing online time and charges by utilizing the host's processing time. All of them store all or parts of an online session in a retrieve buffer, then print or store it in a disc file (such as DIALOGLINK, DIALOG's custom communication software).

For query in chemical data bases, which allow retrieval of structures or substructures, the searcher needs an efficient offline graphic editor. Worth mentioning is the *STN Express* which belongs to the last generation of "front-end" software and MOLKICK, a memory-resident graphics query program.

STN Express was developed for retrieval in technological data bases at STN International. In text mode, it can also be applied to other hosts and data bases. A novice is able to formulate queries by means of a guided search feature without any knowledge of the retrieval language. For some of the more frequently searched topics, *STN Express* contains predefined search strategies. For structure queries in the REGISTRY FILE of CAS and Beilstein-Online there is a built-in structural editor. Structures are drawn with a mouse free-hand, or the mouse is used for selecting predefined components from menus of templates, bonds, elements, and substituent groups. Once drawn, structure diagrams can be directly uploaded for searching, stored for later recall, or printed with high resolution for inclusion in reports.

As mentioned above MOLKICK[6] is a memory-resident program. Once installed, it can be used with quite a number of communication packages. After having finished the structural design offline, MOLKICK automatically complies the required search commands for retrieval, e.g. in CAS (REGISTRY FILE at STN International), and S^4 (Beilstein-Online at STN International and DIALOG). Compared to the graphic module within *STN Express*, the graphic capabilities are more finely tuned.

Nevertheless, one must admit that even the top "user's interface" available at the moment does not make data retrieval more attractive, e.g., in CAS if one is not a chemist.

3.2 Desk-Top Data Bases Versus Online

In a document retrieval system, user's time, not the machine's response time, determines retrieval speed. One often hears the coins dropping during retrieval. The mere thought of retrieving against time to save money frequently hinders the "creative" progress of a retrieval.

This handicap could be avoided if it were possible to install the data bases immediately at the operator's position. Today, the newly created term *Desk-Top Data Base* speaks well for this possibility; an invention over 10 years old has led to a new application in just the last few years: *Desk-Top Data Bases* on optical mass storages.

3.2.1 CD ROM - an Optical Mass Storage

What is CD-ROM? CD-ROM stands for **C**ompact **D**isc-**R**ead-**O**nly-**M**emory. The CD-ROM, on one disc, can store still and/or moving images in black and white and/or color; stereo or two separate sound tracks, integrated with or separated from images; digital program files (e.g. word processors, spread sheets) and digital information flies (e.g. documents, records, catalogues).

At present, the CD-ROM format can store 540 million bytes or ASCII characters in one or more files. This capacity is provided by a disc, which is 120 mm in diameter, and 1.2 mm thick.

---

[6] MOLKICK was jointly developed by the Beilstein-Institut of Frankfurt and SOFTRON GmbH, West Germany.

The CD-ROM inserts easily into the current microcomputer operating environment. If one already possesses a serious, office-type micro computer, one must only add the CD-ROM drive, a controller interface board to link drive and computer, appropriate software, and a supply of discs holding the reference files needed.

3.2.2 Featured CD-ROM Titles

At present, nearly 400 CD-ROMs are included in the third edition of TFPL's CD-ROM DIRECTORY. In the following, only those CD-ROMs are entered which seem to be thematically interesting for the field of environmental protection.

The first mentioned is the »Gefahrgut« CD-ROM (Dangerous Substances on CD-ROM), published and supplied by Springer-Verlag (Heidelberg), containing information on dangerous materials, chemicals and goods; guidelines on handling and safety regulations; information on health hazards; and means of identification. It is widely used in cases of emergency or accidents to obtain critical information on courses of action. There are three sources of data: HOMMEL (Name and Author), a manual of dangerous substances; CHEMDATA, Harwell Library UKAEA; and EINSATZAKTEN, Swiss fire service emergency advice.

The »CHEM-BANK«, published by Silverplatter and supplied by Lange and Springer (Berlin, West Germany), includes a collection of data on hazardous chemicals, containing three major data bases: Registry of Toxic Effects of Chemical Substances (RTECS); Oil and Hazardous Materials-Technical Assistance Data System (OHMTADS); and Chemical Hazard Response Information System (CHRIS).

»Aquatic Science and Fisheries Abstracts« (ASFA), published by Compact Cambridge and »WATER RESOURCES ABSTRACTS«, published by the National Information Service Corporation are both available from Lange and Springer. Both data bases are derived from their online versions and have been drafted above.

Finally, the NTIS (U.S. National Technical Information Service) data base should be mentioned, which includes material from topics of immediate and broad interest, such as environmental pollution and control, energy conservation, natural resources and earth sciences, etc. This data base is available as a DIALOG OnDisc product (also online at DIALOG), or as NTIS-SilverPlatter, published by Silverplatter, supplied by Lange and Springer.

3.2.3 It is Your Choice: CD-ROM and/or ONLINE

During online retrieval, the pressure of hourly charges can often induce stress. Stress can be reduced by planning an online session, but it is often necessary to refine a search strategy interactively, based on intermediate results. It is not necessary to "think on your feet" while searching for information on a CD-ROM; there are no consequences for mistakes, pauses, deadends, or consultations with the manual. All in all, it is a much friendlier situation.

However, some online data bases are typically used in connection with other data bases. Online, it is simple to repeat the same search in a series of data bases since all are available in the same place (e.g., DIALOG's *OneSearch* feature). Offering a single data base on CD-ROM, i.e. one member of such a set, would not make sense, since the user would still have to perform most of the search online in the remaining data bases, negating any advantage of CD-ROM. It would be better to offer either the entire group on CD-ROM or nothing at all. DIALOG at least tried to moderate this problem by establishing DIALOG *OnDisc*. DIALOG *OnDisc* products (e.g., DIALOG *OnDisc* NTIS) make it possible for the searcher to retrieve only the information needed without worrying about connect hour or telecommunications charges. And, as a DIALOG subscriber, one has the opportunity to transfer refined ondisc search strategies from the Command Search mode directly into DIALOG's online files. This feature thus provides an easy link between that portion of a data base which is contained ondisc and the most current updates available online. The most important advantage, therefore, is that one is now able to develop search strategies without any deadline pressure; the more so as the DIALOG *OnDisc Manager* empowers searchers to use the Easy Menu mode, designed to make information access easily available to novice searches. In addition, searchers may opt to use the standard search techniques (the same as online search).

Yet we still have scientists engaged in occasional or varied research projects. Who often cannot predict the data bases they will need to use or how much they will use them. Thus, they will probably continue to use online data bases.

3.2.3.1 Networking CD-ROM. While CD-ROM data bases are more cost-effective than online services, the single-user versions are hampered by access limitations. In order to provide ten people simultaneous access to a data base without a LAN (Local Area Network), not only ten PCs with monitors, but ten CD-ROM drives as well as ten duplicate discs are needed. With a network version, these ten PCs can be attached to a single disc drive, so that all ten people could search the data base at one time. When a multiple-drive system is added to a network configuration, users are granted immediate access to more than one data base from each individual work station.

The increasing use of personal computers and Local Area Networks will make CD-ROM networks a major information delivery system. In particular, universities should consider networked CD-ROMs as a good solution to the problem of providing in-depth research tools at reasonable costs.

4 What comes out?

After having finished your session either on CD-ROM or online, lean back and relax – provided the retrieved buffer has been saved on discs. Depending on the type of data base, either numbers, full texts or, as in our case, bibliographic information are at ones disposal on dics. Figures 1 and 2 show bibliographic documents retrieved from different data bases on different hosts. In addition to bibliographic information, the documents are supplemented with abstracts (in most cases written by the author) and so-called descriptors (terms which classify the article from the view of the data-base producer). Both original papers are written in Chinese, so abstracts and descriptors are the only source materials directly available at present to estimate the relevance of the articles.

4.1 What Happens with the Outcome?

Back from California (or Europe as well) one must now evaluate the data found in particular. Normally, the retrieved information is in a so-called log file. A suggested application of this information is to transfer them to a desktop data-base system, capable of managing all of the project information. It therefore follows almost without saying that the most important facility of

---

DIALOG WATER RESOURCES ABSTRACTS (FILE:117)
 598448 W89-00535
  Hydrological Features of Poyang Lake (II), (in Chinese)
  Zongxian, Y.; Juncai, Z.
  Hydrometeorological Experiment Station of Poyang Lake, Xingzi (China).
  Oceanologia et Limnologia Sinica OECOBX, Vol. 18, No. 2, p 208-214, March
  1987. 7 fig, 3 tab, 1 ref. English summary.,
  Journal Announcement: SWRA2201
  The flow of Poyang Lake is divided into three kinds of flow as follows:
gravity flow, top-lifted flow, and backward flow. A mathematical model is
put forward to reflect the relations of the wind with the wave and the
climbing height of the waves found in Poyang Lake. The quantities of the
suspended matter flowing into and out of the lake were calculated, and
their variations with space and time given. The washing and depositing
regimes are summarized. Although not much sand from the lake's basin flows
into the lake, its influence on some parts of the lake is still serious.
  The contents of common chemical elements, and the variations of the
temperature of the lake water with time, place, and depth were analyzed.
  The lake is polluted slightly, but the water quality of the lake is still
in good condition. (See also W89-00532 and W89-00540) (Author's abstract)
  Descriptors: *Poyang Lake *Lakes ; *Path of Pollutants ; *Limnology ;
*China ; *Mathematical models ; *Sediment transport; Flow ; Hydrology ;
Gravity flow ; Wave action ; Aeolian deposits ; Water temperature ; Lake
basins ; Water quality ; Chemical properties ; China
  Section Heading Codes: 2H (Water Cycle--Lakes); 5B (Water Quality
Management and Protection--Sources of Pollution)

Fig. 1. Sample document from DIALOG's Water Resources Abstracts data base

```
AN  CA107(26):242140r
TI  Modeling study on copper partitioning in sediments - a case study of
    Poyang Lake
AU  Chen, Jingsheng; Dong, Lin; Deng, Baoshan; Wan, Liangbi; Wang, Min;
    Xiong, Zhengliang
CS  Dep. Geogr., Peking Univ.
LO  Beijing, Peop. Rep. China
SO  Huanjing Kexue Xuebao, 7(2), 140-9
SC  61-1 (Water)
DT  J
CO  HKXUDL
IS  0253-2468
PY  1987
LA  Ch
AB  Simulation expts. showed the Cu(II) distribution in geochem. phases
    of lake sediments is related to the content of each phase in the
    sediments. Both measured values and model-calcd. values showed that
    Cu(II) in the sediment of Poyang Lake, China, occurred mainly in the
    org. and Fe(OH)3 phases.
KW  copper partition lake sediment; geochem phase copper distribution
    sediment
IT  Humus and Humic substances
        (copper in lake sediments in relation to)
IT  Process simulation, physicochemical
        (of copper partitioning in lake sediments, geochem. phases in
        relation to)
IT  Geological sediments
        (lake, copper partitioning in, geochem. phases in relation to)
IT  1309-33-7, Iron hydroxide (Fe(OH)3)    1318-74-7, Kaolinite,
    occurrence
        (copper in lake sediments in relation to)
IT  7440-50-8, Copper, occurrence
        (in lake sediments, partitioning of, geochem. phases in relation
        to)
```

Fig. 2. Sample record from CAS's CA FILE data base

such a system: storage and retrieval must be independent of predefined data structures, formats and field sizes. Or an information storage and retrieval system must be found, which allows the use of any text, numbers and even the graphics desired.

Such software is *askSam* 4.2 from North American Software GmbH. This program consists of features additional to the above mentioned facilities. These features allow the establishment of links, built by association rather than by predefined structures (such as mandatory fields). Thus, to find the needle in the haystack, hundreds of paths through the straw can be marked. Some people call this feature *hypertext*.

In one's own data base, the outline of a project is connected to the graphic representation of the research region by a record. At the same time, one can define words at any point of the text as so-called "point and shoot" terms. If such a word is clicked, activation, e. g., of the desired bibliographic file occurs and one can move to a document. In addition, within this file, "point and shoot" terms can be defined which can be linked up with a table of numbers, in the course of which one can read back stored charts of the numbers in short time.

Let us return again to the useful hits from our search for literature. The language of the original papers is specified in the retrieved bibliographic documents as Chinese. Translation from the Chinese to the English language requires time and a benevolent Chinese colleague, so, first of all, we addressed an inquiry to "Technische Informationsbibliothek (TIB)[7] Hannover. We provided this inquiry with *askSam* 4.2 as a report which was transmitted to TIB by telefax. We had to order the copy of one of the original papers directly per online from CAS (via STN International): during an online session, the document accession number of the required document (e. g. in CA FILE: AN, cf. Fig. 2), is submitted in online ordering. Delivery from the USA with the CAS Document Delivery Service takes approx. 4 days.

5 Information Needs Communication

Of course, information is not only found in data bases. Occasionally, scientists should discuss their findings and points of view directly and not in a roundabout way through a paper containing exclusively experiments that turned out well. As a rule, this takes place during symposia and congresses. Yet what about everyday life? How can ideas be exchanged over continents?

One possible solution is to use a *Message Handling System*, simply called *electronic mailbox*. As a user of a mailbox system one is given an address which, together with a password, identifies one as an authorized user of the system and enables one to enter and read out messages.

During "outdoor" research projects, the most important advantage is to be mobile and available at all times. It is possible to link up with the system from nearly any location through the switched network services or the telephone network by means of a portable data terminal (e. g. lap-top PC) with acoustic coupler, which can be carried in a briefcase and used at any given telephone.

Such a mailbox system is provided by the Deutsche Mailbox. The term "system" indicates that it is more complex than just handling incoming and outgoing messages. Furthermore, numerous gateways to worldwide communication standards, such as TELEX/TELETEX, BTX and TELEFAX services are available. With the progressive compatibility of mailbox systems, data exchange within the global mailbox networks is made even quicker and easier. The range of so-called "Value Added Services", such as bulletin boards (e. g. mailbox conferences), teletranslating and access to worldwide data bases (mailbox system as "mini-host") is still increasing.

---

[7] Direct orders come in within 5 days. In cases of urgency express deliveries with telefax transmission are also possible. TIB is "Document Supplier", among others for STN International, DIALOG, ESA-IRS and DIMDI, so online document ordering to TIB is practicable via the mentioned hosts.

In conclusion a brief example of today's practice is given. To establish efficient and cost-effective communications between our institute and several research centers in China, located more than 1000 km from Beijing (e. g. Nanchang, Jingdezhen and even smaller cities), we successfully used the so-called *Interswitch* capability of the Deutsche Mailbox GmbH in March and April 1989: although there is no local mailbox system in China at present, we reached our home mailbox with long-distance calls to a mailbox networks node in Hongkong.

This information exchange between the so-called scientific communities is possible at every scale. The tools for information management adjusted to any information problem are feasible for everyone and for optimal efficiency, one must only use them. The single limiting factor is the daily increase of knowledge, and along with that the possible loss of a general overview of relevant data bases. However, there are already data bases about data bases...

Appendix: Resources

Services

Beilstein-Institut für Literatur der Organischen Chemie Varrentrappstraße 40–42 D-6000 Frankfurt am Main 90, FRG FAX: (+49)69/79 17-321
CAS - Chemical Abstracts Service 2540 Olentangy River Road P.O. Box 3012 Columbus, Ohio 43210, USA FAX: (614) 421-3713
Deutsche Mailbox GmbH Blücherstraße 11 D-2000 Hamburg, FRG FAX: (+49)40/38 02 00 20
DIALOG Information Services, Inc. 3460 Hillview Avenue Palo Alto, California 94304, USA FAX: (415) 858-7069
DIALOG Information Services, Inc. P.O. Box 188 Oxford OX1 5AX, UK FAX: (+44)865/73 63 54
ESA-IRS European space agency's information retrieval service Via Galileo Galilei I-00044 Frascati, Rome, Italy FAX: (+39)6/94180-361
Lange and Springer Wissenschaftliche Buchhandlung Otto-Suhr-Allee 26/28 D-1000 Berlin 10, FRG FAX: (+49)30/3 42 06 11
North American Software GmbH Uhdestraße 40 D-8000 München 71, FRG FAX: (+49)89/79 00 258
SOFTRON Gesellschaft für technisch-wissenschaftliche Software mbH Rudolf-Diesel-Straße 1 D-8032 Gräfelfing, FRG FAX: (+49)89/85 21 70
Springer-Verlag GmbH and Co., KG Dept. New Media/Handbooks Tiergartenstraße 17 P.O. Box 105280 D-6900 Heidelberg 1,FRG FAX: (+49)6221/4 39 82
STN International c/o Fachinformationszentrum Karlsruhe Postfach 2465 D-7500 Karlsruhe 1, FRG FAX: (+49)7247/808 666
STN International P.O. Box 3012 Columbus, Ohio 43210, USA FAX: (614) 447-3713

STN International c/o Japan Information Center of Science and Technology C.P.O. Box 1478 Tokio 100, Japan FAX: (+81)3/581-6446
TIB Universitätsbibliothek Hannover und Technische Informationsbibliothek Welfengarten 1B D-3000 Hannover 1, FRG FAX: (+49)511/71 59 36

Books

CD-ROM Directory 1989, 3rd edn. (1989) TFPL, London 283 pp ISBN 1-870889-11-8
CD-ROM, vol 1: The new papyrus. (1986) Lambert St, Ropiequet S (eds) Microsoft, Redmont, 619 pp ISBN 0-914845-74-8
CD-ROM, vol 2: Optical publishing. (1987) Ropiequet S, Einberger J, Zoellick B (eds) Microsoft, Redmond, 358 pp ISBN 1-55615-000-8
Dirk Gently's Holistic Detective Agency. (1987) Adams D N. Pan, London, 247 pp ISBN 0-330-30162-4
Fachwissen Datenbanken. Die Information als Produktionsfaktor. (1986) Claasen W, et al. (eds) Klaes, Essen, 236 pp ISBN 3-925506-04-7
Fachwissen Online Recherche. Suchstrategien in Online-Datenbanken. (1988) Claasen W, Cornelius P, Ehrmann D, Fischer B, Pichler H, Schwedler E, Tanghe P (eds) Klaes, Essen, 347 pp ISBN 3-925506-13-6
Online Datenbanken. Zugang zum Wissen der Welt mit Personal Computern. (1986) Schubert St SYBEX, Düsseldorf, 199 pp ISBN 3-88745-621-1
Optische Speicher. Fachinformationen auf optischen Massenspeichern. (1986) Schulte-Hillen J, Schwerhoff U. Scientific Consulting Dr. Schulte-Hillen (ed) Klaes, Essen, 122 pp ISBN 3-925506-05-5

Periodicals

cogito Neue Wege zum Wissen der Welt. Zeitschrift für die Nutzung elektronischer Medien. Verlag Hoppenstedt & Co Havelstr. 9 Postfach 4006, D-6100 Darmstadt 1, FRG FAX: (+49)6151/38 03 60
PC MAGAZINE. The independent guide to IBM-standard personal computing. One Park Ave. New York, NY 10016, USA
SOFTWARE TREND. Der Newsletter für die Computergemeinschaft. Software Trend AG Haldenweg 1 CH-8802 Kilchber, Switzerland FAX: (+41) 1/7152441

Microbial Modification of Sedimentary Surface Structures

Hans-Erich Reineck[1], Gisela Gerdes[2], Marianne Claes[2], Katharina Dunajtschik[2], Heike Riege[2], and Wolfgang E. Krumbein[2]

CONTENTS

Abstract		254
1	Introduction	255
2	Study Areas	256
3	Materials in Which Overthrusting Takes Place	258
3.1	Microbial Mats	259
3.2	Mineral Encrustations	261
4	Physical Overthrust Structures and Their Biological Modifications	262
4.1	Description of Structures	262
4.1.1	"Soft Ground" Domes and Folds (alpha-Petees)	262
4.1.2	Encrusted Buckles and Folds (alpha- and beta-Petees)	264
4.1.3	Intermediates Between Petees and Tepees	266
4.2	Classification of Petees	267
4.3	Differentiation of Petees from Tepees	269
5	Discussion and Conclusion	270
5.1	Modes of Petee Genesis	270
5.2	Salterns as a Model	272
References		274

Abstract

In coastal salterns, physical forces cause the upfolding of surface layers. Gel-sticky, microfibrous substrates produced by microbes interfere with these processes. Results are folds and buckles, which differ from abiogenic tepee structures. The modified tepees are termed petees. Basing upon the overthrusting mechanism involved, petees are classified into three types: (1) alpha-petees are buckles or folds which initially derive from subsurficial gas

---

[1] c/o Senckenberg-Institute, Schleusenstraße 39a, D-2940 Wilhelmshaven, FRG.
[2] Institute for Chemistry and Biology of the Marine Environment, University of Oldenburg, P.O.Box 2503, D-2900 Oldenburg, FRG.

pressure, wind or water friction. In alpha-petees, mineral crystallization is either non-existent or initial, crests remain closed and rounded. (2) Beta-petees are advanced stages of alpha-petees with ruptured crests. Mineral crystallization takes place in the whole field of preformed folds, buckles and interspaces. Alpha- and beta-petees can be arranged irregularly or as parallel folds. (3) Gamma-petees are folds around polygons, formed in coherent biogenic surface layers. Their crests are sometimes rounded, but more often ruptured. Gamma-petees are visually closest to tepees and evolve from the same process which is lateral expansion of surface crusts by crystallization pressure.

Within the peritidal zone, petee environments range between subtidal and highest supratidal areas: alpha-petees may already develop in subtidal lagoons and reefs. Gypsum-encrusted beta- and gamma-petees rather indicate intertidal and lower supratidal exposure, and tepees finally the relatively highest topography where microbial mats are excluded.

1 Introduction

Tepees are buckled margins of saucer-like megapolygons in limestone or evaporite crusts. In a vertical two-dimensional exposure, they appear as an inverted "V" (Adams and Frenzel 1950). Characteristic tepee environments experience increased salinity, common changes between wetting and exposure of sedimentary surfaces and the increment of crystals resulting in surface expansion (Assereto and Kendall 1977; Kendall and Warren 1987).

Worldwide, probably even throughout geologic time, microbial mats (Brock 1976; Krumbein et al. 1979; Krumbein 1986) show a typical penchant for such evaporite environments. They coat sedimentary surfaces with flexible hydroplastic layers, interfere with depositional processes and may also react differently from abiogenic materials to physical surface deformations. That microbial mats obviously react differently to shrinkage, was already stressed by Shinn (1983): In the presence of mats, "the rule that thick layers result in large desiccation cracks and thin layers make small cracks generally does not hold true".

Similar structural modifications can be inferred with respect to the shape of tepees. On the occasion of episodic or seasonal wetting, supratidal sabkhas, which are tepee domains, experience invasions of benthic microbes (Gavish et al. 1985; Gerdes et al. 1985). Tepees are known from coastal and continental areas. In continental areas, they are often related to caliche soils (Reeves 1970). Around coasts, they occur in beachrock which is a characteristic strandline indicator (Kendall and Warren 1987; Krumbein 1979; Shinn 1986), and also in expanding gypsum and halite crusts (Gavish et al. 1985). In almost all of these sediments, mat-forming microbes are common. Thus, their tissue-like biogenic matrices may modify tepee morphology in a great variety of evaporite environments. For overthrust structures in initially biogenic surface layers, Gavish et al. (1985) created the term "petee", suggesting that a structure that strikes one as rather modified from a normal tepee may be more likely a petee.

The following text deals with petee examples from salterns. Aims are (1) to document biological overprints of physical deformation structures, (2) to classify petees, and (3) to differentiate petees from abiogenic tepees. Finally, the question will be discussed for which types of depositional environments, salterns may serve as a model.

2 Study Areas

Study areas are salterns on the southern Bretagne coast (France, 45° 20'N), and on the Canary Islands (Lanzarote, Spain, 28° 50'N).

Salterns are man-made evaporite systems. Their main functional principle is to spread laterally apart the precipitation sequence of salts so that calcium and magnesium carbonates, calcium sulphates and finally sodium chloride precipitate selectively in separate basins. This is realized by the regulated horizontal increase of salt concentrations (Fig. 1). Due to this functional principle, all salterns are built almost according to the same engineering design: Tidal flushing is prevented by dykes, and the inflow of seawater regulated by flood gates which hinder also the ponded seawater to flow out at low tides. The seawater, percolating through shallow interconnected ponds, passes through several stages of concentration until it reaches the harvesting

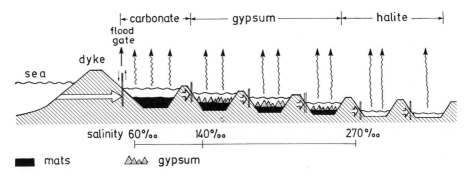

Fig. 1. Idealized transect showing engineering designs of salterns: Series of interconnected ponds in which sodium chloride concentrations increase laterally. Seawater percolates through succeeding ponds into harvesting basins where seawater has evaporated to about one-tenth of original volume. Modifications of the type drawn here occur due to tidal ranges and climates: On coasts with higher tidal range (e.g. Bretagne) salterns are flooded directly through flood gates. On coasts with minor tidal range (e.g. Lanzarote), seawater initially has to be pumped to the highest-lying basins. The full evaporite series (carbonate-gypsum/anhydrite - halite) is only realized in semi-arid and arid climates, while in humid climates, carbonate precipitation is lacking. Drawing by Ms. R. Flügel

basins (Fig. 1). Here, the seawater has evaporated to about one-tenth of the original volume, and sodium chloride precipitates in the rest of the supernatant liquid and on the bottoms where it becomes harvested (Herrmann et al. 1973; Schneider and Herrmann 1980).

The man-made topography of salterns allows for their management even towards higher latitudes. Examples in the temperate climate zone are the salterns at the southern coast of Bretagne (France), which cover about 21 km^2 of the Guerande Bay (Fig. 2a). The salterns are separated from the sea by 3-m-high artificial bars which protect a multitude of geometrical basins against ocean dynamics. In the course of the basins, salinity increases to a maximum of about nine times that of normal seawater (Fig. 1).

On Lanzarote, salterns were studied in the Bay of Janubio (Fig. 2b), which was partially filled by vulcanic eruptions from 1730–1736. The bay is protected against ocean dynamics by a bar upheaved by waves. Salterns cover an area of about 2 km^2. Their water ways and basin walls are built by pyroclastic material, and interspaces are closed with clay to prevent leakage.

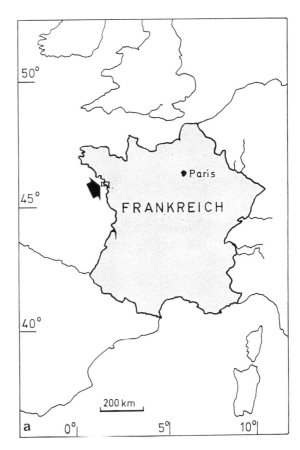

Fig. 2a,b. Study areas. **a** Salt work area in the Bay of Guerande, Bretagne, France. **b** Salt work areas on Lanzarote: Bay of Janubio

Fig. 2b

On the Bretagne coast, the temperate oceanic climate (mean annual temperature is 11° C, annual precipitation 850 mm; Gierloff-Emden 1981) provides changeable weather throughout the year which in the salterns disturbs a long-term duration of evaporation processes. Lanzarote climate tends to semiaridity (mean annual temperature is 20° C, annual precipitation 140 mm; Rothe 1986). Here, longer-lasting periods (at least over seasons) of sun irratiation, low humidity and high evaporation rates support the long-term stability of the horizontal salinity gradient. As in Bretagne salterns, salinity in the course of the basins ranges from seawater salinity upwards to about 300‰ salinity (S).

3 Materials in Which Overthrusting Takes Place

The comparison of the two salterns studied reveals the following differences: As a consequence of climate, massive gypsum crusts are not developed in

Bretagne salterns but form in Lanzarote salterns where crust formation starts at salinity ranges of about 120‰ S. Therefore, Bretagne salterns yield excellent examples of physical deformation structures in soft substrates (*in statu nascendi*), while Lanzarote salterns continue the picture into encrustation of initial soft ground structures. Also, lateral extension of surfaces crystallization pressure can be studied well in the salterns on the Canary Islands. First, some information about microbial mats and evaporative minerals characteristic of the respective environment will be given.

3.1 Microbial Mats

Microbial mats are microfibrous, slime-interwoven, coherent coatings of sediments and rocks (Fig. 3). Creators are microbes of different taxonomic groups (Krumbein et al. 1979). Laminated vertical buildups of microbial mats initiated the term stromatolite for their fossil records (Kalkowsky 1908). Pettijohn and Potter (1964) created the term "growth bedding", visualizing that the laminated patterns deal with the history of living and actively growing benthic systems, controlled in the lateral and vertical extension by the interplay of environment and functional properties of life (Fig. 3).

In both salterns studied, microbial mats are produced by a cyanobacteria-dominated community type that was already described from various hypersaline environments (Cohen et al. 1977a, b; Thomas and Geisler 1982; Holtkamp 1985; Gerdes and Krumbein 1987). Mainly filamentous cyanobacteria (*Microcoleus chthonoplastes*) and coccoid gel-producing types contribute to coherent matrices in surface layers which interact with physical overthrusting forces. In both salterns studied, the coherent filament-dominated microbial mat type proliferates in a range between almost normal seawater salinity and 270‰ S maximum. In this range, the mats interact with the precipitation of carbonates (only Lanzarote salterns) and gypsum. At salinity concentrations exceeding 270‰ S (halite basins), the coherent microbial mat type becomes replaced by flocculous bacterial communities (dominated by flagellates and halobacteria).

The longer periods of undisturbance last, the thicker biolaminated sequences built by microbes are (Gerdes and Krumbein 1987). Within salterns studied, vertical biogenic sequences were observed reaching 5–8 cm (Bretagne) and 10–15 cm (Lanzarote; Fig. 3). Airborne siliciclastic particles rarely contribute to these sequences. The biogenic bedding (Fig. 3a) results from an environmentally controlled periodic dominance change of surface populations of filamentous and coccoid cyanobacteria. By prevailing higher temperature and long-term patterns of stable salinity conditions (characteristic for Lanzarote salterns), coccoid cyanobacteria are prompted to produce large amounts of extracellular polysaccharides. These organisms add units swollen by water-saturated viscous gel to the pile of microbial laminae and contribute to its vertical extension (Fig. 3b). This phenomenon

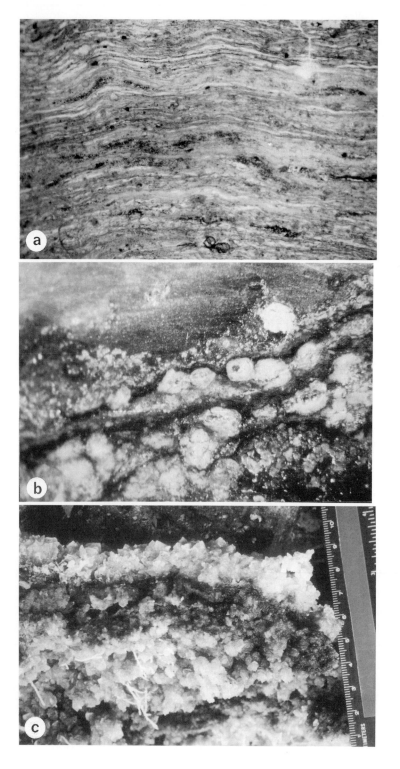

was already reported from the Gavish Sabkha, Sinai coast. Photosynthesis is still active in deeper buried microbial laminae, due to the effect of channelling sun light through viscous bacterial gel. This effect promotes primary production of biomass even in subsurfical, deeper buried layers (Krumbein et al. 1979).

3.2 Mineral Encrustations

Striking differences between both salterns studied are related to mineral formation. In Bretagne salterns, no massive mineral crusts develop except halite. Carbonate formation is completely suppressed, gypsum crusts are ephemeral, thin and fragile, and do not contribute to obvious surface deformations. Reasons for the suppression of carbonates and lack of massive gypsum may be changeable weather and lower mean annual temperatures. Additionally, the salt workers clean out basin bottoms in winter which may also inhibit extensive growth of gypsum crusts. Halite precipitates from about 270‰ S upwards. Crusts form at the bottom of harvesting pans, while smaller crystals precipitate at the water-air interface (termed *"fleur de sel"*). Chemical analyses indicate that sulfate is still concentrated in halite bottom crusts (Métayer 1980).

In Lanzarote salterns, the evaporite series: carbonate – gypsum/anhydrite – halite is evident, and is horizontally spread apart due to the functional principle of salterns (Fig. 1). The carbonates precipitate mainly within the microbial soft substrate mats. Here, foci of nucleation are unicells, cell colonies, gas and liquid bubbles coming from the degradation of organic matter (Krumbein 1986; Gerdes and Krumbein 1987). These nucleation foci determine mainly the particle form (Fig. 3b). Gypsum as the next step in the saline series precipitates in a salinity range between 120 and 270‰ S. Gypsum appears as (1) crusts which typically develop at the sediment-air interface (Cody and Cody 1988), (2) a soft-sediment mush which is preferentially formed subaquatically. Within the salinity range, corresponding in gypsum precipitation, microbial mats are still common. Encrustation proceeds on the mat surface and, as far as these are released from the subsoil, also at subsurfaces (Fig. 3c). Halite as the last step in the salinar series precipitates in correspondence to a salinity exceeding 270‰ S. As in Bretagne salterns, it occurs as crusts and *fleur de sel*.

Fig. 3a–c. Multilayered, completely biogenic piles of microbial mats, dark laminae are caused by the dominance of filamentous cyanobacteria, light laminae by the dominance of coccoid unicells. **a** Vertical section through mats of Lanzarote salterns, vertical length of photo: 1 cm. **b** Biolaminations as in **a** with sand-sized authigenic carbonate particles. Diameter of grains: 100–250µm. **c** Gypsum-encrusted microbial mats from Lanzarote salterns. Encrusted filaments of *Microcoleus chthonoplastes* visible toward the *bottom*

4 Physical Overthrust Structures and Their Biological Modifications

For overthrust structures in initially biogenic surface layers, Gavish et al. (1985) created the term "petee", suggesting that the structure is primarily due to the cohesive character of filamentous cyanobacterial mats.

4.1 Description of Structures

4.1.1 "Soft Ground" Domes and Folds (alpha-Petees)

Structures of this category include single domes, multitudes of buckles, and the transition of buckles into folds. Crests are closed and rounded, due to cohesive microbial mats (Fig. 4a). In Bretagne salterns, where mineral crystallization is not important, the hollow structures remain unconsolidated and fragile.

The following two different processes are considered to be responsible for overthrusting: (1) subsurficial gas concentration which is very common in bioactive sediments (Giani et al. 1984). Gas (H_2S, CH_4, O_2) migrates from deeper buried organic (detrial) or biogenic deposits (microbial buildups) toward the surface where cohesive microbial tissues retard the exchange through the mat-air resp. mat-water interface. By accumulation beyond the surficial biofilm, gas may overthrust the surface into domes and finally elongated folds (Fig. 5). (2) The other processes responsible for overthrusting of the soft biogenic surface tissues is wind or water friction, or slope gravity. These processes are particularly effective where, locally of temporarily, juvenile stages of microbial mats occur, referred to as monolayered mats (Fig. 4b, c). These often represent the initial reestablishment of mats after burial. When these mats establish atop slippery sediments such as gypsum mush, they do not stick well to the subsoil and tend to scour in wind-drifted supernatant liquids (Fig. 4c).

Gas pressure from the subsoil can be so strong that monolayered microbial mats tear and expose the subsoil. Tearing is also supported by moving floods and wind-induced surface waves running through the brine and scouring the folded mats. Upcurlings of mat margins when torn are characteristic. Tears are able to heal, leaving differently colored scars (Fig. 4c).

Fig. 4a–c. Soft ground domes and folds (alpha-petees). **a** Dome generated by subsurficial gas pressure beyond a microbial mat (Lanzarote salterns). **b** Monolayered microbial mat, established on top of a clay mineral layer (Bretagne salterns). Folds were initially produced by subsurficial gas which accumulated beyond the surface mat. Mats tear due to upfoldings by gas pressure and expose the subsoil. **c** Same upfoldings of monolayered mats as in **a**. Mats became reestablished atop subaquatically deposited gypsum mush (Lanzarote salterns). Mats scour on the subsoil due to wind drifting the supernatant liquid, causing folds and tears. Note wide tears and exposure of subsoil due to the non-sticking monolayered mat

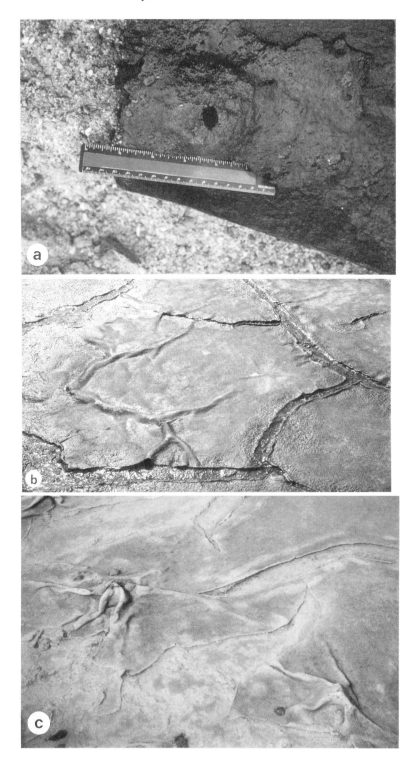

Similar examples of microbial mats undulated by wind and slope gravity were given by Gavish et al. (1985).

4.1.2 Encrusted Buckles and Folds (alpha and beta-Petees)

Subsurficial gas pressure, wind and water friction, and finally slope gravity cause overthrust structures in both Lanzarote and Bretagne salterns. In basins of Lanzarote salterns exceeding 120‰ S, these structures become immediately encrusted by gypsum (Fig. 5). This may increase preservation suc-

Fig. 5. Gypsum-encrusted domes in multilayered microbial mats, some transitions to folds also visible (Lanzarote salterns). Establishment of the "mole-hill"-like surface by gas, which concentrated beyond the formerly non-encrusted, elastic and easily deformable mats (cf. Fig. 4a)

cess of the initially unconsolidated buckles and folds. The formation of massive gypsum is favoured by repeated changes between wetting and exposure of the sediment surfaces as a result of strong evaporation and a low supernatant water table. Gypsum precipitation takes place in the whole field within and around preexisting buckles, folds, and intermediates between both (Figs. 5 and 6). Continued crystallization reveals ruptured crests: beta-petees evolve (Fig. 6).

The internal buckle relief follows the slightly flattened outer curve. The subsoil beyond the buckles and folds is covered by a strongly reduced gypsum mush which is interspersed with bacteria and their extracellular slimes. Outer surfaces are coated by slimes of tan color, produced by numerous coccoid cyanobacteria which protect themselves against phototoxic reactions by carotenoids and slime (Fig. 6). Slickenslides of slimes concentrate within the valleys.

A progressing lateral expansion by crystal increment can press the flanks or folds against each other until one flank overrides the other, rising steeply

Fig. 6a,b. Transitions from alpha- to beta-petees (Lanzarote saltworks): **a** Partially subaerially exposed, gypsum-encrusted microbial mats. Various elongated overthrust structures visible which include both alpha-petees with rounded crests (see *arrow* 1) and their advanced stages (beta-petees) with ruptured crests (see *arrow* 2 and close-up in **b**. **b** Close-up of elongated beta-petee with ruptured crest. Slickensides of slimes across flanks and valleys are produced by coccoid cyanobacteria. The sharp contact at the outer curve indicates the change between covering slime and the overthrusting gypsum crust. Note the rounded form of the crust which is initiated by an elastic microbial mat

Fig. 6b

(Fig. 7). Gas from the subsoil, concentrating beyond mats, may provide zones of weakness against the increase of lateral expansion pressure which finally leads to the overriding of flanks as illustrated in Fig. 7.

4.1.3 Intermediates Between Petees and Tepees

Finally in Lanzarote salterns, overthrust structures occur which are visually closest to tepees and develop from the same sequence of processes: they are folds around polygons, caused by lateral expansion of mineral encrusted microbial mats. Although finally ruptured, crests preserve a rounded appearance due to the mechanical resistance of the mats (Fig. 8a). That means that sharp ridges (Fig. 8b) are made milder by biogenic tissues and binders.

Fig. 7. Distorted beta-petee, due to continuous lateral growth of surface layers by crystal increment. One side of the layer overrides the other producing a steeply rising crust

As in tepee fabrics (Shinn 1969), the rupture of crests may proceed by lateral increment of crystals. A progressing lateral expansion can finally overturn crusts as already mentioned (cf. Fig. 7).

4.2 Classification of Petees

The foregoing description of physical overthrust structures and their biological modifications can be subsumed under the following three categories: structures, materials, mechanisms.

Structures include dome-shaped buckles which occur single and in multitude, buckles with transitions into folds, finally upfoldings around polygons. Crests of these overthrust structures can be rounded or ruptured (Figs. 5–8).

Materials in which these structures develop include (1) soft ground microbial mats (Figs. 3 and 4); (2) evaporite-encrusted microbial mats (Figs. 5–8).

Mechanisms of overthrusting include (1) subsurficial gas pressure (Fig. 4a); (2) wind/water friction or slope gravity which undulate monolayered mats across slippery subsoils (Fig. 4b, c); (3) lateral expansion of surfaces due to mineral increment (Figs. 7 and 8).

By combining these categories, a classification is suggested (Table 1) which includes the following three types of biologically modified overthrust

Fig. 8a,b. Gamma petees and tepees. **a** Gamma-petees: overthrusted, gypsum-encrusted microbial mats in polygonal arrangement with ruptured crests. **b** Tepees: folds surrounding polygons in halite-encrusted terrigenous sediments. Gavish Sabkha, southern Sinai

structures: (1) alpha-petees: dome-shaped buckles and folds generating in soft biogenic substrate, including juvenile (non-encrusted) and encrusted variations with rounded crests or buckles (Fig. 5); (2) beta-petees: modifications of encrusted alpha-petees with ruptured crests (Fig. 6); (3) gamma-petees which evolve from the same processes responsible for tepee formation, but in the presence of microbial mats which influence the form. (Fig. 8a).

As already mentioned, Bretagne salterns are characterized by juvenile alpha-petees only. Lanzarote salterns promote the whole sequence: Juvenile

Table 1. Classification of petees (overthrust structures in biogenic matrices) and tepees (overthrust structures in abiogenic matrices)

Term	Structure	Processes	Ground planes
Alpha-petee	Buckle or fold in biogenic matrix (microbial mats) juvenile or encrusted. Crests rounded, closed	a) Biologically induced: gas accumulation in the subsoil (Figs. 4a, 5). b) Abiogenic: slipping, wind or water friction (Fig. 4b, c)	Dome-shaped, irregular Parallel folds
Beta-petee	Buckle or fold in biogenic matrix, encrusted. Crests broken up	As in alpha-petees (biologically induced or abiogenic), continuation by crystal increment and lateral expansion (Fig. 6)	Dome-shaped, irregular, or as parallel folds
Gamma-petee	Folds in encrusted biogenic matrix surrounding polygons. Crests usually broken up	Lateral expansion due to crystal increment[b] (Fig. 8a)	Hexagonal[a]
Tepee	Folds in encrusted abiogenic materials surrounding polygons. Crest usually broken up	Lateral expansion due to crystal increment (Fig. 8b)	Hexagonal

[a] Hexagonal patterns according to tension minimization.
[b] Lateral expansion of surface layers also possible by hydration, e.g. of anhydrite.

alpha-petees occur in basins of lower hypersalinity (Fig. 4a,c). At salinities exceeding 120‰ S, they become immediately gypsum-encrusted (Fig. 5). With progressive increase in the concentration of brines, petees with ruptured crests appear (beta-petees, Figs. 6 and 7), as well as polygons, surrounded by upfoldings (gamma-petees, Fig. 8a).

4.3 Differentiation of Petees from Tepees

Differences between gamma-petees and tepees are least evident, although visible modifications refer to the rounded shape due to the mechanical resistance of the mats (Fig. 8a). Sharp ridges typical for tepees (Fig. 8b) are made milder by biogenic tissues and binders.

Alpha-petees, primarily soft-substrate structures, experience evaporite encrustation only under appropriate climate conditions (as on Lanzarote). Continuous crystallization may produce massive overthrusted crusts (beta-petee) which are visually similar to gamma-petees and tepees. However, the most important difference to mention is the ground plane which, because of a different primary process, is never polygonal as in gamma-petees and tepees (Table 1).

5 Discussion and Conclusion

Lanzarote and Bretagne salterns represent ecosystems marginal to the sea where engineering designs have effectively limited tide-dominated processes in favour of weather-dominated processes. The latter are responsible for conspicuous deformation structures including overthrust "petee" structures and their transitional stages to tepees.

5.1 Modes of Petee Genesis

Figure 9 concludes the processes involved in the microbially modified physical deformation structures studied: Principally, overthrusting of a continuously horizontal bedding plane is seen to develop from two different starting positions:

1. Biogenic bedding planes become overthrusted prior to crystal increments (Fig. 9a–f). In this case, the tissue-like mats withstand the tension, caused by gas pressure from below, more successfully than non-coherent substrates.
2. Biogenic bedding planes become overthrusted subsequent to their solidification by precipitating minerals (Fig. 9g–k). In this case, the tissue-like mats make otherwise sharp tepee fabrics milder.

Within Lanzarote salterns, soft-substrate petees (Fig. 9a–c) occur at moderate hypersalinity (up to 100–120‰ S). At brine concentrations exceeding 120‰ S, encrustation proceeds (Fig. 9d–f). With continued crystal increment, which is often accelerated by repeated changes between subaerial exposure and wetting, crests break up (beta-petees; Fig. 9e, cf. also Fig. 6). Structural variations are steeply rising gypsum crusts due to overriding of one side of layer across the other (Figs. 7 and 9f).

When gypsum crusts initially form parallel to bedding planes, established either atop (Fig. 9g) or on both sides of microbial mats (Fig. 9h), the lateral expansion may gradually increase with increasing crystal increment, and upfoldings develop (Fig. 9i), arranged in polygonal patterns (see also Fig. 8a). These tepee-like structures can also overturn as in beta-petees (Fig. 9k).

According to Assereto and Kendall (1977), tepees form due to the following sequence of processes: (1) Desiccation and thermal contraction cause fissures. (2) Subsequent wetting causes expansion followed by cracking. (3) Further crystallization continues lateral expansion. Margins of the cracked crusts move against each other, commonly approaching honeycomb patterns and pitched in vertical section into tent form. In this chapter, structures were described that evolve from the same processes as tepees but in the presence of microbial mats (gamma-petees). Although Assereto and Kendall (1977) did not differentiate petees, they might have registered these morphologically similar structures. However, we could not find any argument for the necessity of interaction between thermal contraction and subsequent wetting as

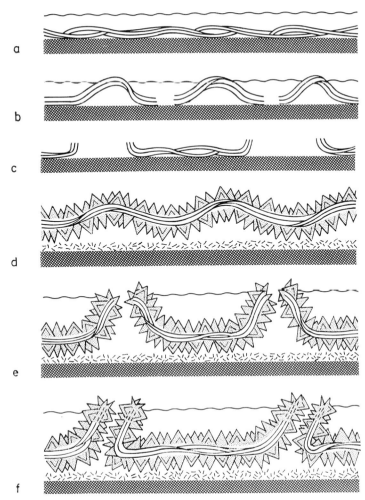

Fig. 9a–f.

Fig. 9a–k. Development of overthrust "petee" structures. **a–c** Soft-substrate structures (alpha-petees). **d–f** Solidification of initially soft-substrate petees (alpha- and beta-petees). **g–i** Petees, genesis in crusts as for tepees (gamma-petees). **k** Tepees. **a** Microbial mat parallel to bedding surface, built by filamentous microbes (microfibrous networks) and coccoid microbes (gelatinous layers; cf. also Fig. 3a). **b** Gas migrating from deeper buried organic or biogenic deposits toward the surface mat produce dome-shaped overthrust structures (cf. Fig. 4a). Tension causes cracks between domes. **c** Cracking of subaerially exposed crests leaves desiccation cracks with curled up margins (cf. Fig. 4c). **d** Dome-shaped overthrust structures (alpha-petees) become gypsum-encrusted (cf. Fig. 5). **e** Gypsum-encrusted domes or folds break up with continued crystal increment. Beta-petees develop (cf. also Fig. 6). **f** Continual gypsum crystal growth produces overriding of one side of the layer across the other. Steeply rising gypsum crusts still contain microbial mats which die off at the ends of the steeply rising crusts (cf. Fig. 7). **g** Gypsum crust established atop microbial mat, parallel to bedding plane. **h** Gypsum crusts forming on both sides enclose the mat. **i** Continual crystal increment initiates fold development around polygons (see also Fig. 8a). Continued lateral expansion forces crests to break up. Tepee-like structures form, capable of overturning as in **f**. **k** For comparison: tepees, overthrust structures in surface crusts without mats. Overturning is also possible as in **f**. Drawing by Ms. R. Flügel

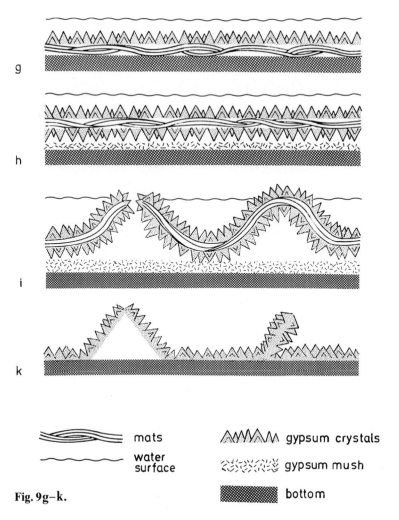

Fig. 9g–k.

mats
water surface
gypsum crystals
gypsum mush
bottom

mentioned above. We are sure that gamma-petees (as well as tepees) can form solely by crystallization pressure. Desiccation in a salt-encrusted material cannot cause contraction, and subsequent wetting never causes lateral expansion.

5.2 Salterns as a Model

1. The two geographic settings of salterns may emphasize the importance of climate overprints on the development of deformation structures. Lower mean annual temperatures and more unpredictable weather conditions are able to suppress the saline sequence even in salterns. Also, the strict horizontal separation of evaporites seems to be only realistic in lower latitude salterns.

2. Microbial growth in salterns is favoured by the continuous supply of water and mineral resources due to the engineering design. Salterns thus serve as models for microbial field cultures.

3. They may furthermore represent recent models for the hydrocarbon accumulation in evaporite salts, due to the intimate association between extensive evaporite precipitation and microbial mats (Warren 1986; Busson 1988).

4. Salterns may provide models for shallow, seasonally subaqueous to subaerially exposed marginal marine environments. Today, various examples of this type exist, e.g. at the Gulf of Mexico (Horodyski 1977), Sinai coast (Friedman 1978; Friedman and Krumbein 1985), Western Australian coast (Arakel 1980) and Gulf of California (Thompson 1968). In Laguna Mormona (Gulf of Mexico), many thousands of square meters are covered by hypersaline surface water, the water table ranging only a few centimeters above microbial mats, and subaerial exposition being possible during periods of stronger evaporation (Horodyski 1977). The center of the Gavish Sabkha (Gavish et al. 1985) is a bar-protected swamp, situated below sea level and fed by percolating seawater. Thompson (1968) described similar low supratidal basins of standing water at Colorado River Delta (Gulf of California).

According to Purser (1985), such coastal swamps have to be distinguished from true coastal sabkhas where the hydrologic regime and authigenic mineral formation are exclusively ruled by capillary evaporation. In salterns and microtidal lagoons, gypsum may also precipitate subaquatically.

Salt swamp situations with prolific mats have been common in earth history (Dunlop et al. 1978; Dean and Anderson 1978; Sarg 1981). Characteristic examples are the Messinian evaporite depositions of the Mediterranean and their association with stromatolites of the marginal marine type (Vai and Lucci 1977; Decima et al. 1988; Rouchy 1988). An ancient model is also the "paralic evaporation basin" type described by Busson and Perthuisot (1977). It describes the stage of a silled coastal area below sea level, ruled by high evaporation intensity. Considerable biogenic accumulations are known to occur within deposits of those basins. Seawater may have been percolating slowly through the salty swamp, initially supplied by surface inflow, later by seepage seawater when surface inflow stopped with retrograding sea level. It is known that gypsum forming in bioactive sediments becomes immediately reduced by bacterial sulfate reduction. This may have been the case particularly in the center of the salty swamps. Perthuisot described gypsum and polyhalite accumulation starting first with decreased bioactivity due to strongly increasing salinity, resulting in the change of the prograding sequence from biogenic layers to evaporites.

Microbe-dominated coastal swamps seem to have been common through earth history. Salterns make possible the comparative study of successional stages in lateral sequences (Fig. 1), namely the decrease of bioactivity with increasing salinity and the resulting change in depositional structures. In this way, the lateral sequence in salterns provides tools for the reconstruction of facies changes which are visible in vertical successions of evaporite sequences.

Acknowledgements. This research has been supported by the Deutsche Forschungsgemeinschaft (German Science Foundation), grant project No. Kr 333/22–1,2).

References

Adams JE, Frenzel HN (1950) Capitan barrier reef, Texas and New Mexico. J Geol 58:289–312
Arakel AV (1980) Genesis and diagenesis of Holocene evaporitic sediments in Hutt and Leeman lagoons, Western Australia. J Sediment Petrol 50:1305–1326
Assereto RLAM, Kendall CGSt (1977) Nature, origin and classification of peritidal tepee structures and related breccias. Sedimentology 24:153–210
Brock TD (1976) Environmental microbiology of living stromatolites. In: Walter MR (ed) Developments in sedimentology, vol 20: Stromatolites. Elsevier, Amsterdam, pp 141–148
Busson G (1988) Evaporites et hydrocarbures. Sci Terre 55:139 pp
Busson G, Perthuisot JP (1977) Interêt de la Sebkha el Melah (Sud-Tunisie) pour l'interprétation des séries évaporitiques anciennes. Sediment Geol 19:139–164
Cody RD, Cody AM (1988) Gypsum nucleation and crystal morphology in analog saline terrestrial environments. J Sediment Petrol 58:247–255
Cohen Y, Krumbein WE, Goldberg M, Shilo M (1977a) Solar Lake (Sinai) 1 physical and chemical limnology. Limnol Oceanogr 22:597–608
Cohen Y, Krumbein WE, Shilo M (1977b) Solar Lake (Sinai) 2 Distribution of photosynthetic microorganisms and primary production. Limnol Oceanogr 22: 609–620
Dean WE, Anderson RY (1978) Salinity cycles: evidence for subaqueous deposition of Castile Formation and lower part of Salado Formation, Delaware Basin, Texas and New Mexico. N Mex Bur Mines Mineral Resourc Circ 159:15–20
Decima A, McKenzie JA, Schreiber BC (1988) The origin of "evaporative" limestones: an example from the Messinian of Sicily (Italy). J Sediment Petrol 58:256–272
Dunlop JSR, Muir MD, Milne VA, Groves DI (1978) A new microfossil assemblage from the Archaean of Western Australia. Nature (London) 274:676–678
Friedman GM (1978) Depositional environments of evaporite deposits. In: Dean WE, Schreiber BC (eds) SEPM short course 4, Oklahoma City 1978, pp 177–188
Friedman GM, Krumbein WE (1985) Hypersaline ecosystemes – the Gavish Sabkha. Ecological studies, vol 53. Springer, Berlin Heidelberg New York Tokyo, pp 484
Gavish E, Krumbein WE, Halevy J (1985) Geomorphology, mineralogy and groundwater geochemistry as factors of the hydrodynamic system of the Gavish Sabkha. In: Friedman GM, Krumbein WE (eds) Hypersaline ecosystemes – the Gavish Sabkha. Ecological studies, vol 53. Springer, Berlin Heidelberg New York Tokyo, pp 186–217
Gerdes G, Krumbein WE (1987) Biolaminated deposits. Lecture notes in earth sciences, vol 9. Springer, Berlin Heidelberg New York Tokyo, 183 pp
Gerdes G, Krumbein WE, Holtkamp EM (1985) Salinity and water activity related zonation of microbial communities and potential stromatolites of the Gavish Sabkha. In: Friedman GM, Krumbein WE (eds) Hypersaline ecosystemes – the Gavish Sabkha. Ecological studies, vol 53. Springer, Berlin Heidelberg New York Tokyo, pp 238–266

Giani D, Giani L, Cohen Y, Krumbein WE (1984) Methanogenesis in the hypersaline Solar Lake (Sinai). FEMS Microbiol Lett 25:219–224

Gierloff-Emden HG (1981) Die Salzgartenlandschaft "Marais Salantes" der Guérande bei Le Croisic. Mitt Geogr Ges München 66:115–139

Herrmann AG, Knake D, Schneider J, Peters H (1973) Geochemistry of modern seawater and brines from salt pans: main components and bromine distribution. Contrib Mineral Petrol 40:1–24

Holtkamp E (1985) The microbial mats of the Gavish Sabkha (Sinai). Diss, Univ Oldenburg, 151 pp

Horodyski RJ (1977) Lyngbya mats at Laguna Mormona, Baja California, Mexico: comparison with Proterozoic stromatolites. J Sediment Petrol 47:1305–1320

Kalkowsky E (1908) Oolith and Stromatolith im norddeutschen Buntsandstein. Z Dtsch Geol Ges 60:68–125

Kendall CGSt, Warren JK (1987) A review of the origin and setting of tepees and their associated fabrics. Sedimentology 34:1007–1027

Krumbein WE (1979) Photolithotrophic and chemoorganotrophic activity of bacteria and algae as related to beachrock formation and degradation (Gulf of Aqaba, Sinai). Geomicrobiol J 1:139–203

Krumbein WE (1986) Biotransfer of minerals by microbes and microbial mats. In: Leadbeater BSC, Riding R (eds) Biomineralization in lower plants and animals. Univ Press, Oxford, pp 55–72

Krumbein WE (1987) Das Farbstreifensandwatt: Bau, Struktur und Erdgeschichte von Mikrobenmatten. In: Gerdes G, Krumbein WE, Reineck HE (eds) Mellum - Portrait einer Insel. Kramer, Frankfurt am Main, pp 170–187

Krumbein WE, Buchholz H, Franke P, Giani D, Giele C, Wonneberger K (1979) O_2 and H_2S coexistence in stromatolites. A model for the origin of mineralogical lamination in stromatolites and banded iron formations. Naturwissenschaften 66:381–389

Métayer C (1980) Le sel guérandais: analyses chimiques et qualités diététiques. In: Société des Sciences Naturelles de L'Ouest de la France (ed): Marais Salants. Soc Muséum d'Histoire Naturelle, Nantes, pp 73–76

Pettijohn FJ, Potter PE (1964) Atlas and glossary of primary sedimentary structures. Springer, Berlin Heidelberg Göttingen, 370 pp

Purser BH (1985) Coastal evaporite systems. In: Friedman GM, Krumbein WE (eds) Hypersaline ecosystems – the Gavish Sabkha. Ecological studies, vol. 53. Springer, Berlin Heidelberg New York Tokyo, pp 72–102

Reeves CC Jr (1970) Origin, classification, and geologic history of caliche on the southern High Plains, Texas, southeastern New Mexico. J Geol 78:352–362

Rothe P (1986) Kanarische Inseln. Sammlung geologische Führer, vol 81. Bornträger, Berlin, 226 pp

Rouchy JM (1988) Relations évaporities-hydrocarbures: l'association laminities-récifs-évaporities dans le Messinien de Méditerranée et ses enseignements. In: Busson G (ed) Evaporites et hydrocarbures. Sci Terre 55:43–70

Sarg JF (1981) Petrology of the carbonate evaporite facies transition of the Seven Rivers Formation (Guadalupian, Permian) southeast New Mexico. J Sediment Petrol 51:73–95

Schneider J, Herrmann AG (1980) Saltworks natural laboratories for microbiological and geochemical investigations during the evaporation of seawater. In: Coogan AH, Hauder L (eds) 5th Symp Salt. North Ohio Geol Soc, pp 371–381

Shinn EA (1969) Submarine lithification of Holocene carbonate sediments in the Persian Gulf. Sedimentology 12:109–144

Shinn EA (1983) Tidal flat environment. In: Scholle PA, Bebout DG, Moore CH (eds) Carbonate depositional environments. Am Assoc Petr Geol Mem 33:172–210

Shinn EA (1986) Modern carbonate tidal flats: their diagnostic features. In: Hardie LA, Shinn EA (eds) Carbonate depositional environments, modern and ancient. Col School Mines Q 81:7–35

Thomas JC, Geisler D (1982) Peuplements benthiques à cyanophcées des marais salants de Salin-de-Giraud (Sud de la France). Geol Méditerr 9:391–411

Thompson RW (1968) Tidal flat sedimentation on the Colorado River Delta, northwestern Gulf of California. Geol Soc Am Mem 107:133

Vai GB, Lucchi FR (1977) Algal crusts, autochthonous and clastic gypsum in a cannibalistic evaporite basin: a case history from the Messinian of northern Apennines. Sedimentology 24:211–244

Warren JK (1982) The hydrological significance of Holocene tepees, stromatolites, and boxwork limestones in coastal salinas in South Australia. J Sediment Petrol 52:1171–1201

Warren JK (1986) Shallow-water evaporitic environments and their source rock potential. J Sediment Petrol 56:442–454

Part II: Environmental Geochemistry

The Pollution of the River Rhine with Heavy Metals

KARL-GEERT MALLE[1]

CONTENTS

Abstract 279
1 Introduction 279
2 Concentrations of Metals 280
3 Dissolved or Undissolved? 284
4 The Metal Load and Its Origin 287
References 289

Abstract

Since the early 1970s the heavy-metal content of the Rhine has gone down considerably. Especially the point sources have become less important. The background pollution counts are around 20% for Pb and Zn, > 20% for Hg and Cd, 40% for Cu, 50% for Cr, and 60% for Ni. As a third source diffuse pollution (rain, surface erosion, etc) plays a role especially with Zn, Cd, Pb. The analytical procedures for evaluating metal concentrations are in some cases no longer sensitive enough to control the very low metal content.

1 Introduction

The large volume and even flow of the River Rhine have been a significant factor in the dense population and heavy industrialization of its catchment area. Today, there are about 50 million people living and working there. The Rhine links five European countries, which use it as a shipping route, a recreational area, and also as a recipient of their wastewater. At the same time, the Rhine provides drinking water for about 8 million people – 20 million if Lake Constance is included. Finally, the Rhine is the largest inlet to the North Sea, and dissolved as well as undissolved substances contained in its water play a major role in its pollution.

---

[1] BASF Aktiengesellschaft DU/A–C 100 D-6700 Ludwigshafen, FRG

For all of these reasons, the quality of the water in the Rhine has been monitored carefully for many years; in fact the Rhine is one of the best monitored rivers in the world. The authorities in the countries bordering the river have installed a range of extremely up-to-date, permanent sampling points, and have carried out complex water analysis. The individual German states publish quality reports[1], the results of which are incorporated in the annual report of the German Commission for the Protection of the Rhine (DKSR)[2]. The German measurements are aligned with those of the other countries and form part of the annual reports of the International Commission for the Protection of the Rhine (IKSR)[3]. Parallel to this, sensitive monitoring is carried out by the water works in the Rhine catchment area, their measurement data also being published in annual reports on the water quality at a national level (ARW, German Water Works[4]; RIWA, Dutch Water Works[5] and at the international level, IAWR[6].

The large number of data show that it is difficult to take representative samples from the Rhine. The great width and shallow depth of the river mean that thorough cross-mixing takes place extremely slowly, and below specific discharge points, trails of increased concentration can be formed, which are many kilometers in length. Official sampling points, which have a fixed location and have been carefully selected, are therefore especially important. When taking samples to determine the metal concentrations, it must also be taken into account that metals are transported in rivers both in dissolved and undissolved form, together with the suspended matter in the river. Comparable samples must, therefore, also include the suspended matter in the river in the same way, which is dependent on the conditions of sampling, e. g. the depth. The time and place of sampling also affect the content of suspended matter, and thus, the metal concentrations found.

At present, the investigating institutions listed all quote the total metal content in *unfiltered* samples. The most important measuring point is at the German/Dutch border, because beyond this the Rhine divides into the Lekl, Waal, and Ijssel without any significant additional discharges of metal occuring beforehand on Dutch territory. The Dutch sampling station on the right bank of the river at Lobith is still within the trail of the Emscher discharge and, as far as some specific parameters are concerned, has higher concentrations than the German samples taken at Bimmen, on the left-hand side.

2 Concentrations of Metals

Figures 1 and 2 show a compilation of the annual average values of the concentrations in unfiltered samples for the eight most important heavy metals, obtained from the various reports on water quality. Metal contents were published for the first time in 1971 by De Groot and co-workers[7]. These are compared in Table 1 with the data[1] recently published for 1988.

Table 1. Content of heavy metals in the Rhine (µg/l), annual average values, German/Dutch border

	Hg	Cd	Pb	As	Cr	Cu	Ni	Zn
1971 [7]	1.4	3.3	36	13	59	31	14	261
1988 [1]	<0.2	<0.3	6	2	6	8	5	36

Table 2. Content of heavy metals 1987 (µg/l); all official German sampling stations

Sampling station	Hg	Cd	Pb	As	Cr	Cu	Ni	Zn
Rhine km 22.9 Öhningen	<0.2	<0.3	< 5	1	< 2	3	< 5	<30
Rhine km 248.9 Weisweil	<0.2	<0.3	< 5	<1	< 2	4	< 5	57
Rhine km 362.3 Maxau	<0.2	<0.3	< 5	<1	< 2	3	< 5	32
Rhine km 498.5 Mainz	<0.2	<0.3	< 5	<1	< 2	7	< 5	28
Rhine km 590.3 Koblenz	<0.2	<0.3	< 5	2	4	7	< 5	36
Rhine km 640.0 Bad Honnef	<0.2	<0.3	< 5	<1	4	6	< 5	24
Rhine km 865.0 Bimmen	<0.2	<0.3	10	2	9	8	5	52
Neckar km 3.2 Mannheim	<0.2	<0.3	< 5	<1	< 2	5	< 5	57
Main km 67.0 Kahl	<0.2	<0.3	< 5	<1	3	6	< 5	24
Main km 3.2 Kostheim	<0.2	0.4	8	2	9	18	9	216
Saar km 91.9 Saarbrücken	<0.2	<0.3	10	<1	< 2	4	< 5	23
Saar km 6.7 Kanzem	<0.2	0.8	9	2	< 2	5	< 5	44
Moselle km 230.0 Palzem	<0.2	<0.3	< 5	2	4	4	< 5	36
Moselle km 2.0 Koblenz	<0.2	<0.3	7	2	4	5	< 5	41
Emscher km 7.8 Duisburg	<0.2	<0.3	6	3	26	12	14	32

It is also interesting to compare Table 1 with the compilation of the measurement data for all the official German sampling stations on the Rhine and its tributaries (Table 2).

All the investigating institutions confirm the noticeable reduction in the heavy metal content of the Rhine, as shown in Figs. 1 and 2 and Table 1. The figures make it clear that over the years the discrepancies in the results, which in some cases were considerable at the beginning, have become smaller in time. Besides general difficulties the possibility of analytic errors in the trace sector, and a different content of suspended matter which may occur during sampling, must be avoided. Moreover, the relationship between concentration and flow rate must be taken into account. In the case of (anthropogenous) discharges at definite points, the load to a large extent does not depend on the flow rate i.e. the concentration is lower when the flow rate is high. In the case of background pollution or diffuse sources (rain, surface erosion), on the other hand, the concentration, at a first approximation, does not depend on the rate of flow. The last year in which the flow rate was noticeably low was 1976 (Bimmen 1390 m^3/s), and years in which the flow

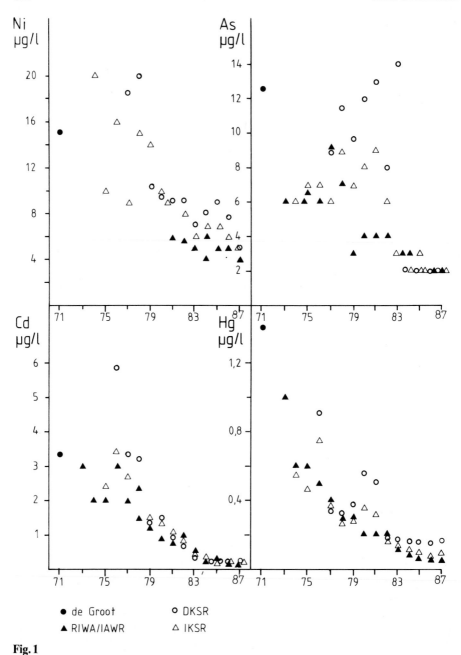

Fig. 1

Figs. 1. and 2. Concentration of metals (annual average, at Bimmen/Lobith, Rhine km 860, German-Dutch border, unfiltered sample). de Groot [7]; DKSR [2];RIWA/IAWR [5,6];IKSR [3]

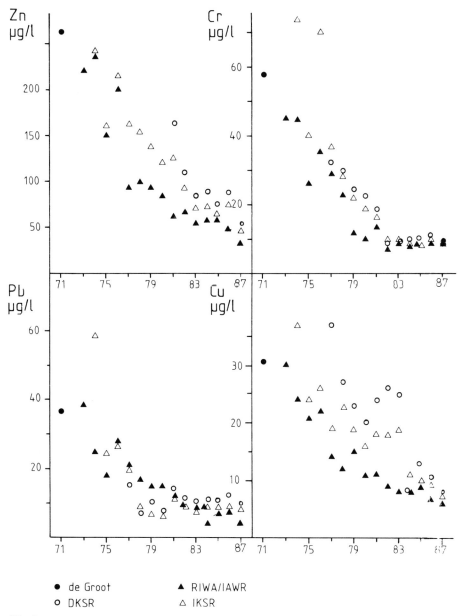

Fig. 2

- ● de Groot
- ○ DKSR
- ▲ RIWA/IAWR
- △ IKSR

rate was high were 1981 (Bimmen 2960 m³/s) and 1987 (Bimmen 2920 m³/s). In 1976, unusually high concentrations of the metals Cr, Cd, and Hg were detected, whereas no particularly low levels were measured in 1981 and 1987. This indicates that discharges at particular points have, in the meantime, become significantly less important, a fact also confirmed by the discharge inventories. Previously, there were considerable discrepancies between the individual investigating institutes, especially in the case of Ni, As, and Cu contents. In the case of Ni and As, the different values can probably be explained by variations resulting from the ubiquitous origins; analysis of copper previously met with procedural difficulties. If, today, the discrepancies between the various investigating institutes are extremely small, then this proves the extent to which it has been possible to standardize the methods of sampling and analysis.

It can generally be stated that metal contents today are often at or below the detection limit of the German standard methods (Deutsche Einheitsverfahren zur Wasser-, Abwasser- und Schlammuntersuchung). This makes it difficult to calculate the metal loads of the Rhine. Only maximum loads can be given for Hg and Cd, as well as Pb, As, Ni, and Cr, at many sampling stations in the catchment area.

The Rhine tributaries are in some cases more heavily polluted; with the Saar, the transport of metals via the Rossel must also be considered. In the case of dammed tributaries such as the Neckar, Main, and Moselle, one must always consider the subsequent discharge of historically polluted sediment when high water occurs.

3 Dissolved or Undissolved?

Rivers transport metals in both dissolved and undissolved form. It is not only the solubility product of defined compounds of low solubility, such as sulfides, which is important, but rather, the transport of the metals absorbed onto the ever present hydroxides of iron and manganese as well as other fine-grained suspended matter with a large surface area. The undissolved proportion of the heavy metals is contained almost without exception in the suspended matter, or in the fine-grained sediment of the < 2 μm fraction. Such fine-grained sediment is found in the undammed part of the Rhine, i.e. in the German section of the river below the last dam level at Baden-Baden not in the entire river bed, but only in relatively few, quite specific locations such as branches of the old Rhine, docks, and the stagnant areas of groynes. For this reason, representative, comparable sampling requires particular experience and local knowledge. It was fortunate that until 1983 the RIWA[5] provided results of analyses both on the filtered and unfiltered samples. When taken with the content of suspended matter, also given, it was thus possible to calculate the metal load of the river, divided according to solution and suspended matter [8,17] (Fig. 3). Recently, the International Commission for the Protection of the Rhine (IKSR) published metal concentrations of the suspended matter, directly centrifugated from the samples taken at the international sampling stations. The data analyzed in 1987, at a rather high water flow are included in Table 3.

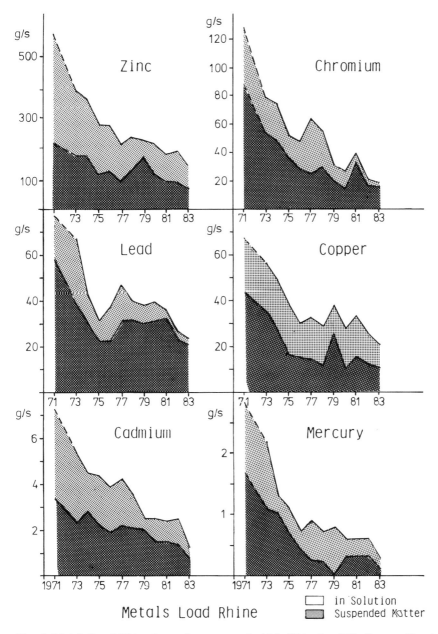

Fig. 3. Metals Load Rhine (annual average, at Lobith, Rhine km 860, German Dutch border, [5])

Table 3. Heavy metal contents of sediment or suspended matter in the Rhine (mg/kg)

		Hg	Cd	Pb	Cr	Cu	Ni	Zn
Sediment	1922 [10]	a	4.4	275	110	68	36	1050
	1958 [10]	a	14	535	640	295	54	2420
	1970 [10]	a	27	445	790	325	62	1855
	1977 [10]	a	27	390	825	285	76	1665
	1979 [9]	5.2	27	440	370	230	a	1640
Suspended matter	1971 [7]	17	34	590	900	439	75	2100
	1973 [8]	13	28	480	650	430	a	2200
	1978 [8]	3.2	29	420	390	160	a	1800
	1983 [8]	(0)	8	200	160	100	50	690
	1987 [3]	1	2	115	80	75	34[b]	490
Port of Rotterdam	1977 [11]	5.4	26	330	480	230	a	1420
Class III	1986 [11]	2.0	7.4	160	130	110	39	690

[a] No data published.
[b] Data from 1986.

The contents in the suspended matter show a surprising conformity with those determined at the same time directly from the $< 2\,\mu m$ of the sediment[9]. It is interesting to compare the data with values obtained on Rhine sediments from the years 1922 and 1958[10], and with the content of dredged material from a particularly polluted part of the port of Rotterdam[11] (Table 3).

When comparing Tables 1 and 3, it is again evident that concentrations of heavy metals in suspended matter or sediment are higher by a factor of approximately 1000 than the values for the total metal content of the unfiltered sample from the river.

The reduction in overall pollution is reflected by the decrease of the pollution of the sediment; a certain "reverberation effect" becomes apparent, especially where the dredged material from the port of Rotterdam is concerned. It takes years for the quality of the sediment to reach that of the water; the sediment has a "memory" for earlier pollution. Nevertheless, the values given for the suspended matter in 1987 are considerably below those for the sediment in 1958, and even in 1922, which gives an indication of how great an improvement has taken place in recent times. As a measure of the level of pollution of a sediment, Müller[12] proposed a six-step geoaccumulation index (Igeo), which takes a geochemical background as reference value. The basis for the reference values is provided by the metal contents of the sediments of aquatic depositional regions, which can hardly suffer human influence, for example, Lake Constance, or unpolluted parts of the North Sea. If the geoaccumulation index (Igeo) is applied, the suspended matter taken from the Rhine in 1987 is classified as "unpolluted" (Igeo = 0), with respect to copper, chromium, and nickel; as "unpolluted to moderately polluted" (Igeo = 0–1), with

Table 4. Undissolved proportion of the total metal content (%)

	Hg	Cd	Pb	As	Cr	Cu	Ni	Zn
1971 [7]	56	46	72	57	69	64	24	37
1983 [8, 17]	>90	60	89	33	86	50	29	50

respect to mercury; and as "moderately polluted" (Igeo = 2), with respect to cadmium, zinc, and lead. Table 4 gives the proportion of the total metal content determined in the suspended matter, i.e. in an undissolved form, in 1971 and 1983.

The metals Pb, Hg, and Cr are mostly transported in undissolved form, i.e. with the suspended matter, whereas Ni and As remain mostly in solution.

4 The Metal Load and Its Origin

If we neglect the specific pollution of such rivers as the Oker, in the Harz region, which flow through ore deposits, and are thus charged with very high geogenous metal loads, every river, regardless of anthropogenous influences, carries a basic load of metals, both dissolved and undissolved. This "background" pollution of sediments and suspended matter was used by Müller [12] to establish his geoaccumulation index. The average content of suspended matter in the Rhine is 36 mg/l[8,17], from which the background pollution of the proportion of suspended matter can be calculated[13]. The background pollution of the solution, i.e. of a filtered sample of riverwater, was recently reported on by Wachs [14]. Table 5 compiles both values and compares their sum with concentrations measured in 1988.

According to this rough calculation, the proportion of background pollution for Pb and Zn is around 20%; for Hg and Cd >20%; for Cu 40%; for Cr 50%, and for Ni 60%. In this respect, very thorough investigations, of ca. 5000 water samples taken at roughly 360 specific discharge points, are of particular interest; they were commissioned by the port of Rotterdam for the entire length of the Rhine, in order to detect the origin of heavy metals in the river (and thus, to a certain extent, also in the dredged material of the

Table 5. Background pollution of the Rhine (µg/l)

	Hg	Cd	Pb	Cr	Cu	Ni	Zn
Undissolved	0.015	0.01	0.7	3.2	1.6	2.4	3.4
Dissolved	0.03	0.05	0.2	0.1	1	0.3	5
Total (Rounded)	0.04	0.06	1	3	3	3	8
Measured in 1988	<0.2	<0.3	6	6	8	5	36

Table 6. Proportions of background pollution and point sources (%)

	Hg	Cd	Pb	Cr	Cu	Ni	Zn
Background	>20	>20	20	50	40	60	20
Point sources[a]	×	22	32	42	18	×	17

[a] × denotes not measured.

port)[15]. Analytical procedures were used, with detection limits well below the values of the German standard methods (Cd 0.02 µg/l; Cr 0.2 µg/l; Cu 0.1 µg/l; Pb 0.2 µg/l; Zn 2 µg/l). Nevertheless, in 1985, only 17% Zn, 18% Cu, 22% Cd, 32% Pb, and 42% Cr could be assigned to definite discharge points, i.e. the trail of certain discharges, waste pipes, or sewers (Table 6).

In this context it should be considered that, especially with Zn, Cd, and Pb, a third source must also play a role: the diffuse pollution of anthropogenous origin (rain, surface erosion, and residual pollution from the outlets of communal sewage treatment plants). According to the current precise annual balances for the total use and discharge of Hg and Cd[16], in 1983, less than 1 t Hg and 9 t Cd from manufacture and commercial use were discharged annually into all German surface waters. The lead load of the waters is, to a great extent, the result of the lead content of gasoline, which means that a reduction can be expected in the future as a result of the use of lead-free gas. Zinc in wastewater is mostly caused by the corrosion of galvanized iron parts, water pipes, sheet metals, and by the decomposition of tires. All these sources lead to diffuse discharge.

When establishing the metal loads carried annually by the Rhine, the proportion of discharges from point sources becomes less and less significant, and the variations in absolute quantities become increasingly dependent on the rate of flow. In this way a high rate of flow has, in the past, already led to the incorrect interpretation that the discharge of heavy metals had again increased. For purposes of longer term overall load calculations, it is probably advisable in the future to use an average flow rate (e.g. Bimmen 1973–1983, 73 billion m^3/a[13]) and concentrations at the time considered. In

Table 7. Metal loads carried by the Rhine with an average flow rate (tons per years, German/Dutch border)

	Hg	Cd	Pb	As	Cr	Cu	Ni	Zn
LWA-NRW 1988[a]	<15	<20	450	150	450	600	400	3000
Dutch values	4.5[5]	8.5[15]	240[15]	180[5]	400[15]	350[15]	360[5]	2800[15]

[a] Landesamt für Wasser und Abfall Nordrhein-Westfalen.

this case, it is rather unsatisfactory that very low concentrations, at or below the limit of detection, must be multiplied by enormous quantities of water. If the concentrations[1] measured in Germany in 1988 are taken as a basis, the result is the rounded loads given in Table 7 with an average flow rate. This table also includes, for purposes of comparsion, the data calculated in recent years by the Dutch authorities, using considerably lower detection limits[5,15].

These loads are only limitedly suitable for the calculation of the North Sea pollution. As regards Cd and Pb, considerable additional pollution[15] can be found at present in the Netherlands, while, as a result of complicated hydraulic engineering measures, dams, and dredging piles, a considerable part of the metal load detected at the German/Dutch border remains in the Netherlands and does not reach the sea[10]. In 1981, it was estimated that around 70% of the metals Cd, Pb, Cr, Cu, and Zn, and 35% Ni, carried by the rivers Rhine and Maas into the Netherlands, also remain there[19], while only the remaining portions reach the North Sea.

Heavy metals have become of particular interest in recent decades within the framework of environmental pollution investigation. This has without doubt been due to the fact that mass mercury poisoning in Japan and elsewhere has given rise to the keen desire for greater precautions. It is, also due, however, to the fact that during this time, relatively simple and highly sensitive analytical procedures have been made available for determining metal content. The metal content in the Rhine has, in the meantime, dropped so low, that these very methods for monitoring particularly relevant metals, such as Hg and Cd, no longer show the required sensitivity.

References

1) Landesamt für Wasser und Abfall, Nordrhein-Westfalen (ed) Rheingütebericht (Annu Rep).
2) Deutsche Kommission zur Reinhaltung des Rheins (ed) Zahlentafeln der physikalisch-chemischen Untersuchungen (Annu Rep).
3) Internationale Kommission zum Schutz des Rheins gegen Verunreinigung (ed) Zahlentafeln. Koblenz (Annu Rep).
4) Arbeitsgemeinschaft Rheinwasserwerke e. V. (ARW) (ed) Jahresberichte. Düsseldorf (Annu Rep)
5) Rhijnkommissie Waterleidingsbedrijven (RIWA) Amsterdam (ed) Berichte (Annu Rep)
6) Internationale Arbeitsgemeinschaft der Wasserwerke im Rheineinzugsgebiet (IAWR) (ed) Rheingüteberichte (Annu Rep)
7) De Groot AJ, Allersma E, Van Driel W Zware Metallen in fluviatiele en marine Ecosystemen. Symp Waterloopskunde, 23/25 05.73 Publ 110 N, Sect 5
8) Malle KG, Müller G (1982) Metallgehalt und Schwebstoffgehalt im Rhein, pt 1. Z Wasser Abwasser Forsch 15:11–15
9) Müller G (1985) Unseren Flüssen geht's wieder besser. Bild Wiss 10:75–97
10) Salomons W, Eysink WD (1981) Pathways of mud and particulate trace metals from rivers to the southern North Sea. Int Assoc Sediment Spec Pap 5:429–450
11) Malle KG (1988a) Baggerschlamm aus Rotterdam. Umwelt 8:143–146

12) Müller G (1979) Schwermetalle in den Sedimenten des Rheins – Veränderungen seit 1971. Umschau 72:192–193
13) Malle KG (1988b) Die Bedeutung der Hintergrundbelastung bei der Bilanzierung der Metallfrachten im Rhein. Z Wasser Abwasser Forsch 21:20
14) Wachs B (1988) Gefährliche Stoffe im Abwasser und Oberflächenwasser, Münchener Beitr Abwasser Fischerei Flußbiol 42:176–243
15) Havenbedrijf Rotterdam (ed) (1989) Project Onderzoek Rijn, Rapportage 2e fase, 15.3.89
16) Balzer D, Rauhut A Quecksilber-Bilanz, Cadmium-Bilanz. In: Landesgewerbeanstalt Bayern (ed) Nürnberg
17) Malle KG (1985) Metallgehalt und Schwebstoffgehalt im Rhein, pt 2. Z Wasser Abwasser Forsch 18:207–209

Interactions of Naturally Occurring Aqueous Solutions with the Lower Toarcian Oil Shale of South Germany

H. PUCHELT AND T. NÖLTNER[1]

CONTENTS

Abstract 291
1 Introduction 291
2 Experimentation 292
3 Results 295
4 Conclusions 307
References 308

Abstract

An oil shale sample of the bituminous Lower Toarcian from South Germany was leached, after different thermal pretreatment, with solutions displaying pH values ranging from 3 to 9. Reactions to the extraction of bulk trace heavy metals dependent on time for water to rock interaction, pH of leaching solutions, solution: solid ratio, and pretreatment of the shale are reported here.

ICP-MS analysis both of bulk rock samples and of extracted fractions permitted the detection of the full trace element spectra below the ng/ml (ppb) range. The ultimate goal of this study was to increase knowledge on the concentration and speciation of trace heavy metals in black shales and, thus, to contribute to better environmental protection.

1 Introduction

Magmatic, metamorphic, and sedimentary rocks react rapidly with interacting solutions to imprint their characteristic major and trace element patterns to the solution. Since certain trace element concentrations are toxic to humans, animals, and plants, it seems important to know the amount of elements which are leached from various substrates and, thus, may enter into the food chain.

---

[1] Institute for Petrography and Geochemistry, Kaiserstr. 12, University of Karlsruhe, D-7500 Karlsruhe, FRG.

Bituminous shales are generally characterized by significant trace element concentrations (Vine and Tourtelot 1970). If soils form from such organic-rich shales, these trace element contents are completely transferred into the new stratum and may even be further enriched by biological processes.

However, little is known about the complete trace element patterns in black shales and the metal speciation, compared to a wealth of organic-geochemical data. Despite high bulk trace element contents (Brumsack and Thurow 1986), these constituents often enter the aqueous solution in minor concentrations only due to the prevailing modes of bonding. The most important bonding types in black shale environments normally are:

1. Carbonate phases (calcite, dolomite, ankerite, siderite);
2. Sulfide phases (pyrite, marcasite, sphalerite);
3. Silicate phases (clay minerals, mica, minor siliciclastics);
4. Organic phases (kerogen, bitumen, organic acids).

All experiments were carried out using a Lower Toarcian oil shale sample (Posidonia shale) collected from an open-pit mine in SW Germany near Balingen, FRG. The Lower Toarcian oil shales attracted attention early due to the European-wide spectacular mass mortality of animals during Upper Liassic times and to the perfect fossil preservation. The shales were studied under different viewpoints because of their economic significance (von Gaertner 1955; Einsele and Mosebach 1955; Hoffmann 1966; Fesser 1968; Joachim 1970; Rieber 1975; Kauffman 1981; Küspert 1982; Seilacher 1982; Walk 1982; Riegraf et al. 1984; Riegraf 1985; Mann 1987; Rullkötter et al. 1987). The sedimentary rocks of the bituminous sequence (thickness 2–20 m in S Germany) were and are industrially processed by carbonizing for distilling of oil and as a component for cement production. Large agricultural areas, particularly in S Germany, have soils with substrates of the Lower Toarcian oil shales. Improved knowledge of metal speciation in black shales is required for better environmental protection.

2 Experimentation

In order to elucidate the importance and the impact of the different mineral phases on the leachability of various elements from the Lower Toarcian oil shales and, thus, their path into the food chain, 5 kg of the material were ground to < 30 µm by treatment in an agate disc mill and then were carefully homogenized. A series of leaching tests was performed with (1) a fresh bituminous shale sample, and the same material pretreated by heating (2) in activated oxygen, (3) at 500° C, and (4) at 1000° C.

Analysis of this material was done by wet chemical procedures and wavelength dispersive XRF (major constituents, only of the fresh sample dried at 105° C), carbon-sulfur analyzer (S + C), atomic absorption, in-

Table 1. XRF, wet chemical, and AAS analysis of the major constituents (in wt %) of the fresh oilshale sample dried at 105° C. Fe_{tot} calculated as Fe_2O_3

SiO_2	23.440
TiO_2	0.450
Al_2O_3	7.840
Fe_2O_3	4.070
MnO	0.052
MgO	1.470
CaO	27.910
K_2O	1.410
Na_2O	0.240
P_2O_5	0.140
S	4.180
H_2O^+, $C_{org.}$, CO_2 } in LOI	
LOI	28.850
Σ	99.78

strumental neutron activation (Kramar and Puchelt 1982), and ICP-MS (Puchelt and Nöltner 1988, 1989). The solutions for ICP-MS analysis were prepared by a microwave pressure digestion method (Nöltner et al. 1989).

The results of the major elements of the bulk material are given in Table 1. Table 2 compiles the full trace element spectra of the fresh and pretreated samples analyzed with ICP-MS.

In order to simulate natural leaching conditions and simultaneously study the kinetics of this process, the following experiments were set up:

1. 4 g and 0.4 g of the oil-shale powder were treated in a 250 ml polypropylene bottle with 200 ml of double distilled water on a shaking table for 0.1, 0.3, 1, 3, 10, and 30 days respectively. Four pH values were used for each reaction time: 3, 5, 7, 9 initially. The pH values were generated by addition of ultrapure sulfuric acid, the basic pH by adding a solution of analytical grade sodium hydroxide. The pH values given are only valid for the beginning of the experiments. After the end of the experiment the solutions were separated from the solids by filtering through microfilters (<0.45 µm) using a carbosulfon vacuum filtering device.

2. In nature, part of the bituminous material is oxidized (weathered). This was simulated by treating the material with activated oxygen in a cold plasma asher (Knapp 1985).

3. During and after World War II, the Posidonia shale in Germany was carbonized at low temperature to distill oil and gasoline. These conditions

Table 2. Trace element concentration in mg/kg (ppm) in the bulk samples after different thermal pretreatment

	Fresh sample	Cold-ashed	500° C	1000° C
Li	19.06	19.86	25.83	33.88
Be	1.39	1.48	1.76	2.45
B	1.42	–.–	1.75	<d.l.[a]
Sc	2.09	7.73	9.17	11.41
Cr	59.94	61.00	63.91	64.57
Mn	402	423	450	562
Co	16.41	16.97	18.62	24.00
Ni	60.39	64.67	73.45	95.03
Cu	59.59	65.83	67.45	88.61
Zn	158	131	142	182
Ga	9.20	10.41	9.06	11.63
As	15.30	16.06	17.48	–.–
Se	1.79	1.92	3.49	<d.l.
Rb	53.13	56.01	59.58	72.31
Sr	2480	2620	2670	3400
Y	16.11	16.79	18.24	23.48
Zr	70.60	73.13	79.23	97.62
Nb	7.22	8.13	8.00	10.09
Mo	19.08	20.40	21.62	30.22
Ag	0.41	0.39	0.46	0.65
Cd	1.29	1.18	1.75	1.19
Sn	1.62	1.90	1.80	1.98
Sb	2.06	2.12	2.14	2.75
Cs	8.69	3.45	3.50	4.16
Ba	161	174	172	220
La	21.57	22.31	23.25	29.72
Ce	40.63	42.82	43.84	56.16
Pr	4.84	5.22	5.27	6.58
Nd	18.97	19.55	19.66	26.21
Sm	3.34	4.06	3.91	5.02
Eu	0.88	0.95	1.06	1.15
Gd	4.05	4.77	4.95	5.32
Tb	0.61	0.64	0.62	0.82
Dy	3.19	3.44	3.85	4.30
Ho	0.70	0.73	0.77	0.95
Er	1.96	2.08	2.10	2.78
Tm	0.28	0.32	0.34	0.43
Yb	2.11	2.01	2.11	2.72
Lu	0.25	0.32	0.33	0.37
Hf	2.03	2.21	2.60	3.11
Ta	0.73	0.65	0.83	0.91
W	0.90	0.92	0.96	1.21
Re	0.23	0.15	0.27	<d.l.
Tl	5.49	5.90	5.92	0.07
Pb	12.92	13.68	20.24	6.25
Bi	0.17	0.18	0.18	0.27
Th	5.88	6.39	6.44	8.15
U	4.86	5.18	5.31	6.69

[a] <d.l. = Below detection limit.

were reproduced by heating the original shale to 500° C in a laboratory muffle furnace until its weight remained stable. According to original documents, temperatures exceeding 500° C were reached during carbonizing of the shale. Along the Suebian Alb (S Germany) several piles of this carbonized material still exist (Maisenbacher 1989, pers. commun.).

4. After 1945, the Lower Toarcian oil shale was locally burned to produce light building blocks. In this process higher temperatures were reached than for the distilling of liquid hydrocarbons. In the present experiments, Posidonia shale was heated to 1000° C in a laboratory muffle furnace to weight constancy.

All the pretreated shale samples were leached in the same way as described under (1). The experiments were performed at the solid:solution ratios of 1:50 and 1:500 respectively.

During oxidizing or heating of the samples a change of the bonding type of many elements occurred, causing either partial loss of some trace constituents or enrichment by different ashing factors. Consequently, the extraction behaviour of the fresh sample differs greatly from the behaviour of the pretreated samples.

3 Results

A qualitative impression of the impact of heating of the Lower Toarcian oil shale is given for two atomic mass unit ranges by the mass spectra from 5–12 amu and from 200–210 amu. In Fig. 1a–d the content of the solid samples is given for lithium, beryllium, and boron. The dilution factor is always 1000. In Fig. 1aa–dd the results of leaching these samples for 30 days at a solution:sample ratio of 50, starting at an initial pH 3 is given. The dilution factor in this case is 1. Trace element analyses have been performed by ICP-MS using a VG Plasmaquad. The analytical procedure is described by Puchelt and Nöltner (1989). The quality of all data was checked by running the NBS reference material SRM 1643b (trace elements in water) at the beginning and at the end of each analytical procedure. In Fig. 1a (fresh sample) the two masses of lithium (6, 7), one for beryllium (9), and two for boron (10, 11) are clearly visible. Figure 1b (sample ashed with activated oxygen) and Fig. 1c (sample ashed at 500° C) exhibit an increase of these elements due to ashing. In Fig. 1d (sample ashed at 1000° C) boron is lost in part, while the other elements are further enriched.

In Fig. 1aa–dd the results of leaching of these materials from the 50:1 solution:solid systems are presented. Berillium is not leached from any of the samples. The extraction of lithium increases more than proportionally with the heating temperature. Boron, only slightly leached from the fresh sample, is strongly activated by heating and is even more leached from the material

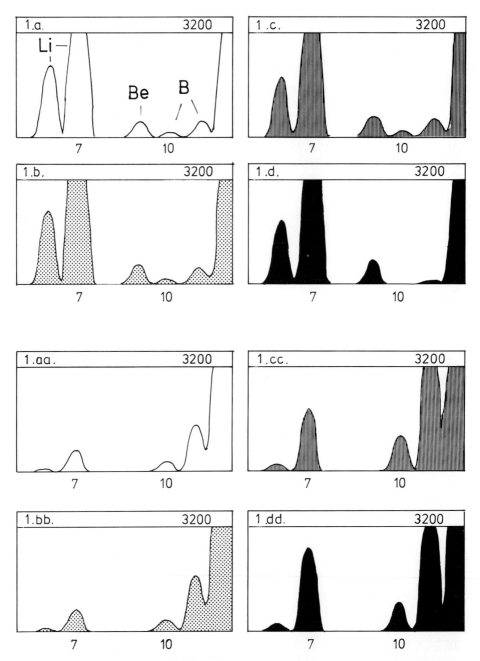

Fig. 1. a–d. Mass spectra from 5–12 amu for fresh Posidonia shale; **b** same material cold ashed with activated oxygen; **c** ashed at 500° C; **d** ashed at 1000° C. Dilution factor for analysis: 1000. **aa–dd** Leaching solutions from materials **a–d** after 30 days of reaction. Solution: sample = 50:1, initial pH 3, dilution factor 1

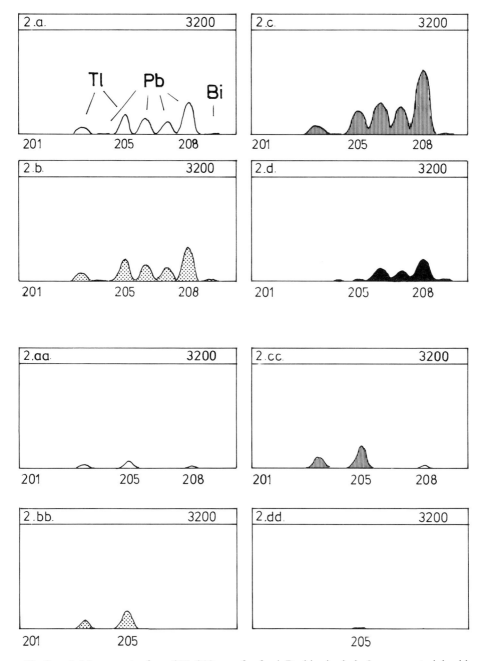

Fig. 2. a–d. Mass spectra from 200–210 amu for fresh Posidonia shale; b same material cold ashed with activated oxygen; c ashed at 500° C; d ashed at 1000° C. Dilution factor for analysis: 1000. aa–dd Leaching solutions from materials a–d after 30 days of reaction. Solution:sample = 50:1, initial pH 3, dilution factor 1

heated to 1000° C than from the fresh shale despite significant loss due to evaporation at this temperature.

The mass range from 200–210 amu is characterized in Fig. 2a–d. Three elements are monitored: thallium (isotopes 203 + 205), lead (204, 206–208), and bismuth (209). During heating, the thallium concentration increases with cold ashing and only slightly during ashing at 500° C. It is almost completely evaporated by heating to 1000° C. Lead is enriched by heating to 500° C but significant loss occurs during heating to 1000° C.

Bismuth is only enriched proportionally to the loss of volatile constituents. The extraction patterns show very limited solubility of lead in the fresh material (Fig. 2aa) and little solubility in the sample ashed at 500° C. No bismuth occurs in any leaching solutions.

Thallium can be slightly leached from the fresh Posidonia shale but much more from the material heated at 500° C. Virtually no thallium is extractable from the sample heated at 1000° C, which has retained a negligible concentration of this element.

In general, trace elements in bituminous shales can be either enriched by heating due to loss of volatile substances or they can evaporate. In addition, the bonding type of elements may change from fresh natural material to the same material heated at various degrees. The final concentrations for most trace elements in the leaching solutions are already reached after a reaction time of 1–3 days. The elements present in the Lower Toarcian oil shale can be grouped as follows:

1. Elements with no considerably different behaviour in fresh and pretreated Posidonia shale material (Co, Hf, Nb, Sn, Zn, Th, Y, Zr, REE).
2. Elements which can be strongly leached from heated samples (As, Cd, Ba, Cu, Cr, Cs, Li, Mo, Ni, Rb, Sr, Ta, V, W).
3. Elements which were hardly or not at all soluble in the fresh and pretreated materials (Be, Ga, Sb, Sc, Th, U).
4. Elements which are at least partially volatilized during ashing of the sample at 1000° C (B, Pb, Re, Se, Tl).

During the leaching experiments all solutions changed to alkaline pH values regardless of the initial (unbuffered) pH. If the experiments started from pH 3, the following values were found after the longest interaction time of 30 days: 7.9 for the fresh sample; 8 for the cold ashed sample; 8 for the 500° C ashed sample; and 12.05 for the 1000° C ashed sample.

Starting at an initial pH 9, most samples produced solutions of pH 8.3–8.5 after 30 days, however, the extraction of the sample ashed at 1000° C reached pH 12.15 (Table 3).

From the group of elements mentioned above except group 3, two examples of their leaching behavior are presented in the following.

Ad 1: For this group, the behavior of cobalt (Fig. 3) and yttrium (Fig. 4) is chosen as examples. Cobalt is only slightly soluble in the solutions used. The cobalt concentrations only slightly increase in accordance to the concentrations in the bulk solid samples (cf. also Fig. 11). Yttrium is significantly leached from the sample ashed at 1000° C after a short reaction time but rapidly disappears from solution (Fig. 4). Yttrium tends to dissolve more easily at higher pH values.

Table 3. Vanadium concentrations versus final pH of the extractions and reaction time, sample ashed at 1000° C, solution:sample = 50:1

Initial pH	3		5		7		9	
Reaction time (d)	Final pH	V conc. (ng/ml)	Final pH	V conc. (ng/ml)	Final pH	V conc. (ng/ml)	Final pH	V conc. (ng/ml)
0.1	11.40	1550	11.50	1464	11.50	1559	11.50	1448
0.3	11.40	1537	11.40	1334	11.50	1320	11.50	1520
1.0	11.05	2008	11.15	1892	11.25	1858	11.30	1825
3.0	11.20	1309	11.30	1204	11.35	1409	11.45	1408
10.0	11.45	1984	12.05	108	12.05	112	12.05	83
30.0	12.05	81	12.10	74	12.10	78	12.15	40

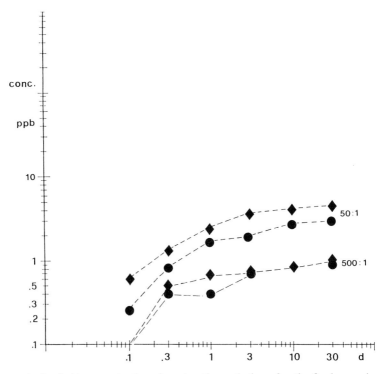

Fig. 3. Development of cobalt concentrations in extraction solutions for the fresh sample and the sample ashed at 1000° C and for different solution:solid ratios; initial pH 3. Symbols indicate concentrations in solutions: ● fresh sample; ◆ sample ashed at 1000° C

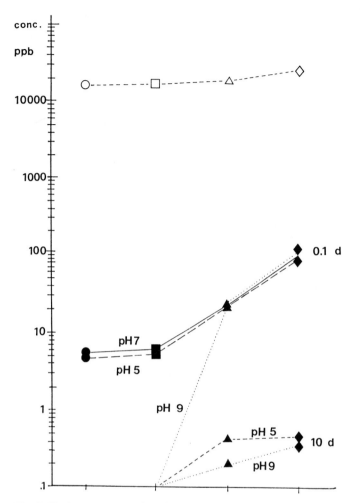

Fig. 4. Yttrium concentrations in extraction solutions from differently pretreated material after two reaction times and bulk yttrium contents. *Open symbols* Concentration in solids; *filled Symbols* concentrations in solutions. ○,● fresh sample; ■,□ cold-ashed sample; △,▲ sample ashed at 500° C; ◇,◆ sample ashed at 1000° C

Ad 2: *Alkali elements* enter into the solution from fresh oil shale to a low final concentration within 1 day (Fig. 5). After heating to 1000° C the leachability of lithium is increased by a factor of 5, of rubidium by a factor of >300. Final concentrations are obtained for rubidium and cesium after 10 days, but lithium concentrations in the solutions increase even after a 30-day reaction time.

Most of the *transition metals* are activated for extraction by heating. For instance, molybdenum is transferred to a more leachable bonding type by thermal treatment. Changes of the extraction pH further in-

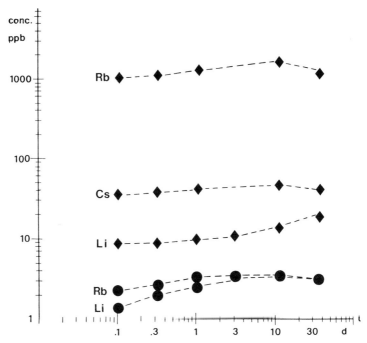

Fig. 5. Development of alkali metal concentration in extraction solutions versus sample pretreatment during 30 days of reaction. Solution:sample = 50:1, initial pH 3. *Symbols* indicate concentrations in solutions: ● fresh sample; ◆ sample ashed at 1000° C

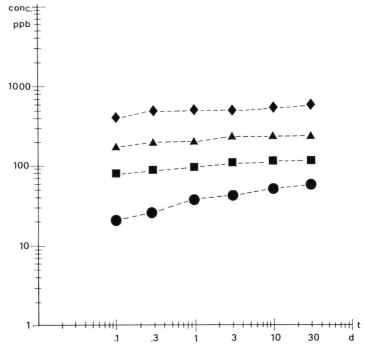

Fig. 6. Development of molybdenum concentrations in extraction solutions versus sample pretreatment during 30 days of reaction. Solution:sample = 50:1, initial pH 3. *Symbols* indicate concentrations in solutions: ● fresh sample; ■ cold-ashed sample; ▲ sample ashed at 500° C; ◆ sample ashed at 1000° C

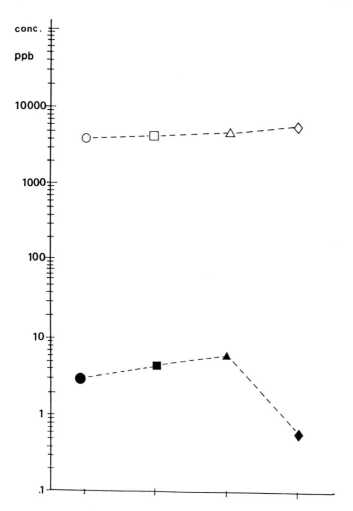

Fig. 7. Leaching of uranium from differently pretreated Posidonia shale samples after 30 days reaction time; initial pH 3. *Open symbols* concentrations in solids; *filled symbols* concentrations in solutions. ○,● fresh sample; ■,□ cold-ashed sample; △,▲ sample ashed at 500° C; ◇,◆ sample ashed at 1000° C

crease the initial solubility. At pH values exceeding 12 the solubility decreases (Fig. 6). A comparable behaviour to that of molybdenum, but extreme pH dependency of the extraction, is found for vanadium (Table 3).

Ad 3: No results can be presented if no solubility (e. g. Be, Th) is visible for any kind of sample pretreatment. In the case of uranium, the fresh, cold-ashed, and 500° C ashed samples show slight solubility. Heating to 1000° C renders this element completely insoluble, although it is enriched in the bulk sample as shown in Table 2 and Fig. 7.

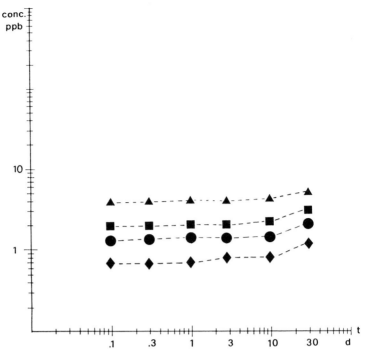

Fig. 8. Development of rhenium concentrations in extraction solutions versus sample pretreatment during 30 days of reaction. Solution:sample = 50:1, initial pH 3. For explanation of symbols, see Fig. 6

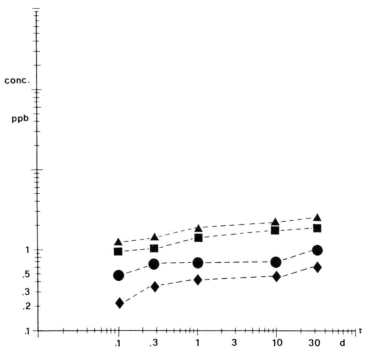

Fig. 9. Development of thallium concentrations in extraction solutions versus sample pretreatment during 30 days of reaction. Solution:sample = 50:1, initial pH 3. For explanation of *symbols,* see Fig. 6

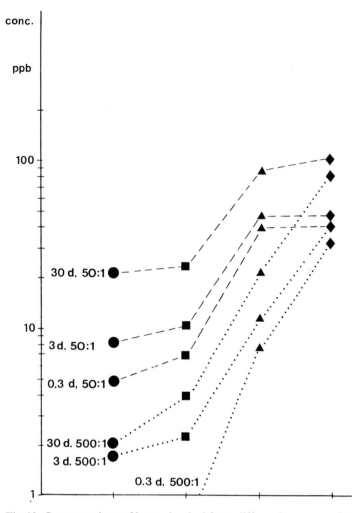

Fig. 10. Concentrations of boron leached from differently pretreated Posidonia shale samples with solution:solid ratios 5:1 and 500:1 after 0.3, 3, 30 day reaction times. For explanation of *symbols,* see Fig. 6

Ad 4: Rhenium is possibly evaporated to a minor portion during cold and 500° C ashing. Significant loss of rhenium occurs during ashing at 1000° C. The loss may be due to the formation of volatile Re_2O_7 which sublimates at ca. 600° C (Nadler and Borchers 1987). Although the amount of rhenium in the bulk sample decreases according to the evaporation, the ratio of $Re_{extract} : Re_{bulk}$ increases in the extract of the material ashed at 1000° C (Fig. 8). The influence of the pH during leaching is the same as for molybdenum (Fig. 6).
Thallium provides another example for volatilization of elements during high temperature heating: although the bulk thallium con-

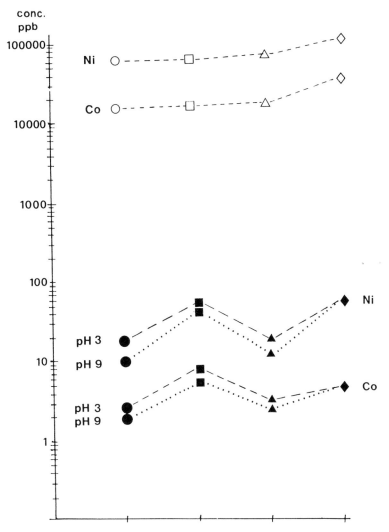

Fig. 11. Leaching of cobalt and nickel from Posidonia shale samples after 30 days reaction time starting from initial pH 3 and 9. For explanation of *symbols*, see Fig. 7

centrations stay in the same range after cold and 500° C ashing, enhanced leaching occurs from the sample ashed at 500° C. Further heating to 1000° C reduced the bulk thallium concentration to almost one tenth of the fresh sample but the extraction (30 days) of the sample ashed at 1000° C still contains about 50% of the concentration in the fresh sample extraction (Fig. 9).

The influence of the solid:solution ratio is given for boron after 0,3 days of reaction. The upper solubility limit for an initial pH 3 appears to be closely approached after 30 days by the 50:1 reaction system (Fig. 10).

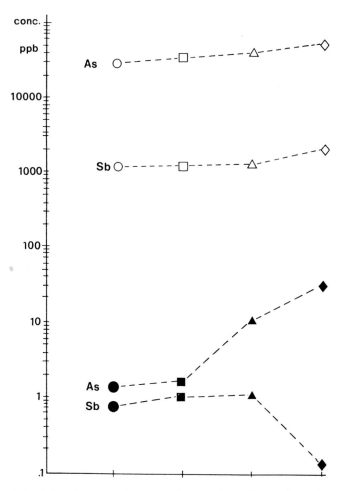

Fig. 12. Leaching of antimony and arsenic from differently pretreated Posidonia shale samples after 0.3 days reaction time, initial pH 5. For explanation of *symbols,* see Fig. 7

Differing initial pH values cause only slight differences of the extraction behaviours of many trace elements. This is demonstrated for cobalt and nickel after 30 days of reaction (Fig. 11).

An analogous increase is observed for arsenic and antimony in the solid during heating. Figure 12 gives the concentration of both elements in solution after 0.3 days, starting with solutions of pH 5; while more arsenic is extracted from the sample heated at 500° C and 1000° C, the solubility of antimony is drastically reduced by extreme heating (Fig. 12).

4 Conclusions

Despite the high concentrations of trace elements in the Posidonia shale, which often surpass the limits set by the Ministry of the Interior of the FRG (Table 4), these elements present no risk to the environment as long as they remain in their original bonding type.

Heating of the material partly or completely removes the organic matter. It also oxidizes the iron sulfides, which contain some of the toxic elements.

Extreme heating causes carbonates to decompose and some elements are volatilized. The mobility of the trace elements is drastically changed by such thermal treatment. Under these extreme conditions, toxic traces may be leached unless they are fixed by another chemical reaction.

Leaching of the heated Posidonia shale results in trace element concentrations of the interacting aqueous solutions which are higher than the critical limits set by the Drinking Water Regulations from 1986 for the FRG or the recommendations of the WHO (Table 5).

Under extreme conditions, the biosphere may be affected (Hock and Elstner 1988) and toxic concentrations of various trace elements can enter into

Table 4. Maximum tolerable concentrations of trace elements in cultural soils for sewage sludge fertilization according to the "Klärschlamm-Verordnung" of the Federal Ministry for Natural Resources and Environment, Baden-Württemberg, FRG

Element	Common concentration in soils (ppm)	Max. tolerable conc. (ppm)
As	0.1 – 20	20
B	5 – 20	25
Be	0.1 – 5	10
Br	1 – 10	10
Cd	0.01 – 1	3
Co	1 – 10	50
Cr	2 – 50	100
Cu	1 – 20	100
Ga	0.1 – 10	10
Hg	0.01 – 1	2
Mo	0.2 – 5	5
Ni	2 – 50	50
Pb	0.1 – 20	100
Sb	0.01 – 0.5	5
Se	0.01 – 5	10
Sn	1 – 20	50
Ti	10 – 5000	5000
Tl	0.01 – 0.5	1
U	0.01 – 1	5
V	10 – 100	50
Zn	3 – 50	300
Zr	1 – 300	300

Table 5. Permitted upper concentrations of inorganic constituents in drinking and mineral water in the FRG (Drinking Water Regulations 1986), WHO recommendations (1984), orientation data according to Quentin (1988), and maximum concentrations found in extractions of the Lower Toarcian oil shale from S Germany; ns: no value set. All values in mg/l (ppm)

	Drinking water Regulations, FRG	WHO Recommendations	Orientation values (Quentin 1988)	Highest conc. in extractions
Ag	0.01	ns	0.01	–.–
As	0.04	0.05	0.04	0.051
B	ns	ns	1.00	0.127
Ba	ns	ns	0.10	0.904
Be	ns	ns	>0.0002	–.–
Cd	0.005	0.005	0.005	0.002
Cu	ns	1.00	1.00	0.015
Cr	0.05	0.05	0.05	1.06
Hg	0.001	0.001	0.001	–.–
Ni	0.05	ns	0.05	0.14
Pb	0.04	0.05	0.04	–.–
Sb	ns	ns	0.01	0.0045
Se	ns	0.01	0.05	–.–
Zn	ns	5.00	5.00	0.025

the food chain (Merian 1984). Under normal natural conditions there is no need that the boundary values of the "Klärschlammverordnung" (Table 4) are to be revised.

Acknowledgements. We thank the project "Water, Waste, and Soil" of Baden-Württemberg for providing the extremely useful ICP-MS method and for supporting the investigations since 1986. Mrs. Doris Baumann and Mrs. Rosemarie Baumann assisted in the experiments and analyses.

References

Brumsack HJ, Thurow J (1986) The geochemical facies of black shales from the Cenomanian/Turonian boundary event (CTBE). Mitt Geol Paläontol Inst Univ Hamburg 60:247–265, 4 figs, 7 tables

Einsele G, Mosebach R (1955) Zur Petrographie, Fossilerhaltung und Entstehung der Gesteine des Posidonienschiefers im Schwäbischen Jura. N Jahrb Geol Paläontol Abh 101:319–430, 5 plates

Fesser H (1968) Zur Geochemie des Posidonienschiefers in Nordwestdeutschland. Beih Geol Jahrb 58:223–286, 17 figs, 9 tables

von Gaertner HR (1955) Petrographische Untersuchungen am Nordwestdeutschen Posidonienschiefer. Geol Rundsch 43:447–463, 2 figs, 1 plate

Hock B, Elstner EF (1988) Schadwirkungen auf Pflanzen. Lehrbuch der Pflanzentoxikologie, 2nd edn. B.I. Wissenschaftsverlag Mannheim-Vienna-Zürich, 348 pp

Hoffmann K (1966) Die Stratigraphie und Paläogeographie der biuminösen Fazies des nordwestdeutschen Oberlias (Toarcium). Beih Geol Jahrb 58:443–498, 4 figs, 1 table, 3 plates

Joachim H (1970) Geochemische, sedimentologische und ökologische Untersuchungen im Grenzbereich Lias delta/epsilon (Domerium/Toarcium) des Schwäbischen Jura. Arb Geol Paläontol Inst, Univ Stuttgart (TH), N F 61, 243 pp, 36 figs, 11 tables, 14 plates, 7 photographs

Kauffman EG (1981) Ecological reappraisal of the German Posidonienschiefer (Toarcian) and the stagnant basin model. In: Gray, Boucot AJ, Berry WBN (eds) Communities of the past. Hutchinson Ross Stroudsburg, PA, pp 311–381, 19 figs

Knapp G (1985) Sample preparation techniques – an important part in trace element analysis for environmental research and monitoring. Int J Environ Anal Chem 22: 71–83, 9 figs, 1 table

Kramar U, Puchelt H (1982) Reproducibility test for INAA determinations with AGV-1, BCR-1, and GSP-1 and new data for 17 geochemical reference materials. Geostand Newslett 6, 2:221–227, 3 figs, 5 tables

Küspert W (1982) Environmental changes during oil shale deposition as deduced from stable isotope ratios. In: Einsele G, Seilacher A (eds) Cyclic and event stratification. Springer Berlin Heidelberg New York, pp 482–501, 5 figs

Mann U (1987) Veränderungen von Porosität und Porengrösse eines Erdölmuttergesteins in Annäherung an einen Intrusivkörper. Facies 17:181–188, 4 figs, 3 tables

Merian E (1984) Metalle in der Umwelt. Verteilung, Analytik und biologische Relevanz. Chemie, Weinheim, 722 pp

Nadler HG, Borchers P (1987) Gewinnung von Rhenium aus den Röstgasen der Molybdänit-Abröstung. Erzmetall 40, 6:293–298, 9 figs, 2 tables

Nöltner T, Maisenbacher P, Puchelt H (1989) Microwave acid digestion of geological and biological reference standard materials for trace element analyses with ICP-MS. Spectrosc Int (in press)

Puchelt H, Nöltner T (1987) Interaction of aqueous solutions with bituminous shales – implications of inductively coupled plasma source mass spectrometry. ICP Inf Newslett 13:139–140

Puchelt H, Nöltner T (1988) Zur Stabilität hochverdünnter Multielement-Eichstandardlösungen – Erfahrungen mit ICP-MS. Fresenius Z Anal Chem 331:216–219, 4 figs, 3 tables

Puchelt H, Nöltner T (1989) ICP-MS analyses of geological materials. VG Elemental, Tech Inf Winsford

Quentin KE (1988) Trinkwasser. Untersuchung und Beurteilung von Trink- und Schwimmbadwasser. Springer Berlin Heidelberg New York London Paris Tokyo 385 pp

Rieber H (1975) Der Posidonienschiefer (oberer Lias) von Holzmaden und die Grenzbitumenzone (mittlere Trias) des Monte San Giorgo (Kt. Tessin, Schweiz). Ein Vergleich zweier Lagerstätten fossiler Wirbeltiere. Jahresh Ges Naturk Württemberg 130:163–190, 8 figs, 1 table

Riegraf W (1985) Mikrofauna, Biostratigraphie und Fazies im Unteren Toarcium Südwestdeutschlands und Vergleiche mit benachbarten Gebieten. Tübinger Mikropaläontol Mitt 3:232 pp, 33 figs, 12 plates

Riegraf W, Werner G, Lörcher F (1984) Der Posidonienschiefer. Biostratigraphie, Fauna und Fazies des südwestdeutschen Untertoarciums (Lias epsilon). Enke, Stuttgart 207 pp, 50 figs, 12 plates

Rullkötter J, Leythaeuser D, Harsfield B, Littke R, Mann U, Müller PJ, Radke M, Schaefer RG, Schenk H-J, Schwochau K, Witte EG, Welte DH (1987) Organic matter maturation under the influence of a deep intrusive heat source: A natural experiment for quantification of hydrocarbon generation and expulsion from a petroleum source rock (Toarcian shale, northern Germany). – Advances in Organic Geochemistry 1987, Org. Geochem., 13, 847–856, 4 figs, 3 tables; Oxford

Seilacher A (1982) Posidonia shales (Toarcian, S. Germany) – stagnant basin model revalidated. In: Montanaro-Gallitelli E (ed) Palaeontology, essential of historical geology. Proc 1st Int Meet Palaeontology, essential of historical geology, Venice, June 2–4, 1981. Mucci Modena, pp 25–55

Vine JD, Tourtelot EB (1970) Geochemistry of black shale deposits – a summary report. Econ Geol 65:253–272, 3 figs, 7 tables

Walk H (1982) Die Gehalte der Schwermetalle Cd, Tl, Pb, Bi und weiterer Spurenelemente in natürlichen Böden und ihren Ausgangsgesteinen Südwestdeutschlands. Diss Fak Bio Geowiss, Univ Karlsruhe, 170 pp, 41 figs, 42 tables

Sediment Criteria Development
Contributions from Environmental Geochemistry to
Water Quality Management

ULRICH FÖRSTNER, WOLFGANG AHLF, WOLFGANG CALMANO,
and MICHAEL KERSTEN[1]

CONTENTS

Abstract		312
1	Introduction	312
2	Identification of Sources and Temporal Developments	313
3	Assessment of Critical Pools of Pollutants in Sediments	315
4	Sediment Quality Criteria	316
4.1	Biological Criteria	317
4.2	Chemical-Numerical Approaches	319
4.2.1	Background Approach	319
4.2.1.1	Standard Values	319
4.2.1.2	Pollution Indices	320
4.2.1.3	Ecological Risk Indices Derived from Enrichment Factors	321
4.2.2	Pore Water Approach	321
4.2.3	Sediment/Water Equilibrium Partitioning	323
4.2.4	Sediment/Biota Equilibrium Partitioning	324
4.2.5	Elution Approach	325
4.2.6	Biological and Chemical Approaches: A Comparison	326
5	Modeling of Sediment Data	327
6	Characterization of Sediment Milieu – Acid-Producing Potential	328
7	Application of Different Criteria Approaches on Sediment Samples from Large Rivers in the Federal Republic of Germany	329
7.1	Index of Geoaccumulation (Müller 1979)	329
7.2	Incorporation of "Toxic Effects" into Accumulation Factors	330
7.3	Numerical Evaluation of the Factor "Element Mobility"	330
7.4	pH-Titrations of Selected Sediment Samples	331
8	Summary and Outlook	333
References		334

---

[1] Arbeitsbereich Umweltschutztechnik, Technische Universität Hamburg-Harburg, Eißendorferstr. 40, D-2100 Hamburg 90, FRG

Abstract

The role of sediments as carriers and potential sources of contaminants is reviewed. A program of sediment studies will normally consist of a series of objectives of increasing complexity, each drawing part of its information from the preceding data base. The study of dated sediment cores has proven particulary useful as it provides a historical record of the various influences on the aquatic system by indicating both the natural background levels and the man-induced accumulation of pollutants over an extended period of time. Since adsorption of pollutants onto particles is a primary factor in determining the transport, deposition, reactivity, and potential toxicity of these materials, analytical methods should be related to the chemistry of the particle's surface and/or to the pollutant species highly enriched on the surface.

New objectives regarding the improvement of water quality as well as problems with the resuspension and land deposition of dredged materials require a standardized assessment of sediment quality. Biological criteria integrate sediment characteristics and pollutant loads, while generally not indicating the cause of effects. With respect to chemical-numerical criteria immediate indications on biological effects are lacking; major advantages lie in their easy application and amendment to modeling approaches. Numerical approaches, on the one hand, are based on (1) accumulation; (2) pore water concentrations; (3) solid/liquid equilibrium partition (sediment/water and organism/water); and (4) elution properties of contaminants. The second component in an assessment scheme would include characteristics of the solid substrate, in particular, buffer capacity against pH-depression. At the present stage of criteria development we propose that the substrate properties should be classified on the basis of the carbonate and sulfide inventory, whereas the pollutant load is advantageously assessed by the accumulation rate multiplied with a toxicity factor for the respective substance.

1 Introduction

Sediments are both carriers and potential sources of contaminants in aquatic systems, and these materials may also affect groundwater quality and agricultural products when disposed on land (Förstner and Müller 1974). Contaminants are not necessarily fixed permanently by the sediment, but may be recycled via biological and chemical agents both within the sedimentary compartment and the water column. Bioaccumulation and food chain transfer may be strongly affected by sediment-associated proportions of pollutants. Benthic organism, in particular, have direct contact with sediment, and the contaminant level in the sediment may have greater impact on their survival than do aqueous concentrations.

In modern sediment research on contaminants four aspects are discussed, which in an overlapping succession also reflect the development of knowledge in particle-associated pollutants during the past 25 years: (1) Identification of sources and distribution; (2) evaluation of solid/solution relations; (3) study of transfer mechanism to biological systems, and (4) assessment of environmental impact. In practice, aspects (1) and (4) are of particular relevance, and recent developments will be treated in the present review.

2 Identification of Sources and Temporal Developments

A program of sediment studies will normally consist of a series of objectives of increasing complexity, each drawing part of its information from the preceding data base (Golterman et al. 1983):

a) *Preliminary site characterization*: Low density sampling with limited analytical requirements, to provide a general characterization of an area for which little or no previous information exists.

b) *Identification of anomalies*: More detailed sampling and analyses, designed to etablish the presence and extent of anomalies.

c) *Establishment of references*: To create reference points, in the form of some measured parameters, for future comparison.

d) *Identification of time changes*: To show trends in variations of sediment data over time, by use of sediment cores or other repeated sediment samplings.

e) *Calculation of mass balances*: To account for the addition and subtraction of sediment-related components with an aquatic environment (a complex study), by means of accurate and representative sampling and analysis.

f) *Process studies*: Specialized sampling to improve state of knowledge about aquatic systems, e. g. by supplementary laboratory experiments.

The principal relationships between *sampling objectives* and *type of activities* for water-related studies are summarized in Table 1.

Program objectives largely control the type, density, and frequency of sediment sampling and associated analyses; whereas the type of environment (rivers, lakes, estuaries, etc.) largely controls the locations and logistics of sampling. Logistic factors include (Golterman et al. 1983):

1. Local availability of sampling platform or vessel;
2. Time available;
3. Access to sampling region;
4. Suitability of survey system to locate sample position;
5. Availability of trained personnel and supportive staff;
6. Availability of equipment;

Table 1. Sampling objectives and type of activities for water-related studies

Type of activity (UNESCO-WHO 1978)	GEMS water objectives (WHO 1978)	Sediment objectives (categories in text)
Monitor		
Continuous standard measurement and observation	Cultural impact on water quality, suitability of water quality for future use	Establish reference point(s); category (c)
Surveillance		
Continuous, specific observation and measurement relative to control and management	Observe sources and pathways of specified hazardous substances	Trace sources (spatial)
Survey		
Series of finite duration; intensive, detailed programs for specific purposes	Determine quality of natural waters	Identify anomalies (category b); calculate mass balances (category e); study process (f)

7. Storage and security;
8. Transport systems;
9. Follow-up capability.

For complex surveys, there are numerous types of sampling patterns from which to choose, e.g., spot samples, square grids (including nested and rotated grids), parallel line grids and traverse line grids (with equal or non-equal sampling), and ray grids or concentric arc sampling, each of which offers some particular advantage (Golterman et a. 1983, pp 76–79).

The suitability of corers and bottom samplers has been tested during equipment trials by Sly (1969). For sources reconnaissance analysis, fine- to medium-grained bottom deposits from a depth of 15–20 cm can be collected, for example, with an Ekman grab sampler. In environments with a relatively uniform sedimentation, for example, in lakes and in marine coastal basins, where the deposits are fine-grained and occur at a rate of 1 to 5 mm/yr, a more favorable procedure involves the taking of vertical profiles with a gravity or valve corer (Jenne et al. 1980).

The study of dated sediment cores has proven particularly useful as it provides a historical record of the various influences on the aquatic system by indicating both the natural background levels and the man-induced accumulation of elements over an extended period of time. Various approaches to the dating of sedimentary profiles have been used but the isotopic techniques, using ^{210}Pb, ^{137}Cs, and $^{239+240}$Pu, have produced the more unambiguous results and therefore have been the most successful. Major contributions to the "Historical Monitoring" by sediment studies (Alderton

1985) have been given by German Müller (1977a, b, 1979, 1981, 1983, 1985) and German Müller et al. (1890) in different parts of the world.

3 Assessment of Critical Pools of Pollutants in Sediments

Since adsorption of pollutants onto air- and waterborne particles is a primary factor in determining the transport, deposition, reactivity, and potential toxicity of these materials, analytical methods should be related to the chemistry of the particle's surface and/or to the metal species highly enriched on the surface. Basically there are three methodological concepts for determining the distribution of an element within or among small particles (Keyser et al. 1978):

1. *Analysis of single particles* by X-ray fluorescence using either a scanning electron microscope (SEM) or an electron microprobe can identify differences in the matrix composition between individual particles. The total concentration of the element can be determined as a function of particle size. Other physical fractionation and preconcentration methods include density and magnetic separations.

2. The *surface of the particles* can be studied directly by the use of electron microprobe X-ray emission spectrometry (EMP), electron spectroscopy for chemical analysis (ESCA), Auger electron spectroscopy (AES), and secondary ion-mass spectrometry. Depth-profile analysis determines the variation of chemical composition below the original surface.

3. *Solvent leaching* – apart from the characterization of the reactivity of specific metals – can provide information on the behaviour of pollutants under typical environmental conditions. Common single reagent leachate tests, e.g. U.S. EPA, ASTM, IAEA and ICES use either distilled water or acetic acid (Theis and Padgett 1983). A large number of test procedures have been designed particularly for soil studies; these partly used organic chelators such as EDTA and DTPA (Sauerbeck and Styperek 1985).
Laboratory techniques for generating leachate from solid materials are generally grouped into batch and column extraction methods. The batch extraction method offers advantages through its greater reproducibility and simplistic design, while the column method is more realistic in simulating leaching processes which occur under field conditions (Jackson et al. 1984). For batch studies best results with respect to the estimation of short-term effects can be attained by "cascade" test procedures at variable solid/solution ratios.

In connection with the problems arising from the disposal of solid wastes, particularly of dredged materials, extraction sequences have been applied which are designed to differentiate between the exchangeable, carbonatic, reducible (hydrous Fe/Mn oxides), oxidizable (sulfides and organic phases),

Table 2. Sequential extraction scheme for partitioning sediments (Calmano and Förstner 1983; Kersten and Förstner 1986, 1987)

Fraction	Extraction	Extracted component
Exchangeable	1 M NH$_4$OAc, pH 7	Exchangeable ions
Carbonatic	1 M NaOAc, pH 5 with HOAc	Carbonates
Easily reducible	0.01 M NH$_2$OH HCl with 0.01 M HNO$_3$	Mn-oxides
Moderately reducible	0.1 M oxalate buffer pH 3	Amorphous Fe-oxides
Sulfidic/organic	30% H$_2$H$_2$ with 0.02 HNO$_3$ pH 2; extracted with 1 M NH$_4$OAc-6% HNO$_3$	Sulfides together with organic matter
Residual	Hot concentrated HNO$_3$	Lithogenic crystalline

and residual fractions (Engler et al. 1977). Despite clear advantage of a differentiated analysis over investigations of total sample – sequential chemical extraction is probably the most useful tool for predicting long-term adverse effects from contaminated solid material – it has become obvious that there are many problems associated with these procedures (Kersten et al. 1985). One of the more widely applied extraction sequences of Tessier and coworkers (1979) has been modified by various authors (Table 2).

4 Sediment Quality Criteria

Three major reasons have been given for the establishment of sediment quality criteria:

1. In contrast to the strong temporal and spatial variability in the aqueous concentrations of contaminants, sediments integrate contaminant concentrations over time, and can, therefore, reduce the *number of samples* in monitoring, surveillance, and survey activities;

2. Long-term perspectives in water resources management involve *integrated strategies*, in which sediment-associated pollutants have to be considered;

3. Wastewater plans will increasingly be based on the *assimilative capacity* of a certain receiving system, which requires knowledge of properties of sedimentary components as the major sink.

Efforts have been undertaken mainly by the United States Environmental Protection Agency to develop standard procedures and criteria for the assessment of environmental impact of sediment-associated pollutants. Initial discussions (US Environmental Protection Agency 1984; Anonymous 1985) suggested five methodological approaches which merit closer consideration: (1) background approach; (2) water quality/pore water approach; (3) sediment/water equilibrium partitioning approach; (4) sediment/organism equilibrium approach; and (5) bioassay approach. Further discussions led to the differentiation of biological and chemical-numerical approaches (G. Chapman et al. 1987):

Biological criteria		*Chemical-numerical criteria*
– Field biological surveys		– Background approach
– Bioassay on original sediments	Sediment-quality-"triad"	– Pore water approach
		– Sediment/water-equilibrium
– Bioassay on spiked sediments		– Sediment/organism-equilibrium
		– Elution tests
		– Substrate composition

4.1 Biological Criteria

Biological criteria have been developed and are already applied in various areas (Anderson et a. 1987; G. Chapman et al. 1987):

1. *Field biological surveys* conduct on-site studies of biota to evaluate possible impact at site (biological response pass/fail).

2. *Bioassay of spiked sediment* estimates effect/no effect sediment concentration for a specific chemical (numerical criterion). May be desired in clearance of new chemicals.

3. *Tissue action level* links sediment concentration to safe tissue concentration (e. g., FDA action level or body burden-response data) through application of equilibrium or kinetic models (numerical criterion).

4. *Aqueous toxicity data* apply toxicity data from typical water-column bioassays to sediments through direct measurement of pore water concentration or estimation of pore water from sediment concentrations through application of equilibrium models. May be desired in evaluation of new chemicals.

Biological approaches on development and application of sediment quality criteria exhibit a common basis in the study of damaging impacts from contaminated sediments on organisms. The biological parameters "bioaccumulation", "toxicity", and "mutagenity" have to be considered separately in any case. Bioassays as well as field surveys are empirical considerations which cannot provide numerical criteria to be transferred to different situations.

Generally, it is difficult to establish clear cause-and-effect relationships between acute or chronic toxic effects on biota and the occurrence of specific pollutants in sediments. One major limitation is that not all sediment-associated chemicals can presently be identified; thus, unidentified compounds cannot be ruled out as principal etiological factors.

Relatively simple and implementable liquid, suspended particulate and solid-phase bioassays have been carried out for assessing the short-term impact of dredging and disposal operations on aquatic organisms (Ahlf and Munawar 1988). Standardized tests are characterized by their lack of variability, but essential information (e.g., lethality, alterations of growth rate) can only be obtained with such a single-species test. The influence of the main environmental variables on the interaction of suspended particulates or in-situ sediment contaminants and organisms should also be determined under simulated field conditions. In particular, benthic bioassay procedures, due to recent developments, are important in evaluating the relationship between laboratory and field impacts (Reynoldson 1987).

With restriction to the effects on benthic communities, the sediment quality "triad" by P.M. Chapman (1986) combines chemistry and sediment bioassay measurements with in-situ studies: Chemistry and bioassay estimates are based on laboratory measurements with field-collected sediments. In-situ studies may include, but are not limited to, measures of resident organism histopathology, benthic community structure, and bioaccumulation/metabolism. Areas where the three facets of the triad show the greatest overlap (in terms of positive or negative results) provide the strongest data for determining numerical sediment criteria.

Studies have been performed in the Puget Sound, Washington by Long and Chapman (1986). In the sediment three dominant and representative chemical groups were distinguished in the analyses and were selected for further study: high molecular weight combustion polyaromatic hydrocarbons (CPAHs), polychlorinated biphenyls (PCBs), and lead. Three types of sediment bioassays were considered: the amphipod *Rhepoxynius abronius* acute lethality test, the oligochaete *Monopylephorus cuticulatus* respiration effects test, and the fish cell anaphase aberration test. Bottom fish histopathology was based on the frequency of selected liver lesions in English sole (*Parophrys vetulus*) from different areas of Puget Sound; liver lesions have been considered most likely to be related to chemical contaminant exposures (Malins et al. 1984). A summary comparison of the data based on effects frequencies of sediment bioassays and in-situ studies (i.e., bottom fish histopathology) indicates that roughly similar sediment contaminant concentrations produce both types of biological responses (Table 3). Bioassay data are divided into those frequencies of effects that contain all rural areas, and those that contain only urban, industrialized areas. Bottom fish histopathology data are divided into frequencies of occurrence of up to 5% and those that are greater than 5%.

The general question, how the borderline concentrations ($\geq$ 130 mg/kg Pb, $\geq$ 6.8 mg/kg PAHs, $\geq$ 0.8 mg/kg PCBs) are related to biological effects, has been studied by G. Chapman et al. (1987) from independent experiments; results from these investigations indicate remarkable coincidence with the

Table 3. Summary comparison of biological effects frequencies with sediment concentrations of selected chemical contaminants in different areas of Puget Sound, Washington (Chapman 1986)

Effects frequency (%)	Chemical contaminants[a] (µg/g)		
	Pb	CPAHs	Total PCBs
Sediment bioassays			
15–50	20– 50	0.2– 5.0	0.01–0.10
55–80	90–800	3.8–24.0	0.10–0.90
Bottom fish histopathology			
0– 5	20– 90	0.2– _3.8_	0.01–0.20
6–40	_130_–800	6.8–24.0	_0.80_–0.90

[a] Borderline concentrations are underlined.

"triad" data. The obvious similarity of results from different areas may be explained by the experience that the spectrum of pollutants in sediments is widely comparable due to the inputs from common sources, such as from atmospheric emissions from combustion of fossil fuels, where PAH and metals exhibit typical associations (Müller 1977).

4.2 Chemical-Numerical Approaches

4.2.1 Background Approach

4.2.1.1 Standard Values. An example of standard values for sediment quality criteria is given by the Dutch sediment quality draft (van Veen and Stortelder 1988). Dutch environmental pollution standards have traditionally been based on contaminant concentrations. The advantages, particularly for monitoring this type of standard, are simplicity and absence of ambiguity. The lack of consideration of the ecological impact is a disadvantage. In a draft for developing sediment standards – aimed at disposal of contaminated sediment on land – the pollution concentrations are normalized to a standard sediment ("underwater soil") consisting of 10% organic matter and 24% clay content (particle size < 2 µm). The level for the target value is based on field observations of sediments in surface waters unaffected by industrial or other discharges. The level for the standard value is based on observations of sediments which are slightly contaminated but with no known ecological effect. The levels for the limit value do not have any ecological background; they are based on existing standardization in the Rotterdam area. These three levels are defined for many toxic compounds, some of which are given in Table 4. Integration of standards for terrestrial and aquatic soils is under discussion and could be of great importance, for example, in the case of disposal of contaminated sediments on land.

Table 4. Draft standards for contaminated sediments (van Veen and Stortelder 1988). Data in mg/kg, except PCB and PAH (µg/kg)

	Cr	Cu	Zn	Cd	Hg	Pb	As	EOX	Oil	PCB	PAH
Target value	100	25	180	0.8	0.3	50	25	–	–	1	50
Standard value	125	70	750	4	1	125	40	5	2000	10	500
Limit value	600	400	2500	30	15	700	100	20	5000	100	3500

4.2.1.2 Pollution Indices. A quantitative measure of metal pollution in aquatic sediments has been introduced by Müller (1979), which is called the Index of Geoaccumulation:

$$I_{geo} = \log_2 C_n/1.5 \times B_n,$$

where C_n is the measured concentration of the element n in the pelitic sediment fraction ($< 2\,\mu m$) and B_n is the geochemical background value in fossil argillaceous sediment ("average shale"); the factor 1.5 is used because of possible variations of the background data due to lithogenic effects. The Index of Geoaccumulation consists of seven grades, whereby the highest grade (6) reflects 100-fold enrichment above background values ($2^6 = 64 \times 1.5$). In Table 5 an example is given for the River Rhine; and a comparison of these sediment indices with the water quality classification of the International Association of Waterworks in the Rhine Catchment (IAWR) has been made. It should be mentioned that – similar to the sediment standards in Table 5 – no further consideration is given to the ecological relevance of the values.

Table 5. Comparison of IAWR water quality indices (based on biochemical data) and Index of Geoaccumulation (I_{geo}) of trace metals in sediments of the Rhine River. (After Müller 1979)

IAWR Index	IAWR water quality (pollution intensity)	Sediment accumulation (I_{geo})	I_{geo}-class	Metal examples Upper Rhine	Metal examples Lower Rhine
4	Very strong pollution	>5	6		Cd
3–4	Strong to very strong	>4–5	5		
3	Strongly polluted	>3–4	4		Pb, Zn
2–3	Moderately to strongly	>2–3	3	Cd, Pb	Hg
2	Moderately polluted	>1–2	2	Zn, Hg	Cu
1–2	Unpolluted to moderately polluted	>0–1	1	Cu	Cr, Co
1	Practically unpolluted	<0	0	Cr, Co	

4.2.1.3 Ecological Risk Indices Derived from Enrichment Factors. A sedimentological approach for an "ecological risk index" was introduced by Hakanson (1980) and tested on 15 Swedish lakes representing a wide range in term of size, pollution status, trophic status, etc. These estimations are based on four "requirements", which are determined in a relatively rapid, inexpensive, and standardized manner from a limited number of sediment samples. Contrary to the afore mentioned approaches, a special term is introduced for estimating the ecotoxicological significance of the individual contaminants. The toxic requirement differentiates the various contaminants according to an "abundance principle", i.e. assuming that a proportionality between toxicity and rarity, and to their "sink effect" exists, i.e. their affinity to solid substrates. After a normalization process the "sedimentological toxic factor" is calculated in the following sequences: $Zn = 1 < Cr = 2 < Cu = Pb = 5 < As = 10 < Cd = 30 < Hg = PCB = 40$.

The "toxic response factor", as formulated by Hakanson (1980) from a complex matrix of assumptions, can possibly be defined much easier from direct measurements of the relative toxicity of typical pollutants in aquatic systems, e.g. from bioassays on water samples. We propose a toxicity factor based on the standardized "Microtox test system", where the individual concentrations are determined from comparable EC_{50} values. According to Walker (1988), the following factors could be used for metallic elements: $Pb = 1; Zn = 5; Cu = 5; Cd = 10; Hg = 35$.

4.2.2 Pore Water Approach

The composition of interstitial waters is the most sensitive indicator of the types and the extent of reactions that take place between pollutants on waste particles and the aqueous phase which contacts them. Particularly for fine-grained material the large surface area related to the small volume of its entrapped interstitial water ensures that minor reactions with the solid phases will be indicated by major changes in the composition of the aqueous phase (Förstner and Kersten 1988).

Interstitial waters are recovered from sediments by dialysis, centrifugation, or squeezing. Oxidation must be prevented during these procedures. Watson et al. (1985) showed that sediments stored prior to the separation of interstitial water yield significant changes on chemical composition compared to samples processed within 24 h of collection. In-situ methods are considered more promising because of their inherent simplicity, and appear to be well adapted to the study of trace metals at the sediment-water interface under field conditions. An in-situ sampler for close-interval pore water studies as presented by Hesslein (1976) can be made from a clear acrylic plastic panel with small compartments predrilled in 1-cm steps or less. This panel can be covered by a nondegradable dialysis membrane or by a polysulfonate membrane filter sheet (Carignan 1984) and displaced into the sediment allowing equilibrium to take place over a period of some days to weeks. An improved sampler of this type has been decribed by Schwedhelm et al. (1988).

While the direct recovery and analysis of waterborne constituents can be seen as a major advantage of this approach, there are several disadvantages,

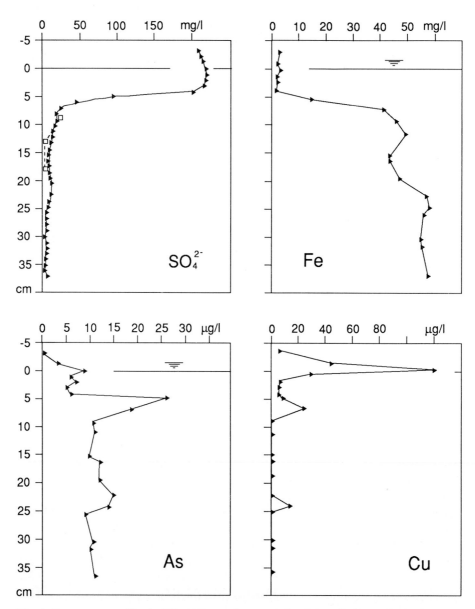

Fig. 1. Pore water profiles in Elbe River sediments (after Schwedhelm et al. 1988). 0-cm-depth contour line reflects the sediment/water interface

particularly arising from the sampling and sample preparation, which are not yet routine procedures, and usually involve considerable precautionary measures such as exclusion of oxygen. In addition, interpretation of profile data may be difficult, as demonstrated from the examples of depth profiles of typical constituents in pore waters from Elbe River sediments (Fig. 1): There are strong gradients for redox-sensitive constituents, such as iron, arsenic, and sulfate; the question is which position in the core profile is the most typical with respect to the uptake by benthic organisms. In this context, the characteristic enrichment of copper at the sediment/water interface, probably due to complexation by organic ligands from degrading organic matter, seems to be particularly relevant.

4.2.3 Sediment/Water Equilibrium Partitioning

This approach is related to a relative broad toxicological basis of water quality data. The distribution coefficient K_D, which is determined from laboratory experiments, is defined as the quotient of equilibrium concentration of a certain compound in sediment (C_s^x, e.g. in mg/kg) and in the aqueous phase (C_w^x; e.g. in mg/l). Since, in particular, water quality management is requesting such simple calculation bases, the problematic nature of these relations – as evidenced from various references (Table 6) – should clearly be indicated. Nonetheless, it seems that sediment quality criteria of the U.S. Environmental Protection Agency will preferentially be based on these approaches.

In practice, three categories of compounds can be distinguished:

1. *Nonpolar organic compounds*, which are dominantly correlated to the content of organic carbon in the sediment sample. The partition coefficient K_D

Table 6. Factors and mechanisms influencing the distribution of pollutants between solid and dissolved phases

Factor/mechanism	Example[a]	References
Sample preparation (e.g. drying)	Metals[a]	Duursma (1984)
Separation (filtration/centrifugation)	Metals[a]	Calmano (1979)
Grain size distribution	Metals[a]	Duursma (1984)
Suspended matter concentration	DDT/Kepone PCBs	Connor and Connolly (1980) Voice et al. (1983)
Kinetics of sorption/desorption	Metals[a]	Schoer and Förstner (1984)
Nonreversibility of sorption	Metals PCBs Chlorophenols	Lion et al. (1982) DiToro and Horzempa (1982) Isaaacson and Frink (1984)
Effect of bioconcentration	1,4-DCB	Oliver and Nicol (1982)

[a] Experiments with artificial radionuclides.

can be normalized from this parameter and the octanol/water coefficient (K_{OW}): $K_D = 0.63\ K_{OW}$/content of organic carbon in total dry sediment (0.63 is an empirical value). For these substances, such as PCB, DDT, and PAH reliable and applicable data can be expected with respect to the development of sediment quality criteria.

2. *K_D-values of metals* are not only correlated to organic substances but also with other sorption-active surfaces. Toxicological effects are often inversely correlated with parameters such as iron oxyhydrate. Quantification of competing effects is difficult, and thus the equilibrium partition approach for sediment quality assessment of metals still exhibits strong limitations.

3. *Polar organic substances* (e. g. phenols, polymers with functional groups, tensids) are widely unexperienced with respect to their specific "sorption" behavior. Partition coefficients are influenced by anion and cation exchange capacity and surface-charge density as a function of pH and other complex properties, so that the K_D-approach at present cannot be taken into consideration.

4.2.4 Sediment/Biota Equilibrium Partitioning

A very important aspect of the assessment of the environmental fate of chemicals is the prediction of the extent to which these substances will achieve concentrations in biotic phases. For organic chemicals, it has been suggested by Mackay (1982) that the bioconcentration factor K_B can be regarded simply as a partition coefficient between an organism consisting of a multiphase system and water; if the dominant concentrating phase is a lipid that has similar solute interaction characteristic to octanol, a proportional relationship between bioconcentration factor K_B and $K_{octanol/water}$ is expected ($K_B = 0.048\ K_{OW}$). This correlation must be used with discretion, particularly for very low K_B, where the amount of solute in nonlipid phases may be appreciable, and for high-K_{OW} compounds (e. g. mirex, octachlorosterene, and higher chlorinated biphenyls). In fact, Oliver (1984) found characteristic dependencies of bioconcentration of oligochaete worms in sediments from K_{OW}-values of the chemicals, where a slow linear increase in bioconcentration factors with K_{OW} is observed for chemicals with K_{OW}'s less than 10^5, and a rapid decrease in bioconcentration occurs for chemicals with very high partition coefficients ($>10^6$). The decrease may be caused by difficulties in chemical transport across worm membranes due to large molecular size or may be affected by strong binding of these chemicals to the sediments making them less bioavailable.

Similar to aqueous equilibria this approach is based on relatively broad experience with food quality data. Here, too, the three categories of compounds – nonpolar organic substances, metals, and polar organic substances – can be distinguished; in practice, again, only for the nonpolar organic compounds sufficient experience is available for developing quality criteria from equilibrium data. As outlined, hydrophobic organic substances show a strong correlation between the bioconcentration factor K_B and the octanol/water coefficient, indicating that the lipid content of the organism constitutes the major concentrating phase in the system water/organism. Simple steady-state correlative models have been used in laboratory studies to predict the bioconcentration from water of organic compounds by fish, mussels, and other aquatic organisms. One major difficulty is that these models were developed from steady-state concentrations and laboratory systems lacking suspended particulate material, and thus cannot deal with the varying concentrations and forms of pollutants found in many environments (Lake et al. 1987). Inclusion of solid particulate matter considers bioaccumulation as a redistribution of contaminants between sources (organic carbon of waste materials) and sinks (dissolved phase and lipids of organisms). For conditions in which the aqueous phase is not important as a sink, e.g. for high solid/aqueous partition coefficients, the bioaccumulation factor will depend on the concentration (C_i = concentration of component i) in the source (oc = organic carbon) and sink (lipids of organisms). Partition factors (PFs) between sediments and organisms have been defined as (Lake et al. 1987):

$$PF = \frac{C_i/g \text{ sediment (dry wt)}/g \text{ oc}/g \text{ sediment (dry wt)}}{C_i/g \text{ organism (dry wt)}/g \text{ lipid}/g \text{ organism (dry wt)}}.$$

PF-values for chlorinated compounds from experiments where exposure concentrations were constant and could be established (i.e., concentrations in sediment were used for infauna; concentrations in suspended particulate matter in dosing systems were used for mussels) were similar to the partition factor of approximately 0.5, which has been calculated by McFarland (1984) from $K_{oc}/BCF_{(lipid)}$ under various assumptions. These findings indicate that modeling bioaccumulation as a redistribution of contaminants between organic carbon of sediments and lipids of organisms is justified for at least some nonpolar, chlorinated organics, organisms, and exposures.

4.2.5 Elution Approach

In Section 3 the significance of surface speciation was indicated with respect to the assessment of the reactivity, mobility, and bioavailability of pollutants in aquatic systems. In practice, the effect of lowering pH-values, either from acid precipitation or from oxidation of sulfidic minerals, plays a dominant role in the mobilization of trace elements from sediments, soils, and solid waste materials. A "mobility test" procedure for soils based on variations of pH-values has been proposed by Kiekens and Cottenie (1985). Application to a large number of polluted and nonpolluted soils indicates that typical mobilization patterns are obtained for the different elements. Besides the nature of the element, the pH-curves reveal typical textural features of different soil substrates. Best results with respect to the estimation of middle-term effects can be attained by cascade test procedures at variable solid/solution ratios: A procedure of the U.S. EPA (Ham et al. 1980) designed for studies on the leachability of waste products consists of a mixture of sodium acetate, acetic acid, glycine, pyrogallol, and iron sulfate. For the study of combustion residues a standard leaching test has been developed by the Netherland Energy Research Centre (van der Sloot et al. 1984). In the column test the material under investigation is percolated by acidulated demineralized water (pH 4; for evaluating most relevant effects of acid precipitation) to assess short-term leaching (< 50 years). In the cascade test the same quantity of material is extracted several times with fresh demineralized water (pH 4) to get an impression of medium-term leaching behavior (50–500 years). As a time scale the liquid/solid ratio (L/S) is used; the maximum leachability is assessed by a shaking experiment at L/S ratio of 100 under mild acid conditions (De Groot et al. 1978). Recent improvements of this method have been achieved by comparing the L/S curves for an individual element with its stability in a wider pH-spectrum; in some cases direct mineralogical evidence can be given for a distinct metal compound (Van der Sloot, personal communication).

Single-extractant procedures are restricted with regard to prediction of long-term effects, e.g. of highly contaminated dredged materials, since these concepts neither involve mechanistic nor kinetic considerations and, therefore, do not allow calculations of release periods. This lack can be avoided by controlled intensification of the relevant parameters, i.e., pH-value, redox potential, and temperature, combined with an extrapolation of the potentially mobilizable "pools", which are estimated from sequential chemical extraction before and after treatment of the solid material. An experimental scheme, which was originally used by Patrick et al. (1973) and Herms and Brümmer (1978) for the study of soil suspensions and municipal waste materials, was modified by inclusion of an ion-exchanger system for extracting the metals released within a certain period of time (Schoer and Förstner 1987). The system can be modified for different intensities of contact between solid materials and solution, by using shakes (e.g., erosion of the depot by rivers) or dialysis bags (flow-by conditions).

4.2.6 Biological and Chemical Approaches: A Comparison

Biological criteria exhibit major advantages in that they integrate effects of multiple factors including sediment characteristics and complex or unknown wastes, and, with respect to field surveys, they are site-specific, requiring minimum extrapolations. On the other hand, field surveys are costly and bioassay organisms may hot represent sensitivity of the natural species assemblage (Anderson et al. 1987).

Major advantages of numerical criteria lie in their easy application and amendment to modeling approaches. However, if criteria do not exist for the chemicals concerned, a biological test may still be required. In addition, some equilibrium modeling approaches will fail if tissue concentration and toxicity are independent (G. Chapman et al. 1987).

5 Modeling of Sediment Data

One of the greatest challenges of environmental chemistry has been to describe the behavior of trace elements in natural aquatic systems based solely on the knowledge of their fundamental physicochemical properties. Initial efforts in applying quantitative models were undertaken for the prediction of metal speciation in solution (Baham 1984). The theoretical foundations for solving the problem of chemical speciation, which is usually not solvable by using experimental analysis, are based upon a model that relates the equilibrium activities of metals and ligands to the formation of complexes in solution. Ionic speciation in multicomponent aqueous systems with hundreds of competing equilibrium reactions can be estimated by computer solution of such geochemical equilibrium models like WATEQB (Arikan 1988). A significant reason for using the geochemical models is that these models yield an estimate of the activity of the metal in solution rather than its total concentration. This is compatible with modern toxicity philosophy because it is the thermodynamic activity of one or more of the aqueous species of the mobile metal in aquatic and terrestrial environments that determines toxicity and bioaccumulation, not the total dissolved concentrations of the metal. These models are also being used increasingly to predict the mobilizable fraction of metals in the sediment (Baes and Sharp 1983) and the transport of toxic metals and radionuclides in ground-waters (Lewis et al. 1987), to assess pollution potential to the ecosystem in general (Luoma and Davis, 1983), and to develop sediment quality criteria (Jenne et al. 1986).

Equilibrium models are, in turn, useful heuristic devices for probing our understanding of the basic physicochemical nature of processes determining metal behavior in natural systems. However, with respect to the modeling of metal partitioning between dissolved and particulate phases in a natural system, e.g. for estuarine sediments (Luoma and Davis 1983), there are still restrictions due to various reasons: (1) adsorption characteristics are related not only to the system conditions (i.e., solid types, concentrations, and adsorbing species), but also to changes in the net system surface properties resulting from particle/particle interactions such as coagulation; (2) the influences of organic ligands in the aqueous phase can rarely be predicted as yet; (3) effects of competition between various sorption sites; and (4) reaction kinetics of the individual constituents cannot be evaluated in a mixture of sedimentary components. These restrictions have been recently discussed in detail by Honeyman and Santschi (1988), who stated that even for aquatic environments of low particle concentration "the non-deterministic and interactive effects described above generally influence the estimation of an apparent partitioning coefficient by 1 to 3 orders of magnitude in either direction". With respect to environments of moderate to high particle concentration such as in soils and sediments they concluded that these theoretical approaches have failed thus far to provide a sound basis for the prediction of trace-element behavior and fate.

6 Characterization of Sediment Milieu – Acid-Producing Potential

Regarding the potential release of contaminants from sediments, changing of pH and redox conditions are of prime importance. In practice, therefore, characterization of sediment substrates with respect to their buffer capacity is a first step in the prognosis of middle- and long-term processes of mobilization, in particular, of toxic chemicals in a certain milieu.

Evaluation of pH-effects can be done relatively easily by titration with acid solutions. For quantifying pH-properties and for better comparison of sediment samples, it is proposed to use the term $\triangle$ pH, which is characterized by the difference of pH-values of 10% sludge suspensions in distilled water (pH_o) and 0.1 N sulfuric acid after 1 h shaking time (Calmano et al. 1986). Three categories of $\triangle$ pH-values can be established, ranging from $\triangle$ pH < 2 (strongly buffered), $\triangle$ pH 2–4 (intermediate) to $\triangle$ pH > 4 (poorly buffered).

Evaluation of the pH-changes resulting from the oxidation of anoxic sediment constituents can be performed by ventilation of sediment suspensions with air or oxygen and subsequent determination of the pH-difference between the original sample and oxidized material. The greater this difference, the higher is the short-term mobilization potential of metals, e.g. during dredging, resuspension, and other processes, by which anoxic sediments contact oxygenated water or – following land deposition of dredged material – atmospheric oxygen.

For a classification of sludges regarding their acid potential, which can be produced by oxidation of sulfidic components, one can preferentially use the data of calcium and sulfur from the sequential extraction scheme as proposed, for example, by Tessier et al. (1979; see Table 2). In anoxic, sulfide-containing sediments the two elements were selectively released during anaerobic experimental procedures (argon or nitrogen atmosphere in glove box) by the Na-acetate step (Ca from carbonates) and peroxide step (S from oxidizable sulfides, mainly iron sulfide). Reaction of oxygen with 1 mol of iron sulfide will produce three $[H^+]$-ions; by reaction with 1 mol of carbonate, two $[H^+]$-ions are buffered. For an initial estimation, one may compare total calcium and sulfur concentrations in the sediment sample.

Experimental approaches for prognosis of the "acid-producing potential" of sulfidic mining residues have been summarized by Ferguson and Erickson (1988). A test described by Sobek et al. (1978) involves the analysis of total or pyritic sulfur; neutralization potential is obtained by adding a known amount of HCl, heating the sample, and titrating with standardized NaOH to pH 7. Potential acidity is subtracted from the neutralization potential; a negative value below 5 t $CaCO_3$/1000 t of rock indicates a potential acid producer. Bruynesteyn and Hackl (1984) calculated acid-producing potential from total sulfur analysis; acid-consuming ability is obtained by titration with standardized sulfuric acid to pH 3.5 (Bruynestein and Duncan 1979). Acid-producing potential is subtracted from acid-consuming ability; a negative value indicates a potential acid producer.

7 Application of Different Criteria Approaches on Sediment Samples from Large Rivers in the Federal Republic of Germany

In the present section examples are given of three potential approaches which could provide direct numerical evaluations of the pollution potential in aquatic sediments; these examples are based on metal data from sediments from five large rivers in the Federal Republic of Germany. In addition, Sect 7.4 will give examples of the different behaviors of sediment substrates with respect to the acid-producing potential, which is a major factor controlling the potential release of critical elements under changing redox conditions (see Sect 6).

7.1 Index of Geoaccumulation (Müller 1979)

The "I_{geo}-approach" of Müller compares advantageously with similar procedures, based on background concentrations, in that it constantly maintains a logarithmic scale. Subsequent to normalization with respect to grain size (which is not discussed here), therefore, pollutant or nutrient concentration data from sediment samples taken from different aquatic systems can be compared. From the examples given in Table 7 the sediment sample from the lower Rhine River (collected from the deeper part of the sediment

Table 7. Index of geoaccumulation (Müller 1979) for sediments from five examples of rivers. ($I_{Geo} = \log_2 A_n/B_n \times 1.5$)

	Neckar	Main	Rhine	Elbe	Weser
Copper	0	1	2	2	0
Lead	1	2	2	2	2
Zinc	1	2	3	4	2
Cadmium	3	2	6	4	4
Mercury	0	1	4	6	1
Average	1.0	1.6	3.4	3.3	1.8

pile) exhibits the highest overall metal accumulation relative to the background concentrations, followed by the sample from the Elbe River (Hamburg harbor); in both examples, mercury and cadmium typically influence the overall factor of enrichment. The sediment samples from the rivers Weser, Main, and Neckar are by far less contaminated than the afore mentioned materials; however, there is still considerable enrichment of cadmium in sediments of both the Weser and Neckar Rivers. From the latter example, it is demonstrated that even with significant recovery, there are still characteristic indications from sediment samples of former situations of extreme pollution (Cd-poisoning of fish in the early 1970s; Müller and Förstner 1973).

7.2 Incorporation of "Toxic Effects" into Accumulation Factors

The somewhat unsatisfying situation regarding the comparison of enrichment factors of elements, which exhibit strongly different impacts on ecosystems, could be overcome by introducing a factor for the relative toxicity of the respective element or compound. As discussed above, this approach again will be controversial, as such arrangements can only consider the relative impact of water constituents at a specific trophic level. Nonetheless, as demonstrated in Table 8, the combination of enrichment factors with a factor of toxicity is advantageous, as such indices more clearly point to the critical compounds in the overall mixture of potential contaminants, and may thus stimulate setting priorities for control and rehabilitation measures. In the case of the extreme mercury pollution of the Elbe River, reduction of a specific source, i.e. emissions from chlorine alkali plants, could significantly contribute to the improvement of the overall sediment quality; for example, with available technology and reasonable costs the concentrations of mercury in the Elbe River could be reduced to such an extent that the overall sediment quality would be at least comparable to the sample from the Weser River.

7.3 Numerical Evaluation of the Factor "Element Mobility"

Without going into detailed discussions, it should be mentioned that there is a possibility for standardizing the data from elution experiments with respect to numerical evaluation. In Table 9 examples are given for an "elution index" based on the metal concentrations exchangeable with 1 N ammonium acetate at pH 7; these metal fractions are considered to be remobilizable from polluted sediments at a relatively short term under more saline conditions, for example, in the estuarine mixing zone. Comparison of the release rates from oxic and anoxic sediments clearly indicates that the oxidation of samples gives rise to a very significant increase of the overall mobilization of the ele-

Sediment Criteria Development

Table 8. Factor of enrichment × toxicity factor ("Microtox")

		Neckar	Main	Rhine	Elbe	Weser
Lead	(× 1)	2	4	4	6	4
Copper	(× 5)	7	10	17	22	6
Zinc	(× 5)	10	15	35	70	30
Cadmium	(× 10)	62	53	500	340	360
Mercury	(× 35)	46	98	805	2520	81
Total		127	180	1381	2958	481

Table 9. Elution index as determined from exchangeable fractions (1 N ammonium acetate solution at pH 7) related to background values from old sediments from the Rhine River: Cu = 51 mg/kg, Pb = 30 mg/kg, Zn = 115 mg/kg, Cd = 0.3 mg/kg (values multiplied by factor 100)

	Neckar	Main	Rhine	Elbe	Weser
Copper	0.2	–	1	1	–
Lead	1	1	2	1	1
Zinc	7	9	28	36	9
Cadmium	30	30	230	30	–
Total oxic	38	40	261	68	10
(Anoxic)	0.5	0.3	8	> 4	4

ment studied here; this effect is particularly important for cadmium. When proceeding further in the extraction sequence, more long-term effects could be estimated (generally with a respective reduction of prognostic accuracy). A major disadvantage of the present approach, however, is that the critical element, mercury, is not yet included in this scheme.

7.4 pH-Titrations of Selected Sediment Samples

Results from titration experiments using 1 M nitric acid on sediment suspensions of 100 g/l are presented in Fig. 2. The titration curve of the Rhine River sediment exhibits a small plateau in the pH-range 5.5 and 6, probably due to a certain fraction of carbonate, which is consumed by addition of 80 nmol of HCl. In contrast, the titration curves of both Elbe River sediments are continuously decreasing due to the low contents of carbonate in these samples. The sediment from the inland harbor basin of Harburg, originally sulfide-rich material which had been stored for 1 year in a closed bottle, has already reached an initial-pH of 4.3; this is probably due to the consumption of the low residual buffer capacity by oxidation of parts of the sulfide fraction.

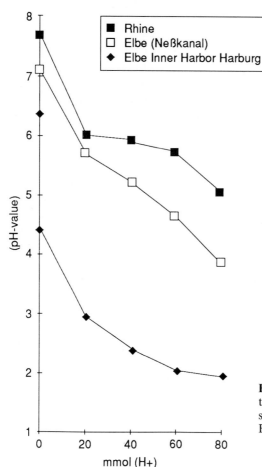

Fig. 2. Variations of pH-values (titration curves) of suspensions (100 g/l) of sediment samples from rivers Rhine and Elbe after addition of 1 M nitric acid

Respective lowering of pH has been found from upland disposal sites of dredged sediments from Hamburg harbor (Tent 1982). Due to the low carbonate content, which is consumed during several months or years, and subsequent lowering of pH, metals are easily transferred to crops, and permissible limits of cadmium have been surpassed in as much as 50% of wheat crops grown on these materials (Herms and Tent 1982). High concentrations of metals have been measured in pore waters from sedimentation polders in the Hamburg harbor area, in the older, oxidized deposits (Maaß et al. 1985). It can be expected that similar effects will occur as well in the aquatic system, particularly in tidal areas affected by periodic drying and wetting (Kersten 1989), and at other high-energy sites exhibiting strong resuspension activities. The situation in the Elbe River estuary is particularly critical since low buffer capacities of the sediments coincide with a relatively long residence time of suspended particles (Müller and Förstner 1975; Tent 1987).

8 Summary and Outlook

Requirements for water quality criteria include (Höpner 1989): (1) The system should be simple; (2) methods should be practicable; (3) criteria should exhibit the ability to register temporal changes; (4) criteria should assess the ecological status of the system in comparison to other situations or sites; (5) the system should sound alarm at critical situations; and (6) ecological status should be described. These requirements reflect a sequence from simpler ones to the most complex systems, i.e. the description of ecological conditions, which can rarely be covered by an easy, practicable approach.

The approaches to sediment quality assessment described so far can be divided into two groups, according to their objectives, i.e., either as a "quality standard" or with regard to an "indicator" function. The first group would comprise, for example, pore water, sediment/water equilibrium, and sediment/organism equilibrium approaches, which are based on toxicologially relevant standards. The approaches of the second group, consisting of the elution and background approaches, will, irregardless of their numerical character, primarily only provide qualitative indications to a certain status of the extent of sediment pollution. However, it seems advantageous to have such a relatively simple initital assessment, which may then be extended, with more complicated procedures, including biological criteria, into an integrated ecological evaluation.

Practical applicability of sediment quality criteria has been proceeding to a different extent. Classifications, on the basis of equilibrium calculations and pore water composition, still seem to require further studies and discussions; on the other hand, interim regulations using modifications of the background approach could well be installed at the present level of knowledge. There are already several examples of statewide water quality evaluations, which use the "I_{geo}"-approach by Müller for the assessment of the pollution status of aquatic sediments. It seems that the next step should be the incorporation of a "toxicity factor", regardless of the underlying test system chosen; there are several initiatives for applying toxicity screening tests, e.g. Microtox, ATP-TOX, and genotoxicity tests, on sediment extracts (e.g. Dutka et al. 1988).

Elution experiments, as demonstrated from the preceding examples, generally confirm the findings of the data from the background approach. However, since elements such as mercury, and organic compounds cannot be incorporated into the presently available scheme, this approach presumably will not be accepted as a primary criterion. On the other hand, recent developments of soil quality criteria should be mentioned, which include a more differentiated approach with respect to metal availability to plants[1].

---

[1] Annex to article 5 of the Swiss regulation on contaminants in soil from June 9, 1986, includes the following standard values for soluble metal concentrations (0.1 M sodium nitrate; weight ratio of soil sample to solution 1:2.5): Pb = 1.0 µg/g; Cd = 0.03 µg/g; Cu = 0.7 µg/g; Ni = 0.2 µg/g; Zn = 0.5 µg/g.

Another aspect of sediment criteria development has been shown for the first time, namely the estimation of potential changes of substrate properties, in particular by oxidation and subsequent lowering of pH-values. This approach again follows the actual developments in describing the soil "filter system", where assessments are performed with respect to the relative bonding strength of heavy metals and to potential hazards for groundwater pollution, by measuring parameters such as "pH", "organic matter content", and "concentrations of iron oxides". Since oxidation processes play a critical role in sediments, we propose that the factor of the "acid producing potential" should be included as a primary parameter in sediment-oriented quality criteria.

Future development of sediment quality criteria will decisively be influenced by the progress of discussions on equilibrium approaches, on which actual efforts are focussed in the U.S.A. (Shea 1988). While sufficient knowledge is already available regarding nonpolar organic compounds, there are still many questions open with respect to the parameters needed for equilibrium calculations involving trace elements and their major substrates. Further basic research is needed in particular for incorporating kinetic data in such models.

References

Ahlf W, Munawar M (1988) Biological assessment of environmental impact of dredged material. In: Salomons W, Förstner U (eds) Chemistry and biology of solid waste – dredged material and mine tailings. Springer, Berlin Heidelberg New York Tokyo, pp 126–142

Alderton DHM (1985) Sediments. In: Historical monitoring, MARC Tech Rep 31. Monitoring and Assessment Research Centre/Univ London, pp 1–95

Anderson J, Birge W, Gentile J, Lake J, Rodgers J, Swartz R (1987) Biological effects, bioaccumulation, and ecotoxicolgy of sediment-associated chemicals. In: Dickson KL, Maki AW, Brungs WA (eds) Fate and effects of sediment-bound chemicals in aquatic systems. Pergamon, New York, pp 267–295

Anon (1985) Sediment quality criteria development workshop, Nov 28–30, 1985. Battelle Washington Operations, Richland

Arikan A (1988) WATEQB – BASICA revision of WATEQF for IBM personal computers. Ground Water 26:222–227

Baes CF, Sharp RD (1983) A proposal for the estimation of soil leaching and leaching constants for use in assessment models. J Environ Qual 12:17–28

Baham J (1984) Prediction of ion activities in soil solutions: computer equilibrium modeling. Soil Sci Soc Am J 48:525–531

Bruynesteyn A, Duncan DW (1979) Determination of acid production potential of waste materials. Met Soc AIME Pap A-79-29, 10 pp

Bruynesteyn A, Hackl RP (1984) Evaluation of acid production potential of mining waste materials. Mineral Environ 4:5–8

Calmano W (1979) Untersuchungen über das Verhalten von Spurenelementen an Rhein- und Mainschwebstoffen mit Hilfe radioanalytischer Methoden. Diss, TH Darmstadt

Calmano W, Förstner U (1983) Chemical extraction of heavy metals in polluted river sediments in Central Europe. Sci Total Environ 28:77–90

Calmano W, Förstner U, Kersten M, Krause D (1986) Behaviour of dredged mud after stabilization with different additives. In: Assink JW, Van Den Brink WJ (eds) Contaminated soil. Nijhoff, Dordrecht, pp 737–746
Carignan R (1984) Interstitial water sampling by dialysis. Methodological notes. Limnol Oceanogr 29:667–670
Chapman G, Adam W, Lee H, Lyman W, Pavlou S, Wilhelm P (1987) Regulatory implications of contaminants associated with sediments. In: Dickson KL, Maki AW, Brungs WA (eds) Fate and effects of sediment-bound chemicals in aquatic systems. Pergamon, New York, pp 413–425
Chapman PM (1986) Sediment quality criteria from the sediment quality triad: an example. Environ Toxicol Chem 5:957–964
Chapman PM, Barrick RC, Neff JM, Swartz RC (1987) Four independent approaches to developing sediment quality criteria yield similar values for model contaminants. Environ Toxicol Chem 6:723–725
Connor DJ, Connolly JP (1980) The effect of concentration of adsorbing solids on the partition coefficient. Water Res 14:1517–1523
Deutscher Verband für Wasserwirtschaft und Kulturbau (ed) (1988) Filtereigenschaften des Bodens gegenüber Schadstoffen, pt 1: Beurteilung der Fähigkeit von Böden zugeführte Schwermetalle zu immobilisieren. DVWK-Merkblätter zur Wasserwirtschaft, vol 212. Parey, Hamburg, 8 pp
DiToro DM, Horzempa LM (1982) Reversible and restistant components of PCB adsorption-desorption: isotherms. Environ Sci Technol 16:594–602
Dutka BJ, Jones K, Kwan KK, Bailey H, McInnis R (1988) Use of microbial and toxicant screening tests for priority site selection of degraded areas in water bodies. Water Res 22:503–510
Duursma EK (1984) Problems of sediment sampling and conservation for radionuklide accumulation studies. In: Sediments and pollution in waterways. IAEA-TECDOC-302, Int Atomic Energy Agency, Vienna, pp 127–135
Engler RM, Brannon JM, Rose J, Bigham G (1977) A practical selective extraction procedure for sediment characterization. In: Yen TF (ed) Chemistry of marine sediments. Ann Arbor Sci Publ, Ann Arbor, MI, pp 163–171
Ferguson KD, Erickson PM (1988) Pre-mine prediction of acid mine drainage. In: Salomons W, Förstner U (eds) Environmental management of solid waste – dredged material and mine tailings. Springer, Berlin Heidelberg New York Tokyo, pp 24–43
Förstner U, Kersten M (1988) Assessment of metal mobility in dredged material and mine waste by pore water chemistry and solid specication. In: Salomons W, Förstner U (eds) Chemistry and biology of solid waste – dredged material and mine tailings. Springer, Berlin Heidelberg New York Tokyo, pp 214–237
Förstner U, Müller G (1974) Schwermetalle in Flüssen und Seen als Ausdruck der Umweltverschmutzung. Springer, Berlin Heidelberg New York
Golterman HL, Sly PG, Thomas RL (1983) Study of the relationship between water quality and sediment transport. Technical papers in hydrology 26. Unesco, Paris, 231 pp
Hakanson L (1980) An ecological risk index for aquatic pollution control. A sedimentological approach. Water Res 14:975–1001
Ham RK, Anderson MA, Stegmann R, Stanforth R (1980) Die Entwicklung eines Auslaugtests für Industrieabfälle. Müll Abfall 12:212–220
Herms U, Brümmer G (1978) Löslichkeit von Schwermetallen in Siedlungsabfällen und Böden in Abhängigkeit von pH-Wert, Redoxbedingungen und Stoffbestand. Mitt Dtsch Bodenkd Ges 27:23–43
Herms U, Tent L (1982) Schwermetallgehalte im Hafenschlick sowie in landwirtschaftlich genutzten Hafenschlickspülfeldern im Raum Hamburg. Geol Jahrb F12:3–11
Hesslein R (1976) An in-situ sampler for close interval pore water studies. Limnol Oceanogr 21:912–914

Höpner T (1989) Statusseminar Gütekriterien für Küstengewässer. Ergebnisse einer Literaturstudie und Kurzfassung der Referate. Arbeitsgruppe für Regionale Struktur- und Umweltforschung, Oldenburg

Honeyman BD, Santschi PH (1988) Metals in aquatic systems. Environ Sci Technol 22:862–871

Isaacson PJ, Frink CR (1984) Nonreversible sorption of phenolic compounds by sediment fractions: the role of sediment organic matter. Environ Sci Technol 18:43–48

Jackson DR, Garrett BC, Bishop TA (1984) Comparison of batch and column methods for assessing leachability of hazardous waste. Environ Sci Technol 18:668–673

Jenne EA, Kennedy VC, Burchard JM, Ball JW (1980) Sediment collection and processing for selective extraction and for total trace element analysis. In: Baker RA (ed) Contaminants and sediments, vol 2. Ann Arbor Sci Publ, Ann Arbor, MI, pp 169–191

Jenne EA, DiToro DM, Allen HE, Zarba CS (1986) An activity-based model for developing sediment criteria for metals. In: Lester JN, Perry R, Sterritt RM (eds) Chemicals in the environment. Selper, London, pp 560–568

Karickhoff SW (1981) Semi-empirical estimation of sorption of hydrophobic pollutants on natural sediments and soils. Chemosphere 10:833–846

Kersten M (1989) Mechanismus und Bilanz der Schwermetallfreisetzung aus einem Süßwasserwatt der Elbe. Diss, T Univ Hamburg-Harburg

Kersten M, Förstner U (1986) Chemical fractionation of heavy metals in anoxic estuarine and coastal sediments. Water Sci Technol 18:121–130

Kersten M, Förstner U (1987) Effect of sample pretreatment on the reliability of solid speciation data of heavy metals – implication for the study of diagenetic processes. Mar Chem 22:299–312

Kersten M, Förstner U, Calmano W, Ahlf W (1985) Freisetzung von Metallen bei der Oxidation von Schlämmen. Wasser 65:21–35

Keyser TR, Natusch DFS, Evans CA Jr, Linton RW (1978) Characterizing the surface of environmental particles. Environ Sci Technol 12:768–773

Kiekens L, Cottenie A (1985) Principles of investigations on the mobility and plant uptake of heavy metals. In: Leschber R, Davis RD, L'Hermite P (eds) Chemical methods for assessing bio-available metals in sludges and soils. Elsevier, London New York Amsterdam, pp 32–47

Lake JL, Rubinstein N, Pavignano S (1987) Predicting bioaccumulation: development of a simple partitioning model for use as a screening tool for regulating ocean disposal of waste. In: Dickson KL, Maki WA, Brungs WA (eds) Fate and effects of sediment-bound chemicals in aquatic systems. Pergamon, New York, pp 151–166

Lewis FM, Voss CI, Rubin J (1987) Solute transport with equilibrium aqueous complexation and either sorption or ion exchange: simulation methodology and applications. J Hydrol 90:81–115

Lion LW, Altman RS, Leckie JO (1982) Trace-metal adsorption characteristics of estuarine particulate matter: evaluation of contribution of Fe/Mn oxide and organic surface coatings. Environ Sci Technol 16:660–666

Long ER, Chapman PM (1985) A sediment quality triad. Measures of sediment contamination, toxicity and infaunal community composition in Puget Sound. Mar Pollut Bull 16:405–415

Luoma SN, Davis JA (1983) Requirements for modeling trace metal partitioning in oxidized estuarine sediments. Mar Chem 12:159–181

Maaß B, Miehlich G, Gröngröft A (1985) Untersuchungen zur Grundwassergefährdung durch Hafenschlick-Spülfelder. II Inhaltsstoffe in Spülfeldsedimenten und Porenwässern. Mitt Dtsch Bodenkd Ges 43/I:253–258

Mackay D (1982) Correlation of bioconcentration factors. Environ Sci Technol 16:274–278

Malins DC, McCain BB, Brown DW, Chan SL, Myers MS, Landahl JT, Prohaska PG, Friedman AJ, Rhodes LD, Burrows DG, Gronlund WD, Hodgins HO (1984) Chemi-

cal pollutants in sediments and diseases of bottom-dwelling fish in Puget Sound, Washington. Environ Sci Technol 18:705–713
McFarland VA (1984) Activity-based evaluation of potential bioaccumulation for sediments. In: Montgomery RL, Leach JW (eds) Dredging and dredged material disposal, vol 1. Soc Civil Eng, New York, pp 461–467
Müller G (1977a) Schadstoff-Untersuchungen an datierten Sedimentkernen aus dem Bodensee. II Historische Entwicklung von Schwermetallen – Beziehung zur Entwicklung polycyclischer aromatischer Kohlenwasserstoffe. Z Naturforsch 32c:913–919
Müller G (1977b) Schadstoff-Untersuchungen an datierten Sedimentkernen aus dem Bodensee. III Historische Entwicklung von N- und P-Verbindungen – Beziehung zur Entwicklung von Schwermetallen und polycyclischen aromatischen Kohlenwasserstoffen. Z Naturforsch 32c:920–925
Müller G (1979) Schwermetalle in den Sedimenten des Rheins – Veränderungen seit 1971. Umschau 79:778–783
Müller G (1981) Heavy metals and other pollutants in the environment. A chronology based on the analysis of dated sediments. In: Ernst WHO (ed) Heavy metals in the environment, Amsterdam. CEP, Edinburgh, pp 12–17
Müller G (1983) Zur Chronologie des Schadstoffeintrags in Gewässer. Geowis Unserer Zeit 1:2–11
Müller G (1985) Unseren Flüssen geht's wieder besser. Bild Wiss 10/1985:76–97
Müller G, Förstner U (1973) Cadmiumanreicherungen in Neckarfischen. Naturwissenschaften 60:258–259
Müller G, Förstner U (1975) Heavy metals in the Elbe and Rhine estuaries: mobilization or mixing effect? Environ Geol 1:33–39
Müller G, Dominik J, Mangini A (1979a) Eutrophication changes sedimentation in part of Lake Constance. Naturwissenschaften 66:261–262
Müller G, Kanazawa A, Teshima S (1979b) Sedimentary record of fecal pollution in part of Lake Constance by coprostanol determination. Naturwissenschaften 66:520–521
Müller G, Dominik J, Reuther R, Malisch R, Schulte E, Acker L, Irion G (1980) Sedimentary record of environmental pollution in the western Baltic Sea. Naturwissenschaften 67:595–600
Oliver BG, Nicol KD (1982) Chlorobenzenes in sediments, water, and selected fish from Lakes Superior, Huron, Erie, and Ontario. Environ Sci Technol 16:532–536
Oliver BG (1984) Uptake of chlorinated contaminants from anthropogenically contaminated sediments by oligochaete worms. Can J Fish Aquat Sci 41:878–883
Patrick WH, Williams BG, Moraghan JT (1973) A simple system for contolling redox potential and pH in soil suspensions. Soil Sci Soc Am Proc 37:331–332
Reynoldson TB (1987) Interactions between sediment contaminants and benthic organisms. In: Thomas RL, Evans R, Hamilton A, Munawar M, Reynoldson TB, Sadar H (eds) Ecological effects of in situ sediment contaminants. Hydrobiologia 149:53–66
Sauerbeck D, Styperek P (1985) Evaluation of chemical methods for assessing the Cd and Zn availability from different soils and sources. In: Leschber R, Davis RD, L'Hermite P (eds) Chemical methods for assessing bio-available metals in sludges and soils. Elsevier, London New York Amsterdam, pp 49–66
Schoer J, Förstner U (1984) Chemical forms of artificial radionuclides and their stable counterparts in sediments. Proc Int Conf Environmental contamination, London. CEP, Edinburgh, pp 738–745
Schoer J, Förstner U (1987) Abschätzung der Langzeitbelastung von Grundwasser durch die Ablagerung metallhaltiger Feststoffe. Wasser 69:23–32
Schwedhelm E, Vollmer M, Kersten M (1988) Bestimmung von Konzentrationsgradienten gelöster Schwermetalle an der Sediment/-Wasser-Grenzfläche mit Hilfe der Dialysetechnik. Fres Z Anal Chem 332:756–763

Shea D (1988) Developing national sediment quality criteria – equilibrium partitioning of contaminants as a means of evaluating sediment quality criteria. Environ Sci Technol 22:1256–1261

Sobek AA, Schuller WA, Freeman JR, Smith RM (1978) Field and laboratory methods applicable to overburden and mine soils. US Environ Protec Ag Rep EPA-600/2-78-054

Tent L (1982) Auswirkungen der Schwermetallbelastung von Tidegewässern am Beispiel der Elbe. Wasserwirtschaft 72:60–62

Tent L (1987) Contaminated sediments in the Elbe estuary: ecological and economic problems for the Port of Hamburg. In: Thomas RL et al. (eds) Ecological effects of in-situ sediment contaminants. Hydrobiologia 149:189–199

Tessier A, Campbell PGC, Bisson M (1979) Sequential extraction procedure for the speciation of particulate trace metals. Anal Chem 51:844–851

Theis TL, Padgett LE (1983) Factors affecting the release of trace metals from municipal sludge ashes. J Water Pollut Control Fed 55:1271–1279

UNESCO (ed) (1978) Water quality survey. A guide for the collection and interpretation of water quality data. Studies and reports in hydrology, vol 23. Unesco, Paris, 350 pp

US Environmental Protection Agency (ed) (1984) Background and review document on the development of sediment criteria. EPA Contract 68-01-6388. JRB Associates, McLean/Virg

Van der Sloot HA, Piepers O, Kok A (1984) A standard leaching test for combustion residues. Shell BEOP-31. Studiegroep Ontwikkeling Standaard Uitloogtesten Verbrandingsresiduen, Petten/Netherlands.

Van Veen HJ, Stortelder PBM (1988) Research on contaminated sediments in the Netherlands. In: Wolf K, Van Den Brink WJ, Colon FJ (eds) Contaminated soil 88. Kluwer, Dordrecht, pp 1263–1275

Voice TC, Rice CP, Weber WJ Jr (1983) Effects of solids concentrations on the sorptive partitioning of hydrophobic pollutants in aquatic systems. Environ Sci Technol 17:513–518

Walker JD (1988) Effects on microorganisms. J Water Pollut Control Fed 60:1106–1121

Watson PG, Frickers PE, Goodchild CM (1985) Spatial and seasonal variations in the chemistry of sediment interstitial waters in the Tamar estuary. Estuar Coast Shelf Sci 21:105–119

WHO (ed) (1978) GEMS – global environmental monitoring system. WHO, Geneva, 313 pp

Transport of Matter in Sediments: A Discussion

DIETRICH HELING[1] and RALF GISKOW[2]

CONTENTS

Abstract 339
1 Geologic Evidence 340
2 Transport Processes 340
2.1 Advective Transport 340
2.2 Diffusive Transport 341
3 Model Calculations 343
4 Short-Range Diffusive Transport 348
5 Conclusions 350
References 350

Abstract

Time intervals necessary for the cementation of some selected sandstones were calculated on the basis of experimentally determined diffusion coefficients, assuming that the transport of matter was accomplished solely by diffusion. The time intervals needed to transfer the enormous quantities needed for cementations under natural conditions proved to be unrealistically long even if geologic time periods are considered. The calculations suggest that for the long-range transport (exceeding dm- to m-distances) advection is the only possible mode of transport.

Considering the small concentrations differences in the pore solutions occurring in natural sediments, diffusion is likely to be the only possibility for the short-range transport, i.e. within distances of less than 1 dm. In fine-grained sediments advection fails to transfer matter due to small pore openings.

A constraining example of diffusive transport is the diagenetic replacement of solids by pore cements, effected by slight and locally constricted concentration differences. The solution and precipitation of dissolved and replacing compounds, respectively, appears to be controlled dominantly by

---

[1] Institut für Sedimentforschung, Im Neuenheimer Feld 236, D-6900 Heidelberg, FRG
[2] Hauptstraße 187, D-6500 Mainz, FRG

the pH. The causes for the implied changes of the pH in time and space can be presumed presently as unproven.

1 Geologic Evidence

One of the most important diagenetic processes in porous sediments is cementation, which means postsedimentary precipitation of minerals in the pore space of sediments. Often, the chemical components for such precipitation are not contained in the detrital composition of the sediment, but have to be conveyed from an external source by aqueous solutions. Cements derived from matter supplied from outside of the sediment are called "allochthonous" (von Engelhardt 1960). Wherever conditions of oversaturation prevail the dissolved matter will precipitate as cementing minerals.

Well-known examples for allochthonous pore cements introduced from an external source are calcite or gypsum in a primarily pure siliclastic sandstone. Silica cement, on the contrary, is mostly autochthonous, i.e. the silica is derived from the sediment itself, but may sometimes be supplied from an outside source as well; this being likely, in particular, in sandstones lacking "pressure solution" and displaying pore spaces totally filled by silica cement devoid of syntaxial overgrowth rims.

The substances necessary to build up the pore cements of allochthonous, as well as autochthonous, origin are supplied by aqueous solutions. The following rough estimation may help to determine the quantities of pore solutions needed to carry the matter for a pore space cementation.

In order to reduce the porosity of a sandstone from 0.2 to 0.1 by cementation of calcite 260 kg $CaCO_3$ per m^3 of rock are necessary. The solubility of $CaCO_3$ is about 0.1 g/l. Hence, volumes of pore solution of no less than 2.6×10^3 m^3 per m^3 of rock are needed for the cited minor cementation. This means that enormous quantities of pore solutions have to pass through the sediment in order to supply the cementing compounds. Pore water currents of this dimension have been described by Lemcke and Tunn (1956) in the Molasse strata of southern Germany.

Further proof of pore solution currents of regional extension in range and volume is the accumulation of oil and gas in trapping structures of reservoir rocks, following long-distance migration through porous sediments. With the formation of oil and gas reservoirs, migration distances of up to 10 km often have to be postulated (Levorsen 1967).

2 Transport Processes

2.1 Advective Transport

Pore water currents of regional migration distances in permeable sediments, i.e. rocks with communicating pore spaces, are caused by pressure gradients.

Transport of Matter in Sediments: A Discussion

The flow rate per area unit of such currents, called advective flows, is described by the Darcy law:

$J = k \, grad \, p$.

In addition to the pressure gradient (grad p), the pore water current depends on the coefficient of permeability k, which is controlled by the structure of the rock being passed by the solution; in particular, by its pore widths.

Pressure gradients in sediments may be caused by:

1. The compaction of clayey sediments, where the interstitial water is squeezed out by the weight of the overburden. According to the pressure gradient, such currents are generally directed to the surface.

2. The expansion of the pore water due to locally or regionally increasing temperatures, for instance, adjacent to rising salt diapirs, magmatic intrusions, or hydrothermal vents. Temperature-induced pressure gradients are directed parallel to the pressure gradients.

3. Tectonic or halokinetic movements exerting increased pressure on parts of compressible sediments.

4. The decrease in pressure due to surficial erosion in combination with reexpansion of strata. Pressure gradients caused in this way are directed towards the site of the pressure minimum.

5. The dehydration of smectitic clays. By exchange for potassium ions interlayer water of smectite is released, which contributes to a pressure increase by an increase of its volume.

6. The release of gas originating from the diagenesis of organic substances included in the detritus.

7. Artesian water, which is overpressurized by communicating with elevated source areas.

The pressure gradients caused by these effects may produce extensive pore water movements lasting for geologic time intervals. The higher the permeability of the rock, the more readily the pressure gradient is reduced and eventually levelled off by water flow in the direction of the gradient. A gradient can be produced only if the pressure is built up at a higher rate than it is diminished by the discharge of the pore water. If the permeability is high enough to let the water bleed off without delay, than the buildup of a pressure gradient is impossible.

2.2 Diffusive Transport

As a different mode of transport, matter can be transferred in porous and permeable sediments by diffusion. Diffusion is evoked by concentration gradients. The linear, single-phase diffusion current, as described by Fick's first law, is formally analogous with the Darcy-law:

$$J = D \text{ grad } c,$$

where D is the diffusion coefficient which depends on the diffusing ion or compound, its concentration, and temperature. Grad c is the concentration difference over the distance 1.

Besides rock salt, and, under favourable conditions, carbonates, the solubilities of the cementing compounds are generally very low, resulting in concentration gradients too small to cause major diffusion currents. Rock salt plays only a minor role as a pore cement since the saturation concentration necessary for precipitation is seldom reached. Carbonates are transported extensively by advective currents, as is proven by the readily accomplished dissolution and cementation processes in porous carbonatic rocks.

For the bulk of potential cement-forming matter, however, the solubilities are so low that no considerable concentration differences can

Table 1. Parameters of grain size distribution of the investigated sandstones

Sandstone sample	Clay fraction (%)	Mean size mm	Mean size $\bar{x}(\Phi)$	Size class[a]	Coefficient of sorting (δ)	Coefficient of skewness (α_3)
Braun-grauer Sandstein	3.83	0.09	3.541	vfs	0.882	−0.169
Dogger β-Sandstein	2.53	0.14	2.865	fs-vfs	0.657	0.823
'Feinsandstein'	3.09	0.12	3.067	fs-vfs	0.874	0.488
Gipskeuper-Sandstein	1.79	0.40	1.351	ms	0.957	0.234
Glaukonit-Sandstein	0.96	0.15	2.721	fs	0.931	0.287
'Grobsandstein'	–	0.66	0.658	cs	0.859	0.878
Hilssandstein plattiger Teil	5.31	0.06	3.971	vfs-csi	0.699	0.282
Hilssandstein massiger Teil	2.83	0.08	3.551	vfs	0.642	0.072
Lettenkeupersandstein	1.79	0.13	2.961	fs-vfs	0.871	0.115
'Mittelsandstein'	2.09	0.20	2.306	fs	0.658	0.477
Muschelkalksandstein	4.53	0.12	3.051	vfs-fs	0.827	0.052
Roter Sandstein	2.27	0.11	3.145	fs-vfs	0.937	0.186
Rotliegend Sandstein 4	3.74	0.12	3.091	fs-vfs	0.761	0.291
Rotliegend Sandstein 5	4.59	0.23	2.121	ms-fs	0.861	0.227
Schilfsandstein, rot	4.04	0.09	3.501	vfs	0.661	0.424
Schilfsandstein, hell	3.37	0.10	3.426	vfs	0.711	0.107
'Siltstein'	3.47	0.07	3.820	vfs	1.413	0.229
St. Peter-Sandstein	1.67	0.20	2.341	fs	0.891	−0.317

[a] Size classes: csi = coarse silt; vfs = very fine sand; fs = fine sand; ms = medium sand; cs = coarse sand.

develop. Therefore, the diffusive transport for these compounds is insignificant, if only long-distance transport is considered.

3 Model Calculations

In order to approach realistic geologic conditions more closely, the diffusion coefficients of a number of natural sandstones were evaluated experimentally (Giskow 1987). For the experiments, sandstones with the greatest available differences in texture and composition were selected. Tables 1 and 2 show the major petrographic data of the investigated sandstones. Their grain size distributions range from coarse silt to coarse sandstone with clay contents from 1 to over 5%, porosities between 4 and 30%, permeabilities between 0.4 and about 1000 mD, and BET surfaces ranging from 1 up to little more than 10 m^2/g.

The diffusion currents were measured on clyindric plugs 3 cm in diameter and about 1 cm height, which were inserted in a measuring cell separating two vessels, which contained the same electrolyte solution, but in different concentrations. The concentrations (c) were determined by means of electric conductivity; the diffusion coefficients, by the time-lag method (Golubev and

Table 2. Petrophysical data of the investigated sandstones

Sandstone sample	Porosity (%)	Formation resistivity factor F	Specific surface area m^2/g (BET)	Tortuosity	Permeability (mD)
Braun-grauer Sandstein	26.3	20.0	4.11	5.26	105
Dogger β-Sandstein	29.8	9.9	10.62	2.95	1000
'Feinsandstein'	25.6	10.8		2.76	–
Gipskeupersandstein	24.0	17.6	4.74	4.22	256
'Grobsandstein'	9.5	49.3		4.68	–
Hilssandstein	17.5	35.7	7.06	6.25	–
Hilssandstein	18.9	35.0		6.62	2
Hilssandstein, plattiger Teil	26.8	19.9	8.41	5.33	2
Lettenkeupersandstein	19.9	35.0	2.57	6.97	2
'Mittelsandstein'	13.6	35.4		4.81	–
Roter Sandstein	17.3	35.6	1.59	6.16	91
Roter Sandstein	16.7	29.6		4.94	–
Schilfsandstein, rot	26.3	18.1	5.04	4.76	–
Schilfsandstein, rot	25.3	16.9		4.28	154
Schilfsandstein, hell	22.7	16.3	5.01	3.70	113
Schilfsandstein, hell	24.1	15.4		3.71	–
Schilfsandstein, grau	24.6	24.8		6.10	–
'Siltstein'	3.8	224.3		8.52	–
St. Peter Sandstein	8.8	–	–	–	0.4
Unterer Muschelkalksandstein	23.7	–	–	–	240

Table 3. Measured coefficients of diffusion (D*) and rates of diffusive transport (Q/t)' of the investigated sandstones (d = day)

Sandstone sample	0.5 N NaCl		0.2–0.4 N CaCl$_2$	
	D*cm^2/d	µS/(Q/t)' d/cm^2	D*cm^2/d	µS/(Q/t)' d/cm^2
Grau-grauer Sandstein	0.056	3.622		
Dogger β-Sandstein	1.824	17.949		
'Feinsandstein'	0.009	1.300	0.006	1.723
Gipskeupersandstein	0.057	0.835		
Glaukonitsandstein	0.561	16.284		
Hilssandstein, plattiger Teil	0.117	2.037		
Hilssandstein	0.137	0.805	0.017	0.329
Hilssandstein	0.260	0.622		
Lettenkeuper-Sandstein	0.288	0.509		
'Mittelsandstein'	0.034	0.581	0.027	1.159
'Mittelsandstein'	0.070	0.536	0.025	1.295
'Mittelsandstein'	0.068	0.390		
'Mittelsandstein'	0.058	0.238		
Roter Sandstein	0.080	0.882	0.006	0.712
Roter Sandstein	0.035	2.101		
Roter Sandstein	0.032	3.040		
Rotliegend Sandstein 4	0.375	0.965		
Rotliegend Sandstein 5	0.343	0.762		
Rotliegend Sandstein 28	0.055	1.642		
Schilfsandstein, rot	0.313	6.588	0.027	1.267
Schilfsandstein, rot	0.141	5.490		
Schilfsandstein, rot	0.004	4.738		
Schilfsandstein, rot	0.103	7.218		
Schilfsandstein, hell	0.055	1.569	0.013	1.215
Schilfsandstein, hell	0.023	3.111		
Schilfsandstein, hell	0.067	3.402		
Schilfsandstein, hell	0.054	1.388		
Schilfsandstein, hell	0.044	3.939		
St. Petersandstein	0.078	2.026		
Stubensandstein	0.151	0.998		
Unterer Muschelkalk-Sandstein	0.121	3.658		
Unterer Muschelkalk-Sandstein	0.149	2.289		
'Siltstein'			0.007	0.290
'Grobsandstein'			0.141	0.866

Garibyants 1971), which is based on the time that elapses from the beginning of the diffusion process until the stationary diffusion flow commences (tx). The diffusion coefficient is then calculated from $D = 1^2/6\, t_x$, where 1 is the height of the sample plug. To obtain reliable results, it is essential to keep the temperature during the experiment precisely constant. Moreover 0.5 N NaCl and 0.2 or 0.4 N CaCl$_2$ solutions were selected for the diffusing electrolyte on the basis of aptitude tests. The diffusion coefficients of the investigated sandstones resulting from these measurements are shown in Table 3.

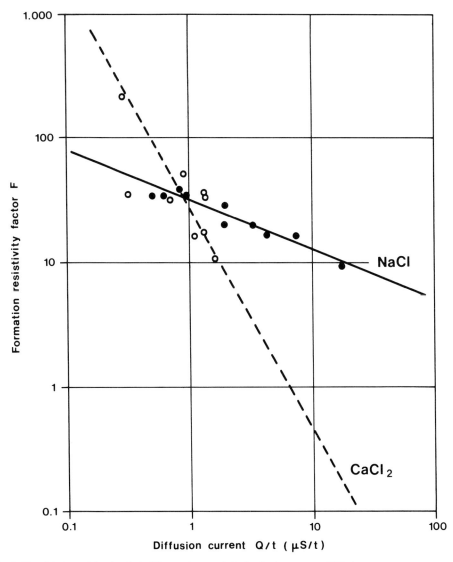

Fig. 1. Double logarithmic plot of formation resistivity factor versus diffusion current

The diffusion rate of $CaCl_2$ is up to 15 times greater than that for NaCl, all other conditions being equal. Within the concentrations, ranging from 0.1 to 1.0 mol/l, the diffusion coefficients are nearly independent of the concentration of the diffusing electrolyte. The current, however, rises with the concentration within the same concentration interval exponentially.

Diffusion coefficients and diffusion current do not correlate significantly with the petrographic parameters of the investigated sandstones. For instance, no significant correlation between the diffusion coefficient and the mean grain size or the specific surface area can be detected.

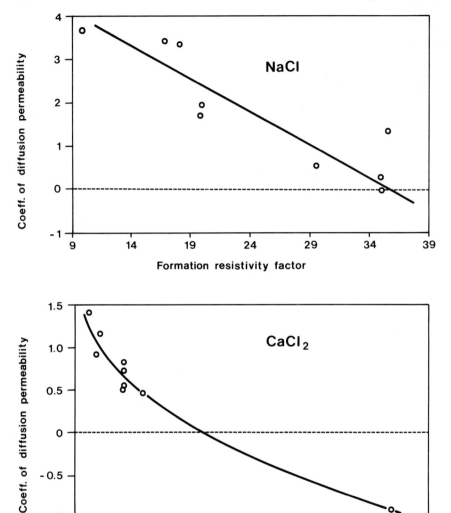

Fig. 2. Plot of logarithm of coefficient of diffusion permeability versus formation resistivity factor

Since the diffusion parameters depend on a variety of different petrographic properties, a significant correlation with a single petrographic factor is not to be expected. According to the opposing effects of different influences, any correlation with a single factor becomes insignificant. In evaluating the diffusion experiments, a system of interdependent influences has to be considered.

A more significant correlation exists between the diffusion current (Q/t) and the formation resistivity factor (F). Plotted in a double-logarithmic net,

F versus Q/t yields a linear function (Fig. 1). The formation resistivity factor is the quotient of the electric conductivity of the pore water and that of the porous rock totally saturated with that pore water. Hence, F depends on porosity, tortuosity (anisotropy of fabric), and the clay content, which may exchange cations with the pore solution.

Obviously, these are the dominant factors influencing the diffusion current. Therefore, the formation resistivity factor appears to be the main indicator for the diffusion properties of a permeable rock. Migration of electric charge by moving ions in an electric field, and diffusion through a rock appear to follow similar rules. The higher F, the lower the diffusion current Q/t. Similarly, a reciprocal relation exists between F and the coefficient of diffusion permeability Pk (Fig. 2). (Pk = $\beta \cdot D$, β = ionic capacity = coefficient of solubility.)

In the attempt to apply the laboratory results to geologic reality, the question arises as to which time intervals must be considered as realistic for the diagenetic transport of matter in porous rocks if the transport depends exclusively on diffusion. To answer this question, calculations were done for the time needed to allow the diffusion of the quantity of NaCl or $CaCl_2$ necessary to completely fill the pore space of a sandstone prism of 1000 × 10 × 10 m. The area through which the solutions enter the sandstone was assumed to be 10 × 10 m, in other words, the face area of the prism was open to diffusion, while the body was sealed on all remaining surface planes. The concentrations in the solution-spending reservoir were assumed to be 0.5 N for NaCl, and 0.2/0.4 N for $CaCl_2$. Further assumptions were made regarding: (1) the homogeneity of the rock and (2) a constant temperature of 20° C. The calculations were based on three different models:

Model 1: Estimation of the time interval assuming a stationary diffusion current throughout, i.e. time calculations without consideration of the time interval of the nonstationary introductory phase.

Porosity and coefficients of diffusion as evaluated for the investigated sandstones, are listed in Tables 2 and 3.

Model 2: Estimation of time interval assuming the porosity of all sandstones to be constant (0.2). Diffusion coefficients as evaluated are listed in Table 3.

Model 3: Estimation of time interval and amount of diffusing matter until commencement of the stationary stage of diffusion.

The calculations for the investigated sandstones yielded the following results:

	NaCl	$CaCl_2$
Model 1	8 –200 × 10^3 a	70–370 × 10^3 a
Model 2	8 –260 × 10^3 a	75–450 × 10^3 a
Model 3	0.06– 9 × 10^3 a	5– 27 × 10^3 a

The calculated time intervals are within the scope of quite realistic geologic processes. However, the calculations start from concentration gradients that are much greater than those found in natural sediments, as proven by numerous pore water analyses (v. Engelhardt 1970). Moreover, due to advective pore water currents interfering with the diffusion currents, the concentration gradients cannot be considered stable, neither over distances of some tens of meters nor throughout time intervals of some 10000 to 100000 years.

In order to maintain a sufficiently high concentration gradient over a sufficiently long time, the pore solution in the spending area has to be undersaturated, the dissolved ions have to be removed continuously, and, in the receiving area, the solution has to be oversaturated. By increasing the concentration of dissolved ions, the pH, and, consequently, the saturation concentration, are changed. Furthermore, the saturation concentration depends on the temperature, being itself a function of the depth of burial and of tectonic dislocations. The saturation concentration is influenced by the concentration of the accompanying ions, almost ubiquitously present in pore solutions.

By the precipitation of pore cements in the receiving rock, the volume of pore water equal to the volume of cement is replaced, causing an advective pore water current which deviates, or even opposes the diffusion current. Additionally, the tortuosity of the receiving rock is increased by the cementation.

By each of these processes – with the exception of the termperature increase – the diffusive transport is diminished under natural geologic conditions. Although for the quantitative estimation of all the effects influencing diffusion, the specific conditions prevailing in the particular case have to be considered, it is evident that the roughly estimated time intervals necessary for a diffusion over 100 to 1000 m distance under natural conditions are too short by orders of magnitude. This result implies that, considering the actual influences of diffusive transport, time intervals will result, over which the postulated constancy of the diffusion coefficient becomes improbable, if not impossible.

4 Short-Range Diffusive Transport

As a result of the foregoing evaluation, it appears unlikely that diffusive transport ranges will exceed distances of the order of magnitude of a maximum of 10 m even during geologic time intervals. However, what statements can be made on the diffusive transport within the mm to cm range?

In extensively cemented sandstones, replacements by pore cements can often be observed. When the pore space of a sandstone is totally filled by a pore cement, for instance, calcite which fills the pores of a sandstone consisting predominantly of quartz grains, the process of cementation is by no means completed. More cement can be precipitated in the space where detrital components or an earlier cement have been removed by dissolution. Such diagenetic processes are called "replacements".

Replacements, therefore, include the dissolution of a solid phase of a rock and the subsequent precipitation of the replacing cement. To accomplish these processes, the pore solution must be undersaturated on behalf of the solid phase in order to be dissolved, and oversaturated with regard to the cementing phase in order to be precipitated.

If quartz is the phase to be replaced and calcite the replacing cement, the pH favorable for a replacement is considered to be 8–9 (Krauskopf 1967). Replacements can proceed to such an extent that, for instance, a quartz sandstone is converted into a rock of pure calcite.

The transport processes implied by replacements, i.e. the removal of the dissolved, and the supply of the replacing compounds, can be accomplished only by diffusion. Due to the extremely small paths of transport on the grain or crystal surface planes, advective transport is impossible, since advective currents along the submicroscopically small pores would demand extremely high, and therefore, unrealistic pressure gradients.

How is it possible that different ions can diffuse in opposite directions along the narrow pore spaces of less than 1 µm in width? The gradients producing the diffusion must be caused by locally confined concentration differences. A concentration gradient may be caused by a change in the pH. For example, the dissolution of quartz and the precipitation of calcite can be a consequence of an increased pH. Since at the very contact between the replacing calcite and the quartz being replaced calcite precipitation occurs, an increased pH (>7) must be postulated.

On the other hand, for the supply of calcite in dissolution, a lower pH (>7) would be necessary. However, checking favorable pH conditions, it has to be considered that actually both the removal of dissolved SiO_2 and the supply of dissolved $CaCO_3$ occur on the same paths by diffusion. The fact that both compounds diffuse in opposing directions excludes advective currents as a possible mode of transport.

At the pH prevailing along the path of diffusion, both compounds participating in the processes of replacement must be kept in solution. The replacing mineral precipitates at the site of the replacement. Therefore, here the solution must be oversaturated by the replacing compound. These considerations allow the presumption that the subtle changes of saturation concentrations causing replacements are controlled by differentations of the pH in constricted spaces. PH inhomogeneities produce dissolution and precipitation processes leading to the diffusion of matter.

Possible causes for the spatial pH inhomogeneities can only be supposed. They are difficult to measure, due to their small dimensions. A possible change in pH for the example of calcite replacing quartz could be conceived in the following way: Quartz is dissolved, forming monomeric H_4SiO_4, which is dissociated to a minor extent (Siever 1962). Through the hydrogen needed for the dissociation, the equilibrium between H^+, HCO_3^-, and H_2CO_3 is shifted in such a way that the pH immediately adjacent to the surface of contact between replacing calcite and replaced quartz is increased, and $CaCO_3$ is deposited.

The pore space is not reduced by replacements, since the replaced volume is substituted by the same volume of replacing cement. Therefore, no pore water movement is produced, as it occurs by normal cementation. At replacements, the pore water remains stagnant.

5 Conclusions

Diagenetic transport of matter in porous and permeable sediments above dm dimensions is brought about by advective flow of pore solutions. Within microscopically small ranges, however, advective transport fails, due to small pore widths. Ions dissolved here can be moved by diffusion only. Hence, with diagenetic replacements by pore cements, the removal of the dissolved, and the supply of the replacing minerals, is accomplished by diffusion. The cause of the concentration gradients generating diffusion remains hidden. Only presumptions can be made.

Certainly, under natural conditions, the concentration gradients are small. However, in relation to short ranges, they are considered sufficient to effect the migrations observable in cemented sandstones of geologic time intervals. In summary, it can be stated: *Long-range* transport of matter in sediments can be accomplished by advective pore flows only; the *short-range* infiltration on a microscopically small scale, however, is produced by diffusive transport.

References

von Engelhardt W (1960) Der Porenraum der Sedimente. Springer, Berlin Heidelberg New York
von Engelhardt W (1973) Die Bildung von Sedimenten und Sedimentgesteinen. Sedimentpetrologie, pt 3. Schweizerbart, Stuttgart
Giskow R (1987) Messungen von Diffusionsparametern an Sandsteinen: Ein Beitrag zur Frage des Stofftransportes in Sedimenten. Heidelberger Geowiss Abh, vol 7
Golubev VS, Garibyants AA (1971) Heterogeneous processes of geochemical migration. Consultants Bur, New York, London
Krauskopf K B (1967) Introduction to geochemistry. McGraw-Hill, New York
Lemcke K, Tunn T (1956) Tiefenwasser in der süddeutschen Molasse und in ihrer verkarsteten Malmunterlage. Bull Ver Schweiz Petrol Geol Ing 23:35
Levorsen AI (1967) Geology of petroleum, 2nd edn. Freeman, San Francisco London
Siever R (1962) Silica solubilitiy, 0–200° C, and the diagenesis of siliceous sediments. J Geol 70:127–150

Pathways of Fine-Grained Clastic Sediments – Examples from the Amazon, the Weser Estuary, and the North Sea

GEORG IRION and VOLKER ZÖLLMER[1]

CONTENTS

Abstract	351
1 Introduction	352
2 Transport and Composition of Fine-Grained Particles in Rivers and in the Sea	353
3 Method of Tracing Pathways of Suspended Matter	353
4 Examples	354
4.1 Quaternary Sediments in Western Amazonian Lowlands	354
4.2 Aspects of Sediment Distribution in the Recent Floodplains of Rio Amazon	356
4.3 Sediments in Estuaries of the Inner German Bight	360
4.4 Sediments of the Near-Coastal Area of the Southern and Eastern North Sea	362
References	365

Abstract

The analysis of the clay-mineral association of mainly illite, chlorite, kaolinite and smectite in sediments makes it possible to trace pathways of fine-grained sediments. Along the transport routes, mixing effects, deposition and resuspension of the sediments can be deduced.

In Amazonia, the source of a huge Pleistocene sediment deposit is traced back to different source areas and the buildup and history is explained in connection with sea-level changes. In the recent Amazonian floodplains, mixing of different sediment types takes place.

In the SE corner of the North Sea the landward transport of muddy sediments within estuaries was determined and in the North Sea itself, the transport of fine-grained particles along the coast from SW to NE seems to be obvious. Human influence on these sedimentational processes is assumed.

---

[1] Forschungsinstitut Senckenberg, Institut für Meeresgeologie und Meeresbiologie, Schleusenstr. 39 a, D-2940 Wilhelmshaven, FRG

1 Introduction

Among the geological dynamics of the earth, transport of clastic sediments is an essential part. To know their pathways is one key to the unterstanding of processes on the earth's surface. The erosion tends to level elevations and the eroded material is washed into the sea. The delivery of river sediments to the world oceans is estimated to be 15×10^9 t/a (Milliman and Meade 1983). Aeolian transport delivers only about 10^7 t/a of sediments, and a similar amount results from volcanic activities (Gorsline 1984). Most of the river-transported material is a fine-grained, clastic, suspended load. Only about 10% coarse-grained sediments comprise the bed load. Figure 1 gives a typical example of grain-size distributions of suspended loads and bed material taken from Rio Amazon. Generally, more than 50% of the suspended load consists of grain sizes < 63 µm. This may be taken as a definition for "fine-grained sediments", and corresponds to the common definition for "mud".

River sediments may reach the sea without a major time lag. But, generally, they are stored in floodplains for longer periods before finally reaching the sea (Meade 1988; Irion 1989).

Most river sediments delivered to the sea are deposited in estuaries and in near-shore areas. For example, the Amazonian sediments are in part transported northwards along the coast up to the estuary of Rio Orinoco (Eisma and van der Marel 1971). But the Amazon is as well one of the few examples where river sediments are transported into the deep sea. Its submarine cone (Damuth and Flood 1984) reaches a depth of 4800 m at a distance of 800 km from the river mouth.

In total, a very small proportion of the world's river sediments finally reaches the deep sea, where aeolian material is of relatively greater importance.

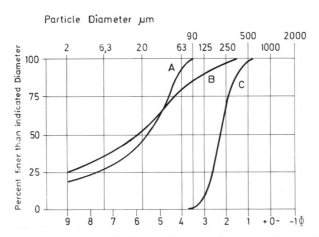

Fig. 1. Particle-size distribution in suspended matter (*A* and *B*) and bottom load (*C*) of Rio Amazon. *A* and *C* from author *B* from Meade et al. (1979)

2 Transport and Composition of Fine-Grained Particles in Rivers and in the Sea

In most rivers the particles seem to be transported as individual mineral grains. This is not the case for rivers draining industrialized regions. In the highly polluted river Rhine Eisma et al. (1980) showed that aggregates are formed as a result of its anthropogenic salt content (0.06% salinity) and due to the organic pollutants present. From experiments with natural sediments in artificial salt waters, Gibbs (1983) showed that already at salinities between 0.05 to 0.1% coagulation of particles takes place. At the transition from the terrestrial to the marine environment a very intensive flocculation is observed in the estuaries. This flocculation may have various causes: glueing together of the particles by organic material, high turbulence and high concentration of suspended sediments, and the increasing electrolyte content in the mixing zone (Lüneburg 1953; Meade 1972; Wellershaus 1981; Eisma 1986).

In the marine environment the settling velocity is accelerated by the formation of flocs (Wunderlich 1969; McCave 1971). Eisma (1987) showed that suspended material of the North Sea has a settling velocity of a hydraulical equivalent to quartz spheres of about 50 µm diameter.

In the fraction finer than about 6 µm, clay minerals, mainly illite, chlorite, smectite and kaolinite are dominant. The first two minerals originate primarily from different magmatic and metamorphic rocks, whereas smectite and kaolinite are predominantly products of weathering at the earth's surface. During geologic time the clay minerals may have been reworked many times and hence the circumstances of their formation are often unknown.

3 Method of Tracing Pathways of Suspended Matter

The given clay mineral association of a source area (e.g. the upper reach of a river system) is taken as a tracer for studying the pathways of sediments. Together with other fine-grained particles, some clay minerals are released into bottom sediments during transport. This allows one to trace the pathways. On its way mixing with other sediments, deposition and resuspension and even some alteration due to weathering processes may be observed. It is possible as well to trace the pathway back from the depositional area to the source area.

The method of tracing the pathways of sediment transport by means of clay mineralogical studies is demonstrated best by some examples. Referring to sedimentological programs carried out by the Department of Sedimentary Petrography of the Senckenberg-Institute in Wilhelmshaven, four studies have been selected. The first two examples deal with riverine systems, and the third and fourth with the marine environment of the North Sea. The studies comprise the following:

1. Quaternary sediments in western Amazonian lowlands,
2. Aspects of sediment distribution in the recent floodplains of Rio Amazon.
3. Sediments in estuaries of the inner German Bight;
4. Sediments of the near-coastal area of the southern and eastern North Sea between the Rhine estuary and the Skagerrak.

4 Examples

4.1 Quaternary Sediments in Western Amazonian Lowlands

Little is known about the geology of the central area of the Amazonian lowlands. Due to its large extension and the difficult access to the tropical rain forest, field studies are restricted to only a few sites. The datings of the sedimentary deposits have been quite controverse. According to the geological map of 1964 (Carte géologique de L'Amérique du sud): "they belong to the Tertiary, in 1978", thus they were termed Cenozoic. Results of our own field trips in 1972 and later demonstrate, however, a Pleistocene age.

For our considerations it is of great benefit that the distribution of the clay mineral composition in recent river sediments of the Amazon basin shows fundamental differences between the single rivers (Irion 1987). The sediments of Rio Amazon coming from the Andes are rich in smectite, illite and chlorite. Rivers draining the south-western lowland yield sediments which contain more than 80% smectite in the <2 μm fraction.

We studied large areas in the lowlands (Irion 1976a, 1984, 1987). However, for demonstrating the principles of our studies, the following description focuses on one place only. We selected the area of Lago Aiapuá near Rio Purús about 120 km above its confluence with Rio Amazon (Fig. 2).

Figure 3 shows a section of a radar map of this area. The eastern part of the section shows the Holocene alluvial plain of Rio Purús with scroll bars and lake sediments. The surroundings of Lago Aiapuá and areas west of it show a pattern, which allows the assumption that older scroll bars were present, supposingly formed by Paleo-Purús. We tested this assumption by analyzing samples from a 12-m-long sediment core taken from one of these structures. The grain-size distribution shows a fining upward cycle as expected in a scroll bar deposited by rivers. Most importantly, the clay mineralogy of the lower samples corresponds to the clay mineral association of Rio Purús (Fig. 4). In both sediments a low-charged type of smectite is dominant, and typically chlorite is missing. In an upward direction of the sediment core, the smectite content decreases due to weathering.

From mineralogical and sedimentological investigations, together with the interpretation of the radar map, it can be concluded that two generations

Pathways of Fine-Grained Clastic Sediments 355

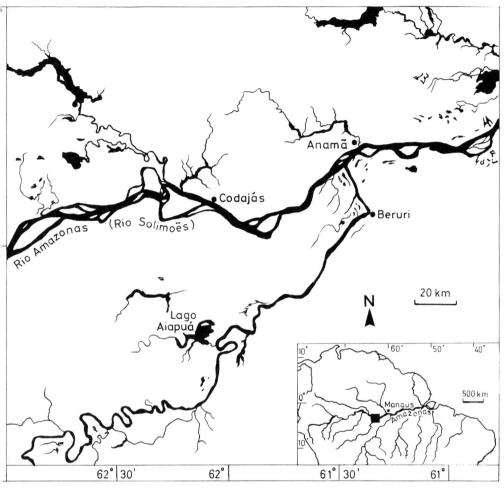

Fig. 2. Rio Purus with Lago Aiapuá

of large flood plains exist in the analyzed areas of Lago Aiapuá, whereby the old floodplain is situated 5 to 15 m higher than the recent one. The source area of both floodplains was the same – the area of the Brazilian state of Acre. The surface of the higher floodplain is altered mineralogically and chemically during weathering processes.

Situations like this, where two generations of floodplains formed by sediments of the same source, are found in many places of the western Amazonian lowland. It reveals that the pathways of sediments have not changed since the time of formation of the older floodplains. In some places, three or even four generations of floodplains can be recognized. The causes of the different floodplain levels, which can be traced over a large area, must be linked to fluctuations of the Pleistocene sea level (Irion 1976b). This seems

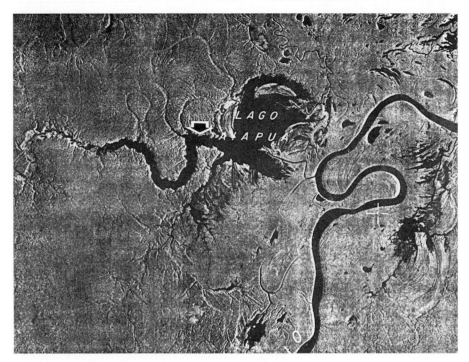

Fig. 3. Section of the radar map of Lago Aiapuá and Rio Purus with the location of a sediment core (*white arrow*). The width of the section is 55 km

to be possible since the inclination of the slope of the Amazonian water system is very low. In the lower reach of Rio Amazon, between Manaus and the Atlantic, it amounts to about 1 cm/km (Mississippi 11 cm/km). Sea-level changes may therefore influence the Amazonian water tables far from the river mouth (Irion 1989).

The similarity of the mineralogical composition of the pre-Holocene land surfaces with the recent riverine sediments shows that source areas and pathways as documented in the floodplains did not change over a long period in the eastern Amazon lowlands. The degree of the surface weathering allows the conclusion that the age of the landscape does not exceed Early Pleistocene or Late Pliocene. A formerly unexpected large sediment body has developed since that time.

Our general assumptions have to be supported by more detailed studies in the very remote areas of the western Amazonian lowlands.

4.2 Aspects of Sediment Distribution in the Recent Floodplains of Rio Amazon

Sediment distribution in major lowland rivers, like the Amazon, may be relatively complicated. Water masses with different suspended loads meet in the

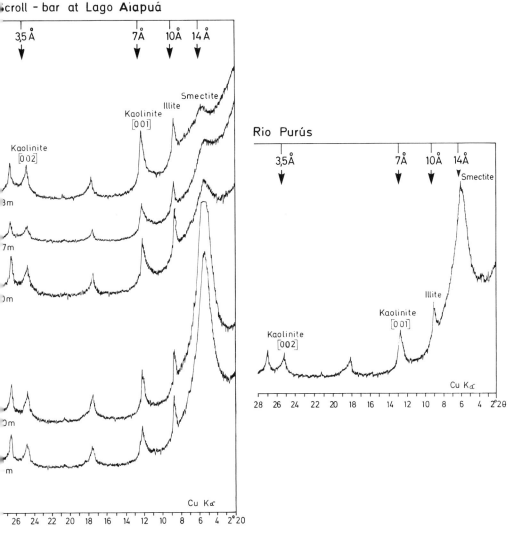

Fig. 4. X-ray diffraction patterns of the <2 μm fraction of samples from the sediment core from the location indicated in Fig. 3 *Left:* Depths below surface; *right:* for comparison one X-ray diffraction of the recent Rio Purus

floodplains. Seasonal changes of water levels are not synchronous in the different flow regimes and, furthermore, in valleys overflowed during the Flandrian transgression, lakes were formed (Sioli 1957, 1984). But in most cases, differences in the clay mineralogy of the suspended loads allows us to trace the pathways of the fine-grained sediment, as shown above.

First, clay mineral analysis may help to estimate the contribution of the various source areas to the sediments of Rio Amazon itself. Samples from

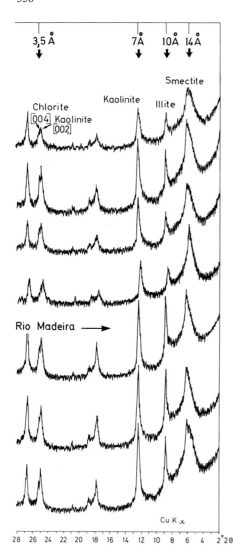

Fig. 5. X-ray diffraction patterns of the <2 µm fraction of riverbank sediments from Rio Amazon taken between the Columbian/Brazilian border (*upper pattern*) and the upper reach of the delta. The samples are taken at roughly equal distances. The location of the confluence with Rio Madeira is indicated. Please observe that after this location, the illite content increases

more than 100 locations from riverbank sediments of Rio Amazon and from its major tributaries were collected (Irion 1976b, 1987), and the < 2 µm fraction was separated for clay mineral analysis. The results of Rio Amazon riverbank samples collected on the whole length of its Brazilian section (Fig. 5) does not show any change in composition with the exception of the lower river section after the confluence with Rio Madeira. The contribution of all other lowland rivers to the sediment of the Rio Amazon is obviously negligible. This means that for the sediment delivery of Rio Amazon, only the Andes and the Sub-Andean regions (Irion 1976b) are of importance. Yet it should be emphasized that the suspended load of western Amazonian lowland rivers is nevertheless high, their contribution to the main river ap-

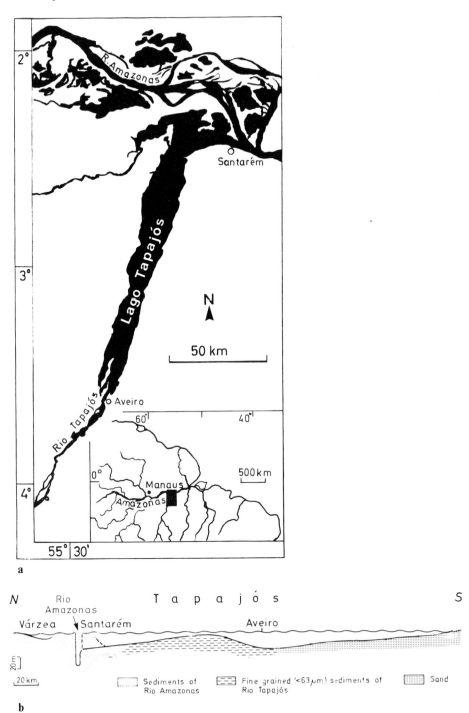

Fig. 6. a Location map of Lago Tapajós; **b** longitudinal section of the lower course of the ria-lake of Rio Tapajós

pears only to be low because the sediment transport of the Rio Amazon itself is extraordinarily high.

The reliability of estimating sediment balances by comparing clay mineral compositions can be tested with the results of Meade (1988). He calculated the sediment transport by measuring the suspended load directly from riverwaters and found that 90% of the sediments of the Lower Amazon is derived from the Andes. This is in good agreement with our findings.

One of the most striking phenomena of the aquatic systems of Amazonia are the above-mentioned lakes. The lakes – in Amazonia they are named "ria-lakes" – are very widespread in the central Amazon lowlands. One major example of such a ria-lake is the lower reach of Rio Tapajós. The lake has a length of 170 km and a maximal width of 20 km. Figure 6a, b shows a profile along its length axis. In the suspended load of Rio Tapajós only kaolinite occurs, whereas in Rio Amazon smectite, illite and chlorite are dominant. The ria-lake is a trap for sediments and has an estuary-like shape. In contrast to estuaries (Sect. 4.3), where fine-grained particles coagulate to flocs of higher settling velocities, in the fresh-water regime of the ria-lakes the mineral particles are transported separately. From coagulation in estuaries a large-scale deposition of sediments results, whereas in ria-lakes fine-grained sediments are deposited only when the water velocity is very low. This explains that in the upper section of the Tapajós ria-lake only quartz sands are deposited, whereas in the lower section fine-grained, kaolinite-rich sediments delivered by Rio Tapajós are deposited. In the lowermost 30 km only fine-grained sediments with the clay mineral association of Rio Amazon, smectite, mainly illite and chlorite, are found. This can be explained by the fact that the waters of Rio Amazon enter this part of the ria-lake section during seasonal floods. The upstream immigrating Amazon riverwaters strongly influence the sedimentation processes and may change the biological activity in the affected area since the sediments of Rio Amazon are richer in nutrients than those of Rio Tapajós.

4.3 Sediments in Estuaries of the Inner German Bight

Most estuaries along the world's coastlines were developed during postglacial sea-level rise. They are depositional areas. Meade (1969) demonstrated a landward transport of sediments in estuaries of the Atlantic coast of the USA. Landward transport is due to tide effects and low river discharge in comparison to the size of the estuary. Another prerequisite for sedimentation in estuaries is the formation of large sediment flocs with relatively high settling velocities as described by Eisma (1987) and Wellershaus (1981).

We (Irion et al. 1987; Schwedhelm 1988) studied the clay mineral distribution and some geochemical parameters in three estuaries of the inner German Bight: the Ems, Weser and Elbe estuaries. Annually rivers carry together only 1.4 million tons of suspended matter into their estuaries. The

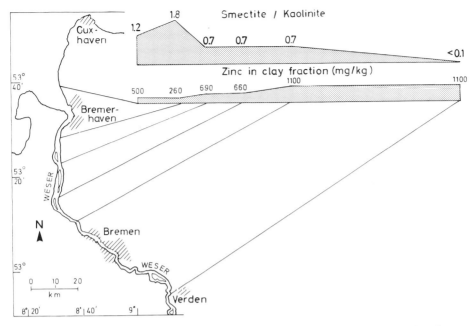

Fig. 7. Smectite/kaolinite ratio of 0.6–2 µm fraction in samples from the Weser river from locations indicated. The smectite/kaolinite ratio is very low in Weser river sediments. First mixing with marine sediments takes place below Bremen, and no river sediments could be found below a few km south of Bremerhaven. (Irion et al. 1987)

high anthropogenic pollution level of these river sediments makes it possible to distinguish them from the less polluted marine sediments. Müller and Förstner (1975) used the anthropogenic heavy metal content to trace river sediments of the Elbe into its estuary and they detected intense mixing with marine sediments. Salomons and Mook (1977) carried out similar studies in the Ems estuary. Van Straaten (1960) concluded from his more general studies of the Ems estuarine sediments, that "the influence of sediment supply by river Ems proves to be negligible (in the estuary)".

The method of using heavy metals as tracers is restricted to about the past 100 years since they are regarded of industrial origin. Also, their behaviour in estuaries has been discussed controversely. We therefore used the clay mineral associations of the sediments themselves to trace the sediment transport in the estuarine environment.

We analyzed the clay mineral composition and some heavy metals (Zn, Pb, and Cu) in sediment samples taken along the estuaries. The results of all three estuaries Ems, Weser, Elbe showed a similar pattern. This will be discussed for the Weser estuary:

The sediments of the Weser river are low in smectite. The smectite content is lower than that of the North Sea sediments (Fig. 7). The mixing is first observed a few kilometers north of Bremen and a clay mineral association of "pure" marine sediments is obtained below the last sampling location, south of Bremerhaven. The heavy metal distribution in the <2 μm fraction follows closely the patterns established for the mixing of the two sediment types. From the patterns of the clay mineral association and as well from heavy metal contents it follows that marine components are dominant far upstream in the estuary. Thus it may be concluded that the net input of sediment to the estuary from the sea is by far larger than that by the river.

Over the last decades very intensive dredging, in order to deepen the fairways of the estuaries, has influenced the sedimentation patterns. In the Elbe estuary about 10 million m^3 of sediments are dredged each year, whilst in the past between 1974 and 1978 50 million m^3 were dredged annually. Most of the dredging is done by hopper-suction dredgers. These dredgers lose suspended sediments with concentrations of up to 300 g l^{-1} over their weirs during operation. If only a small proportion of the dredged sediments are lost by this outflow, and additionally sediments are eroded and kept in suspension by turbulence caused by ship propellers, it may be considered that the quantity of sediment discharge from these man-made suspensions reaching the sea may greatly exceed the 1.4 million tons of sediments delivered to the estuaries by the rivers. Since the fairways of the three estuaries are situated to their largest part in areas where marine derived sediments are deposited, mainly marine and not fluvial sediments to terrigenous origin are resuspended by the described processes. To conclude, in the net transport from the estuaries into the sea, the proportion of marine sediments exeeds today by far that of terrestrial origin.

4.4 Sediments of the Near-Coastal Area of the Southern and Eastern North Sea

Whereas the origin of the North Sea sands can be related predominantly to coarse-grained glacial deposits, little is known about the source areas of the fine-grained sediments. The fine-grained sediments are restricted to the inner zone of the tidal flats, to the Skagerrak and to some other local mud deposits. But the fine fraction is also present in coarse-grained material which occupies the largest part of the North Sea. The overall distribution of fine-grained particles in the bed material of the North Sea allows one to trace the pathways of the fine-grained sediments in the whole area.

In coarse-grained sediments the <2 μm fraction generally makes up less than 1% of the sediment. The fine-grained particles may stay only temporarily in the sediments after having been deposited during calm periods; during storm events with surfs they will, at least in part, be resuspended.

First investigations were carried out in the confined area of the inner German Bight (Irion et al. 1987). Expecting a relative homogeneity in the clay mineral distribution, a minimized grain-size spectrum with an interval of 0.6–2 μm was analyzed in order to obtain even the smallest differences in

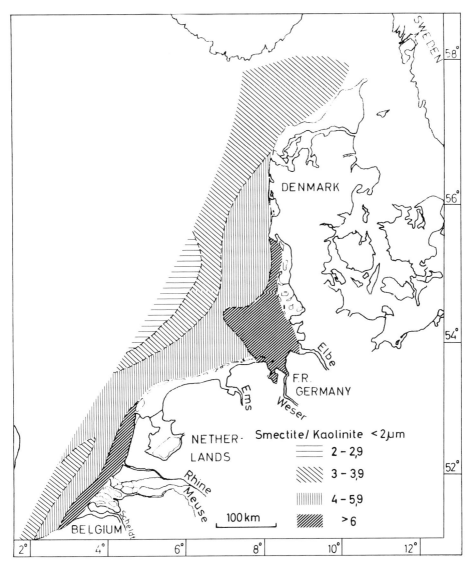

Fig. 8. Smectite/kaolinite ratios (weighted peak areas) of the clay fraction in surface sediments of the southern to northeastern parts of the North Sea

composition. During later investigations a modified method, using the total < 2 μm fraction was applied (Zöllmer and Irion in preparation). In this case the mineral ratios were determined by weighing the peak areas of the X-ray pattern. With this method almost 400 surface samples covering the entire North Sea area have been analyzed.

By depicting some of the results of our investigations, we confine the area from the mouths of the Rhine and Scheldt to Skagen (Northjutland). In Fig. 8 smectite/kaolinite ratios are shown and classified into four groups

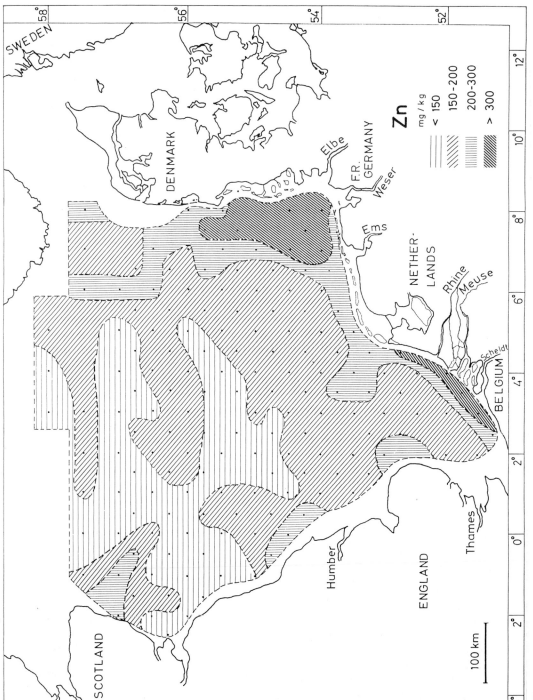

from 2 to >6 (<2 occurs only off the coast of the United Kingdom; not shown here). From this it can be deduced that next to the coast, the highest smectite/kaolinite ratios occur, with decreasing values towards the center of the North Sea. The maximum values are found near the mouths of the Rhine and Scheldt and in the inner German Bight off the estuaries of the Ems, Weser and Elbe rivers.

The source area of the sediment with high smectite/kaolinite ratios must be located in the coastal zone and not in the rivers since, as it was mentioned above, the river sediments are transported only in part, if at all, to the North Sea. In addition, the smectite/kaolinite ratios of the rivers, with the exception of the Ems river, are much smaller than in the areas outside the estuaries.

The zinc content has increased over the past 100 years due to industrialization and is transported together with the fine-grained fraction of sediments; this can be used as a (good) tracer of polluted sediments. Its distribution (Fig. 9; Irion and Müller 1987) shows a similar pattern to the smectite/kaolinite ratio. This supports the idea that sediments with a high zinc content (Irion et al. 1987) along with high smectite/kaolinite ratios are mobilized by human activities (dredging) in the estuaries, and influence the sedimentary processes in the coastal area.

From the smectite/kaolinite pattern (Fig. 8), along with the distribution of zinc, a coastal transport of fine-grained material from the south, continuing along the eastern coast of the North Sea to the Skagerrak seems to result. This is in accordance with the general current distribution in this part of the North Sea.

Acknowledgements. We thank B. Krause, R. Flügel and T. Holzlehner for their assistance.

References

Damuth JE, Flood RD (1984) Morphology, sedimentation processes, and growth pattern of the Amazon deep-sea fan. Geo Mar Lett 3:109–117

Eisma D (1986) Flocculation and deflocculation of suspended matter in estuaries. Neth J Sea Res 20/2–3:183–199

Eisma D (1987) Flocculation of suspended matter in coastal waters. Mitt Geol Paläontol Inst Univ Hamburg, SCOPE/UNEP Spec Vol 62:259–268

Eisma D, Van Der Marel HW (1971) Marine muds along the Guyana Coast and their origin. Contrib Mineral Petrol 31:321–334

Eisma D, Kalf J, Veenhuis M (1980) The formation of small particles and aggregates in the Rhine estuary. Neth J Sea Res 14/2:172–191

Gibbs RJ (1983) Coagulation rates of clay minerals and natural sediments. J Sediment Petrol 53/4:1193–1203

Gorsline DS (1984) A review of fine-grained sediment origins, characteristics, transport and deposition. In: Stow DAV, Piper DJW (eds) Fine-grained sediments: deep-water processes and facies. Geol Soc, Spec Publ 15:0–659

Irion G (1976a) Quaternary sediments of the upper Amazon lowlands of Brazil. Biogeographica 7:163–167

Irion G (1976b) Mineralogisch-geochemische Untersuchungen an der pelitischen Fraktion amazonischer Oberböden und Sedimente. Biogeographica 7:7–25

Irion G (1984) Sedimentation and sediments of Amazon rivers and evolution of the Amazon landscape since Pliocene times. In: Sioli H (ed) The Amazon – limnology and landscape ecology of a mighty tropical river and its basin. Junk, The Hague, pp 201–214

Irion G (1987) Die Tonmineralvergesellschaftung in Flußsedimenten der feuchten Tropen (Amazonas-Becken, West-Papua Neuguinea) als Ausdruck der Verwitterung im Einzugsgebiet. Habilschr, Univ Heidelberg, pp 1–268

Irion G (1989) Quaternary geological history of the Amazon lowlands. In: Holm-Nielsen LB, Nielsen IC, Balslev H (eds) Tropical forests. Academic Press, New York London, pp 23–34

Irion G, Müller G (1987) Heavy metals in surficial sediments of the North Sea. Heavy metals Environ 2:38–41

Irion G, Wunderlich F, Schwedhelm E (1987) Transport of clay minerals and anthropogenic compounds into the German Bight and the provenance of fine-grained sediments SE of Helgoland. J Geol Soc 144:153–160

Lüneburg H (1953) Die Probleme der Sinkstoffverteilung in der Wesermündung. Veröff Inst Meeresforsch Bremerhaven 2:15–51

McCave IN (1971) Wave effectiveness at the sea bed and its relationship to bed forms and deposition mud. J Sediment Petrol 41:89–96

Meade RH (1969) Landward transport of bottom sediments in sediment to the oceans. J Geol 91:1–21

Meade RH (1972) Transport and deposition of sediments in estuaries. In: Nelson BW (ed) Environmental framework of coastal plain estuaries. Geol Soc Am Mem 133:

Meade RH (1988) Movement and storage of sediment in river systems. In: Lerman A, Meybeck M (eds) Physical and chemical weathering in geochemical cycles. Kluwer, Dordrecht, pp 165–179

Meade RH, Nordin CF Jr, Curtis WF, Mahoney HA, Delany BM (1979) Suspended-sediment data, Amazon River and its tributaries, June–July 1976 and May–June 1977. US Geol Surv Open File Rep 79–512:42 pp

Milliman JD, Meade RH (1983) World-wide delivery of river sediment to the oceans. J Geol 91:1–21

Müller G, Förstner U (1975) Heavy metals in sediments of the Rhine and Elbe estuaries: mobilization or mixing effect. Environ Geol 1:33–39

Salomons W, Mook WG (1977) Trace metal concentrations in estuarine sediments: mobilization, mixing or precipitation. Neth J Sea Res 11:119–129

Schwedhelm E (1988) Determination of sediment and suspended matter origins in the Elbe estuary using natural tracers. In: Schwedhelm E, Salomons W, Schoer J, Knauth H-D (eds) Provenance of the sediments and the suspended matter of the Elbe estuary. GKSS 88/E/20:0–76

Sioli H (1957) Sedimentation im Amazonasgebiet. Geol Rdsch 45:608–633

Sioli H (1984) The Amazon and its main affluents. Hydrogr Rundsch 45:608–633

van Straaten LMJU (1960) Transport and composition of sediments. Verh Kon Ned Geol Mijnbkd. Gen/Geol Ser, D1 XIC Symp Ems-Estuarium (Nordsee), pp 270–292

Wellershaus S (1981) Turbidity maximum and mud shoaling in the Weser estuary. Arch Hydrobiol 92/2, 161–198

Wunderlich F (1969) Georgia coastal region, Sapelo Island, U.S.A.: sedimentology and biology. III. Beach dynamics and beach development. Senckenbergiana Mar 4:47–79

Zöllmer V, Irion G (in preparation) Clay mineral distribution in North Sea surface sediments.

Subject Index

abstract 249
acid-producing potential 328, 329, 334
algae
  decaying 236
  floating 220
  remains 220
algal
  blooms 235
  laminations 101
  mats 118
alluvial fans 129
analcite 226
analytical methods
  for determining heavy metal concentrations 279, 289
anhydrite/gypsum 186–189, 191, 211, 215, 216, 218
  beds 213
  idioblasts 214–216, 218
  intercalation 212
  kieseritic 214
  layer 212, 213
  layered 194
  residual beds 191
ankerite 236
anthracite 16–23
Apalachian
  Basin 22, 24, 26
  Mountain Chain 20
aragonite 211
  $\delta^{13}C$-concentration 91, 92
  $\delta^{18}O$-concentration 89, 92
ash layers
  basaltic 52, 70, 72
  deep sea 72
  discrete 52
  dissaminated 52
  rhyolitic 52, 63, 70
  silicic 72, 74
  thickness 52–55
  frequency 72–77
authigenic feldspar
Azores Islands 31

basalt
  tholeiitic, pillowed 35
basin modeling 137, 146, 147
basins
  foreland 125, 131
  intramontane 125, 130, 131
  sedimentary, hydrocarbon prospective 134, 146
beach-rock 30, 255
bentonite 61, 63, 65, 66, 68, 69
bioaccumulation 312, 317, 325, 327
bioassy 317, 318
biocalcarenites 35
bioconcentration factor 324, 325
bioproduction 235, 236
bioturbation 12
birds-eyes 101, 119, 121
bituminous shale 292
  trace elements in 298
buffer capacity 328, 331, 332

calcarenite 36
calcite
  associated with evaporites 211
  $\delta^{13}C$-concentration 89, 91
  $\delta^{18}O$-concentration 89, 91
  dental Mg-calcite 109, 119
  high-Mg 30, 31, 38, 88, 89, 92
  primary Mg-calcite 114, 115
  twisted 118
caliche 43, 255
  nodules 235
Canary Islands 30, 31
Cape Verde Islands 31, 36
cap-rock
  gypsum/anhydrite 180–191
  sequence 183, 184, strata 191
  strata 191
  erosion 191
carbon
  total organic (C_{org}) 228, 233
carbonate
  fabric

carbonate
  fabric
    grainstone   103, 105
    packstone   103, 105
    stromatolitic   119
  platform   143
  Dinaric   97
  sand dunes   40
  sediments
    C-isotope composition   88–92
  cements
    drusy mosaic   38
    dog-tooth   38
    meniscus   38
    micritic envelopes   38
    micritic linings   39
    interparticle   39
    intraparticle   39
    granular gravitational   115–117, 119
    marine vadose   115, 116, 119
carnallite   186, 215, 217, 218
  fibrous   213, 214
cascade test   326
cathodoluminescence   21
Catskill Mountains   17
chalcophile elements   233, 234, 236
chlorite   206
clastic sediments
  transport   352, 353
    pathways   351–365
coagulation/flocculation   327, 353, 360
coccoliths   166
  $\delta^{13}C$-concentration   89
communication network   240, 242, 243
compaction
  effect on primary petroleum migration
    166, 175
concentration
  gradient   348, 349
  saturation   348, 349
contaminants   321
  concentration   316, 318
    borderline   318
CPAH (combustion polyaromatic
  hydrocarbons)   318, 319, 325
cyanobacteria   259, 265

data bases   248, 250–252
  on-line   241–244
  desk-top   246, 247, 249
  electronic   242
  numeric   242
  full-text   242
  search-in   242

  bibliographic   245
depth of burial   22
descriptor   249
desert varnish   206
detection limit   284, 288, 289
diagenesis of carbonate sediments
  submarine   37
  subaerial   37
  regressive   38
diaprism   218
diffusion   10, 341, 349, 350
  coefficient   342, 343, 345, 347, 348
  current   342, 345–348
    stationary   347
  water film   7
dissolution
  selective   10
dissolution-molecular diffusion
  precipitation sequence   5, 8
dissolution-seam   6
dolomite   90
  diagenetic   118
  $\delta^{18}O$-values   88–92
  poorly ordered   88, 92
  feldspathic
    laminated   18
    mottled   18
  saddle   25
dolomitization   39, 143

easy menue mode   248
ecology   245
eluation
  behaviour   295
  experiments   298, 304
  index   330, 331
  pattern   298
  solution   291
elutriation
  analysis   198
environment
  aquatic   325, 327
  estuarine   361
  fluvial   223, 225
  intertidal (marine vadose)   101, 107, 119
  lacustrine (lake-)   221, 223
  marine   353, 354
  reducing   236
  savanna   235
  sedimentary   328
  subtidal (marine phreatic)   101, 107–115, 119
  supra-tidal (meteoric vadose)   101, 107, 115, 119

Subject Index

terrestrial   327, 353
environmental pollution (heavy metals)
  anthropogenous   287, 288
  geogenous   287
  North Sea   289, 365
environmental
  protection   245
  risk   307
epeirogeny   25
epilimnion   236
estuary sedimentation   352, 360, 362, 365
  smectite/kaolinite ratio   363–365
evaporite
  cycle   211
  facies   211
exploration risk
  economic   148
  environmental   148
  geologic   148
  reduction by probabilistic methods   147–150
facies
  alluvial red   221
  fluvio-lacustrine, grey   221
  marsh   17
fitted-fabric   6
flocculation
  see coagulation/flocculation
flood plane   355–357
flower structure   138, 140
fluid-homogenization temperature   22, 25
food chain
  transfer   291, 292, 312
foraminifera
  benthic   103
  $\delta^{18}O$-concentration   89, 90
formation resistivity factor   343, 346, 347
fossil dunes
  lithified   36

geothermal gradient   236
glaciation   198, 206
glass shards   55, 57–59, 62, 70
grain size distribution   197–208
gypsum   273
see also anhydrite/gypsum
  crusts   258, 261, 269

halite   218
  cubes   192
  encrustations   261
  fibres   192–213
  pseudomorphs   183
  secondary   194

translucent   194, 195
types
  "Liniensalz"   182
  "Rosensalz"   182
  "Schneesalz"   182
  "Schwadensalz"   212, 213
  "Tonbanksalz"   182
  "Tonbrockensalz"   182
heavy metals
  concentration   280–284, 361
    annual average   280
    related to discharge   281, 282
  distribution
hematite   216
hypertext   250

ichnofabric   3
  tiering   3, 13
ignimbrite   55
Illawara reversal (magneto-stratigraphic marker horizon)   128, 129, 132
index of geoaccumulation (I_{geo})   286, 287, 320, 329, 330, 333
information
  bibliographic   242, 249
  exchange   251, 252
  problem   241
in-situ sampler
  for pore water sampling   321
interstitial water   321
isostatic unroofing   25, 26
isopachous rims   111–114, 119

karstification   184, 189
kerogen   154, 158, 159, 292
  network   159, 174, 175
  types   168
kieserite   212, 214–218
koenenite   206

land sat imagery   138, 139
land turtle
  eggs   40
level of organic metamorphism (LOM)   18, 22
lithophile elements   234, 235
loess   206, 208
log file   249

magma discharge rates   72, 76
magnesite   206, 211
mail box
  electronic   251

marine terraces 40, 42
  elevated 41
microbial mats 259–261, 266–268, 270–273
  monolayered 262, 267
micro fractures
  horizontal 169
  vertical 169
miliolids 101, 103
molluscs
  δ^{13}C-concentration 89

network
  local area (LAN) 248

oil shale 221, 236
oil window 155
ostracods 103
ostrich eggs 139
overthrust structures 262

palagonite 52
paper shales 225–237
  lacustrine (Saar-Nahe Basin) 221, 223, 225
    bedding 225
  "siliclastic" 228, 232, 233
  containing carbonates 228, 232
partition coefficient 323–325, 328
PCB (polychlorinated biphenyls) 318, 319, 324
permeability 11
  alterations 137
  of HC-reservoirs 140
  reduction by illite formation 147
  coefficient of 341, 341
  relative 159
  of source rocks 161, 164, 166, 168, 169, 174, 175
petee 254, 256, 262, 265–267, 269–272
petroleum
  formation 153–155
  migration 152–178
    primary 153, 155–161, 174
      pathways 174
    secondary 155
photic zone 226
phreatomagnetic eruption 55
pollutants 312
  adsorption 315
  bioavailability 325
  concentration 325, 329
  organic 353
  particle-associated 313
  sediment-associated 316

toxic 321
Zn in estuaries 364, 365
pollution
  background 287
  level 361
  potential 329
polyhalite 213–215, 217, 218, 273
  acicular 215
polyhalitisation 218
pore cement 348
  allochthonous 340
  autochthonous 340, 342
  replacement 349
pore size 156, 169
  distribution 153, 161, 163, 164, 168
    curves 166
      diagenetic effects on 164, 175
      lithofacies effects on 163, 175
pore solution current 340, 341
porosity 11, 343, 347
  minus cement 115
  secondary 166
Posidonia Shale 291–298, 304–307
potash
  seam 186, 212
  layer 191
pressure dissolution 3, 7
pressure gradient 340, 341
pyroclast
  essential (juvenile) 50
  accessory (cognate) 50
  accidental 50

radioactive waste 212
rare earth elements (REE) 232, 233
  chondrite normalized 228
reef
  patch 98
rhodochrosite
  δ^{18}O-concentration 92
Rosin-Rammler equation 200
Rotliegend
  sedimentary (Upper) 126, 128–132
  volcanic (Lower) 126–128
    K/Ar- and ^{40}Ar/^{39}Ar-dating 128

Saalian unconformity 126
sabkha 255, 273
saltern 254, 256, 264, 272–274
salt plug 140
sampling objectives 313
sand
  aeolian 131, 132
  beach 131

Subject Index

scroll bar 354
sea level
  eustatic changes 91
search technique (EDP)
  standard 248
  on-line 248
sediment quality-"triad" 317–319
seismic resolution 137
Selvagens Islands 31, 36
siderite 236
sideromelane 52
single-extractant procedure 326
soft ware 245, 247, 250
source rock 153–156, 162, 168, 174
  pore network 153, 159
  petroleum potential 155
  organo facies 175
subrosion
  of rock salt 179–196
suspended matter/load 280, 284, 286, 287, 353, 358, 360
sylvite 186, 214–216, 218
  seam 212, 213
synaeresis cracks 105
stromatolite 225, 259, 273
stylobreccia 4
stylolites
  stratiform 3
  growth of 9
  flow in active 9
  distribution of 13

tachyhydrite 186
tachylite 250

tepee 254–256, 266, 269–272
tephra 21, 48–82
  particles 55
  fall out 59
    subaerial 50
    subaquatic 50
    distal 60
    near source (proximal) 60
tephrostratigraphy 76
tidal flat 362
tortuosity 347, 348
toxicity factor 330, 331, 333
  sedimentological 321
toxic response factor 321
trace elements 228, 233, 293, 327
  analysis 295
  behaviour 328
  bituminous shales 292
  concentration 291, 292, 294, 298, 307
  metals 221
  mobility 307
  toxic 307
transport of matter
  advective 340, 348–350
  diagenetic 347, 350
  diffusive 341, 348
  long-distant 342, 350
  short-range 339, 350

Variscan foldbelt 125, 126
vertical uplift 25
vitrinite reflectivity 18
volcanic episodicity 72–77

zeolites 21